MUSHROOMS AS FUNCTIONAL FOODS

MUSHROOMS AS FUNCTIONAL FOODS

Edited by

Peter C. K. Cheung

The Chinese University of Hong Kong

WILEY

A JOHN WILEY & SONS, INC., PUBLICATION

Published by John Wiley & Sons, Inc., Hoboken, New Jersey
Published simultaneously in Canada

Library of Congress Cataloging-in-Publication Data:

Cheung, Peter C. K.
Mushrooms as fuctional foods / Peter C.K. Cheung.
p. cm.
Includes bibliographical references and index.
ISBN 978-0-470-05406-2 (cloth)
1. Mushrooms—Therapeutic use. 2. Functional foods. I. Title.
[DNLM: 1. Agaricales—chemistry. 2. Food Analysis. 3. Food Technology.
QW 180.5.B2. c526m 2008]
RM666.M87C45 2008
615'.3296—dc22

2007050403

10 9 8 7 6 5 4 3 2 1

This book is dedicated to my family members:
Carmen, Timothy, Rebekah, and Anthony Cheung

CONTENTS

FOREWORD

It has been over twenty years since the concept of "functional foods" was first introduced as a factor in the analysis of foods after nutrients. Consumers are now deeply interested in food bioactives that provide beneficial effects to humans in terms of health promotion and disease risk reduction. They also demand more detailed information about food factors in order to obtain appropriate functional food products.

In Asian countries, like China and Japan, mushrooms have been collected and cultivated for hundred of years. They have a long history of use for their health promotion benefits. In recent years reports on the chemistry, and the nutritional and functional properties of mushroom have been overwhelming. In the *Journal of Agricultural and Food Chemistry* alone there have been more than 300 articles related to mushrooms published since 1990. However, there is no in-depth comprehensive reference book of mushrooms as functional food available. The current book of Professor C. K. Cheung, *Mushrooms as Functional Foods*, is a timely and well welcomed book for scientists and students working in functional food research.

Besides covering the agricultural production, nutritional values, and health benefits of mushrooms, this book also introduces emerging molecular analysis and functional genomics to the study of mushroom. Health benefits of mushrooms, such as, antioxidative, hypocholesterolemic, and hypoglycemic effects are discussed in depth. Polysaccharides are the best known and potent mushroom-derived substances with immunomodulating and antitumor activities and this topic has been treated extensively in a separate chapter. Included also is a unique and useful chapter on regulatory issues of mushrooms as functional foods in different countries.

Scientists and students who research mushrooms will certainly benefit from reading this comprehensive monograph to gain in-depth knowledge for the development of mushrooms into functional foods.

CHI-TANG HO

Rutgers University

PREFACE

Mushrooms have been known for their nutritional and culinary values as well as viewed as tonics and used as medicines by humans for ages. In modern terms, they can be considered as functional foods which can provide health benefits beyond the traditional nutrients they contain. There are monographs that cover the medicinal and healing properties of some individual traditional mushrooms and fungi such as the *Ganoderma*, Shiitake mushroom, and *Cordyceps* for the general public. However, there are very few in-depth and up-to-date comprehensive reference books in the scientific literature of both the basic and applied aspects of mushrooms as functional foods.

This book is an integration of the recent research conducted on the biological and chemical aspects of mushrooms when being utilized as a functional food. Topics that are covered in this book range from the agricultural production of mushrooms to the use of molecular biological techniques like functional genomics, from nutritional values of newly cultivated mushroom species to the multifunctional effects of the unconventional form of the mushroom (sclerotium), and from the mechanistic actions of the physiological benefits and pharmacological properties of bioactive components in mushrooms to the regulations of their uses as functional foods and dietary supplements in different parts of the world.

This comprehensive book should serve as a reference for scientists; chemists; biologists; food manufacturers; students majoring in food science, nutrition, biology, and biochemistry, to name a few; and all those who are interested in obtaining a stronger background in the development of mushrooms and edible fungi into functional foods.

PETER C. K. CHEUNG

The Chinese University of Hong Kong

ACKNOWLEDGMENTS

I want to thank the many colleagues and research collaborators who have graciously given support and advice in the development of this book: Professor Shu-ting Chang and Marilyn M. L. Yu, The Chinese University of Hong Kong; Professor Lina Zhang and Dr. Mei Zhang, Wuhan University; and Nian-Lai Huang, Sanming Mycological Institute. The clerical support provided by Yuk-fan Ng and Kit-fong Tong is deeply appreciated.

P. C. K. C.

CONTRIBUTORS

Eden Akavia, International Center of Biotechnology and Biodiversity of Fungi, Institute of Evolution, University of Haifa, Haifa, Israel

Shu-Ting Chang, Department of Biology, The Chinese University of Hong Kong, Hong Kong, China

Peter C. K. Cheung, Food and Nutritional Sciences Programme, Department of Biology, The Chinese University of Hong Kong, Hong Kong, China

Winnie W. Y. Chum, Department of Biology, The Chinese University of Hong Kong, Hong Kong, China

Hoi-Shan Kwan, Department of Biology, The Chinese University of Hong Kong, Hong Kong, China

Vincent E. C. Ooi, Department of Biology and Institute of Chinese Medicine, The Chinese University of Hong Kong, Hong Kong, China

Solomon P. Wasser, International Center of Biotechnology and Biodiversity of Fungi, Institute of Evolution, University of Haifa, Haifa, Israel

Ka-Hing Wong, Department of Biology, The Chinese University of Hong Kong, Hong Kong, China

CHAPTER 1

Overview of Mushroom Cultivation and Utilization as Functional Foods

Shu-Ting Chang

Department of Biology, The Chinese University of Hong Kong, Hong Kong, China

CONTENTS

1.1 INTRODUCTION

In 1957, R. Gordon Wasson, a world known amateur mycologist, proposed the division of people into two classes for which he coined the following terms:

Mycophiles—Those who love and know their mushrooms intimately.
Mycophobes—Those who fear, dislike, and do not know their mushrooms.

I think all readers of this book belong to the former and not the latter.

Knowledge of numerous new mushroom species has accumulated through time. The number of recognized mushroom species has been reported to be 14,000, which is about 10% of the total estimated mushroom species on the earth (Hawksworth, 2001). China is estimated to have about 1500–2000 edible mushroom species with 981 species identified. By 2002, 92 species have been domesticated while 60 of these have been commercially cultivated (Mau et al.,

Mushrooms as Functional Foods, Edited by Peter C. K. Cheung

2004). However, mushrooms have nearly always been around, with a very long and interesting history. Mushrooms have been found in fossilized wood that is estimated to be 300 million years old, and almost certainly prehistoric man used mushrooms collected in the wild as food. Recently, the importance of the role of mushrooms in history was evidenced by the fact that the desert truffle, *Terfezia arnenari*, was described in the Bible as "bread from heaven" and also "manna of the Israelites" (Pegler, 2002).

It may be interesting to have a charming mushroom poem as a beginning for this chapter: "Without leaves, without buds, without flowers, yet, they form fruit; as a food, as a tonic, as a medicine, the entire creation is precious" (Chang and Miles, 1989, p. 345). The first part describes the morphological and physiological characteristics of mushrooms while the second states the nutritional and medicinal properties of mushrooms.

Our attitudes to the phenomenon of nature are seldom based on simple observation. There are, however, examples throughout history where certain living things have inspired fear and loathing simply because they are regarded as ugly species with peculiar behavior and supposedly evil. For example, in some communities, bats, snakes, spiders, toads, and owls have all been associated with the devil or regarded as harbingers of illness and even of death. This is one of the reasons why some refer to the poisonous mushroom as a "toadstool." Actually, the name has no scientific basis at all and should not be used in any situations, although it is possible to find a toad sitting beside or even on top of a mushroom. Mushrooms attract toads, not due to the mushroom itself, but because of the various insects which are harbored in them. Insects certainly are interested in mushrooms as a source of food (Chang, 2005).

It cannot be denied that some mushrooms, even though they represent less than about 1% of the world's known mushrooms, are dangerous if eaten. Some are deadly poisonous. But perhaps a more likely explanation for the widespread abhorrence of wild mushrooms in communities is that they are by nature a rather strange and mysterious group of organisms, quite unlike the green plants. In some ancient communities, the seemingly miraculous manner of its growth without seed, without leaf, and without bud, its fruiting body's sudden appearance after rain, especially after lightning and thunderstorms, its equally rapid disappearance, and its curious umbrellalike shape gave rise to a wealth of illusions and mythologies.

Fungi are found just about everywhere. Mushrooms, a special group of macrofungi, are rather more selective than other fungi in that the size of the fruiting body requires the availability of more nutrients than are required for the production of asexual spores by microfungi. Nevertheless, in damp places, such as tree-fern gullies and areas of rain forest, plentiful moisture leads to mushroom formation and mushrooms can be collected during most of the year. There may be a particular flora of mushroom species associated with the seasons of autumn, summer, and spring. Relatively few mushrooms are produced during the cold winter months, although there are perennial fruiting bodies that persist during the winter. But in drier regions mushrooms occur only after seasonal rains. Formation of mushroom fruiting bodies depends very much on the pattern of rain and, in some years, there may be virtually a complete lack of fruiting.

There has been a recent upsurge of interest in mushrooms not only as a health vegetable (food) which is rich in protein but also as a source of biologically active compounds of medicinal value. Uses include complementary medicine/dietary supplements for anticancer, antiviral, immunopotentiating, hypocholesterolemic, and hepatoprotective agents. This new class of compounds, termed *mushroom nutriceuticals* (Chang and Buswell, 1996), are extractable from either the mushroom mycelium or fruiting body and represent an important component of the expanding mushroom biotechnology industry. It has been shown that constant intake of either mushrooms or mushroom nutriceuticals (dietary supplements) can make people fitter and healthier. In addition, mushroom cultivation can also help to convert agricultural and forest wastes into useful matter and reduce pollution in the environment. Therefore, mushroom cultivation can make three important contributions: production of health food, manufacture of nutriceuticals, and reduction of environmental pollution.

1.2 WHAT ARE MUSHROOMS?

1.2.1 Definition of a Mushroom

Mushrooms along with other fungi are something special in the living world, being neither plant nor animal. They have been placed in a kingdom of their own, called Myceteae (Miles and Chang, 1997). But what are mushrooms? The word *mushroom* may mean different things to different people and countries. It was reported (Chang and Miles, 1992) that specialized studies and the economic value of mushrooms and their products had reached a point where a clear definition of the term mushroom was warranted. In a more broad sense "mushroom is a macrofungus with a distinctive fruiting body, which can be either epigeous or hypogeous and large enough to be seen with naked eye and to be picked by hand" (Chang and Miles, 1992). Thus, mushrooms need not be Basidiomycetes, or aerial, or fleshy, or edible. Mushrooms can be Ascomycetes, grow underground, have a nonfleshy texture, and need not be edible. This definition is not a perfect one but can be accepted as a workable term to estimate the number of mushrooms on the earth (Hawksworth, 2001). The most common type of mushrooms is umbrella shaped with a pileus (cap) and a stipe (stem), that is, *Lentinula edodes*. Other species additionally have a volva (cup), that is, *Volvariella volvacea*, or an annulus (ring), that is, *Agaricus campestris*, or both, that is, *Amanita muscaria*. Furthermore, some mushrooms are in the form of pliable cups; others are round like golf balls. Some are in the shape of small clubs; some resemble coral; others are yellow or orange jellylike globs; and some even very much resemble the human ear. In fact, there is a countless variety of forms.

The structure that we call a mushroom is in reality only the fruiting body of the fungus. The vegetative part of the fungus, called the mycelium, comprises a system of branching threads and cordlike strands that branch out through soil, compost, wood log, or other lignocellulosic material on which the fungus may be growing. After a period of growth and under favorable conditions, the established (matured) mycelium could produce the fruit structure which we call the mushroom. Accordingly mushrooms can be grouped into four categories: (1) those which are fleshy and edible fall into the edible mushroom category (e.g., *Agaricus bisporus*); (2) mushrooms which are considered to have medicinal applications are referred to as medicinal mushrooms (e.g., *Ganoderma lucidum*); (3) those which are proven to be or suspected of being poisonous are named poisonous mushrooms (e.g., *Amanita phalloides*); and (4) a miscellaneous category, which includes a large number of mushrooms whose properties remain less well defined, may tentatively be grouped together as "other mushrooms." Certainly, this approach of classifying mushrooms is not absolute and not mutually exclusive. Many kinds of mushrooms are not only edible but also possess tonic and medicinal qualities.

1.2.2 Ecological Classification of Mushrooms

Mushrooms can be ecologically classified into three categories: saprophytes, parasites, and mycorrhiza.

There are only a few parasitic mushrooms. Most of the cultivated gourmet mushrooms are saprophytic fungi. Some are mycorrhizal mushrooms, for example, Perigold black truffle (*Tuber melanosporum*) and matsutake mushroom (*Tricholoma matsutake*). It is difficult to bring these pricey wild gourmet species into cultivation because they are mycorrhiza. These mushroom species have a symbiotic relationship with some vegetation, particularly trees, that is, there is a relationship of mutual need. Therefore, the substratum (host) should be carefully recorded, as this can be an important feature in identification and in classification, for example, whether the mushroom is growing on dung, wood, bark, living trees, litter, or soil. If the mushroom is growing on a living plant or on dead parts recognizable as belonging to a nearby plant, flowers, fruits, or other parts of the plant, these should be collected for identification of the host or substrate if its name is not known.

Saprophytes obtain nutrients from dead organic materials; parasites derive food substances from living plants and animals, causing harm to the hosts; and mycorrhiza live in a close physiological association with host plants and animals, thereby forming a special partnership where each partner enjoys some vital benefits from the other.

However, some mushrooms do not fall neatly within these man-made categories and can share two of these categories (Figure 1.1). For example, some *Ganoderma* spp., including *G. lucidum*, are common saprophytes but can be pathogenic too; also *T. matsutake*, while initially appearing to be mycorrhizal on young roots, soon becomes pathogenic and finally exhibits some saprophytic ability.

1.2.3 Identification of Mushrooms

Successful identification of wild mushrooms requires a basic knowledge of the structure of fungi and of the way in which they live. To identify a given mushroom,

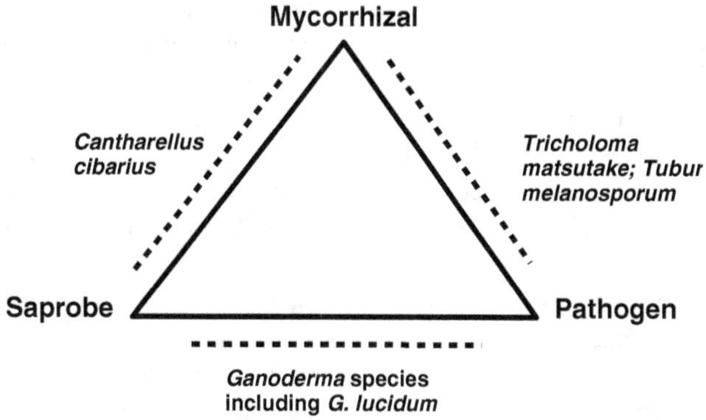

Figure 1.1 Modified triangular model for ecological classification of mushrooms (Hall et al., 2003b).

it is necessary to examine the fruiting bodies with the utmost care. A fresh fruiting body is much easier to identify than a pickled (preserved in formalin) or a dried one. A good reference material, usually a book with color, pictures of the different mushrooms known, is a basic requirement. A key is usually provided to simplify identification in most reference texts (Arora, 1986; Carlucccio, 2003; Chang and Mao, 1995; Fuhrer, 2005; Shepherd and Totterdell, 1988; van der Westhuizen and Eicker, 1994).

In using the reference, it is essential that one knows some specific characteristics of the mushroom being identified. These characteristics are (1) size, color, and consistency of the cap and the stalk; (2) mode of attachment of the gills to the stalk; (3) spore color in mass; and (4) chemical tests or reactions.

Although the color of the gills is a good indication of the spore color, there are instances when the experienced mycologist will have to resort to what is called "spore print" examination to determine the real color of the spores. For specimens with a distinct cap and stem, the cap is removed and placed fertile-side down, preferably on a microscope slide, but in the absence of such, on white paper, black paper, or cellophane. Then it is covered with a bowl or similar object to prevent air currents. A thin spore print is often visible after as little as a half hour, but a useful deposit usually requires longer time (up to 2 hours or more). The print is necessary to determine overall spore color. It is also a source of mature spores for microscopic examination and measurements.

The mode of gill attachment to the stem indicates the genus of the mushroom and should be carefully noted. To determine the mode of attachment, the mushroom is cut longitudinally through the cap, exposing the point of attachment of the gills to the stem.

The environment in which the mushroom was picked should also be noted. It is important to know whether the mushroom grows directly on the ground, on decaying wood, on a living tree trunk, or on compost. One should not overlook the species of those on which the mushrooms are found growing or the type of grasses or moss present in the area where the mushrooms are collected.

There is no single reference work in which all mushrooms are illustrated or described. In most cases, mushroom species in publications are grouped by region or locality, for example, North American mushrooms, mushrooms of the Western Hemisphere, and mushrooms of South Africa. While certain mushrooms are easy to identify, many are not. In fact, there is a great number of look-alikes. To avoid any unpleasant experiences, especially when identifying mushrooms for the purpose of determining edibility, experts should always be consulted (Quimio et al., 1990).

Collectors should always remember when using keys that the mushroom they have in hand might not be included in the book they are consulting (or in any other book, for that matter). Once they have obtained a name with a key, they must read the detailed description provided for the mushroom and compare it with the one they are trying to identify. If the description does not fit the specimen, then they must go back to the key and try again, following a different route. If they exhaust all of the possible routes and still cannot find a description that fits, they should assume that the fungus in hand is not in the books being consulted. Using the information gained, they may then consult other appropriate references that may be available or they may seek the assistance of specialists working with the group in question. They should never attempt to force a specimen into a category where it does not fit.

Some mushrooms are very palatable due to their exotic taste, but some mushrooms are very poisonous. Unfortunately, there are no general guidelines for distinguishing between the poisonous and edible species. The only means by which a nonspecialist can determine the edibility or toxicity of a given mushroom is to carry out an accurate identification of the specimen. Such identification may be obtained by consulting the relevant literature, preferably with illustrations, or experts in the subject. Identification of a mushroom at the generic level is inadequate since, within a given genus (e.g., *Lepiota*) some species are edible while other species are highly poisonous.

Several species of *Amanita* are extremely poisonous, but obvious symptoms do not appear until 8–12 hours after ingestion. The poisonous compound, amatoxin, is not destroyed by boiling or processing. Some less poisonous mushrooms produce only nausea or gastric upset within 30–60 minutes of ingestion (Hall et al., 2003a; Quimio et al., 1990). Mushrooms partially eaten by animals or insects are not necessarily fit for human consumption. When the mushroom is in doubt, throw it out. *If you are not absolutely sure whether a given mushroom is edible or otherwise, do not touch it. Leave the strange mushroom alone.*

1.3 CONCEPT OF MUSHROOM BIOLOGY AND APPLIED MUSHROOM BIOLOGY

1.3.1 Mushroom Biology

The biological science that is concerned with fungi is called mycology. Mushroom biology is the branch of mycology that deals with mushrooms in many disciplines.

When knowledge increases and areas of specialization develop within the discipline, it is convenient to indicate that area of specialization with a self-explanatory name. In biology, there are such specializations as neurobiology, bacteriology, plant pathology, pomology, molecular biology, virology, fungal physiology, embryology, endocrinology, phycology, and entomology. These names indicate either a group of organisms (e.g., bacteria, algae, and insects) and/or an approach to the study (e.g., disease, development, and physiology).

Although several terms for this important branch of mycology that deals with mushroom have been used, and each of these has its merit, when we get down to the matter of definitions, it seems that there is a place for a new term—*mushroom biology* (Chang and Miles, 1992). Mushroom biology is a new discipline concerned with any aspect of the scientific study of mushrooms, such as taxonomy; physiology, and genetics.

1.3.2 Applied Mushroom Biology

Applied mushroom biology is concerned with all aspects of the application of mushroom biology. It consists of three main components: mushroom science, mushroom biotechnology (Chang, 1993), and mushroom bioremediation (Figure 1.2). *Mushroom science* deals with mushroom cultivation and production (mushrooms themselves) and encompasses the principles of mushroom biology/microbiology, bioconversion/composting technology, and environmental engineering (Figure 1.3); *Mushroom biotechnology* is concerned with mushroom products (mushroom derivatives) and encompasses the principles of mushroom biology/microbiology, fermentation technology, and bioprocess (Figure 1.4). Mushroom biotechnology, both as a technology and as the basis for new mushroom products, requires industrial development. It, like many bioscience

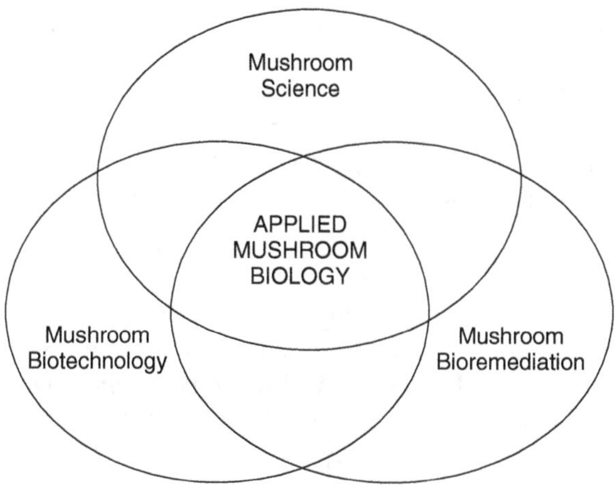

Figure 1.2 Applied mushroom biology consists of three components: mushroom science, mushroom biotechnology, and mushroom bioremediation.

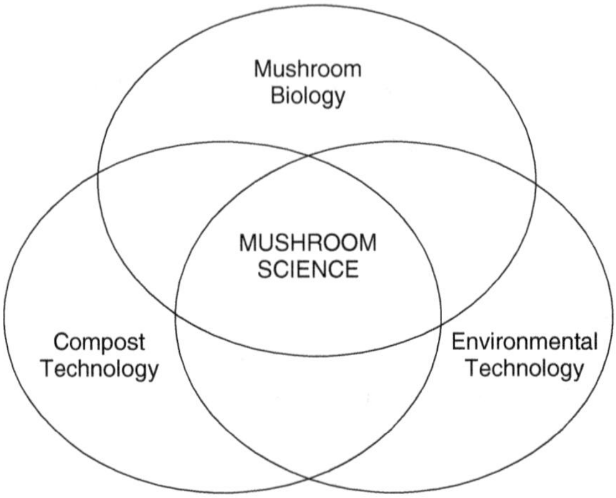

Figure 1.3 Mushroom science: concerned with mushroom cultivation and production.

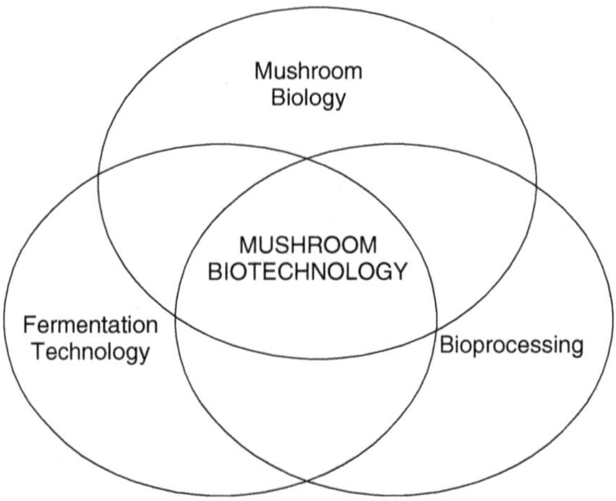

Figure 1.4 Mushroom biotechnology: concerned with mushroom products (mushroom nutriceuticals/dietary supplements).

industries, operates at the cutting edge of science and involves numerous regulatory issues. The third component of applied mushroom biology has been developed in recent years. This is *mushroom bioremediation* and is concerned with the beneficial impacts of mushrooms on the environment (from mushroom mycelia) and encompasses principles of mushroom biology/microbiology, ecology, and bioconversion technology (Figure 1.5).

Therefore, the aims of the discipline of applied mushroom biology (Figure 1.2) are to tackle the three basic problems—shortage of food, diminishing quality of

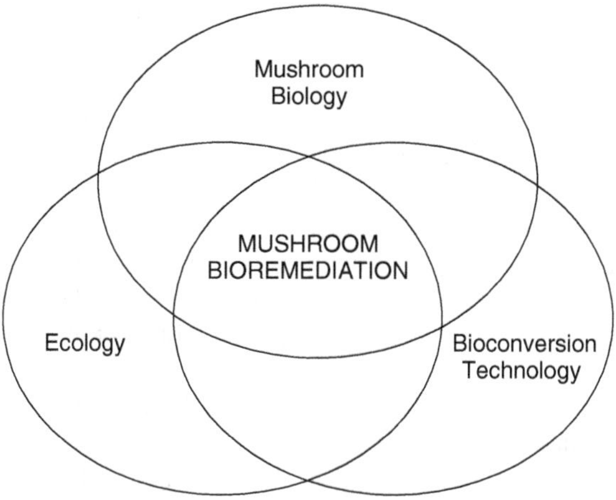

Figure 1.5 Mushroom bioremediation: concerned with beneficial impacts of mushrooms on environment.

human health, and pollution of the environment—which human beings still face, and will continue to face, due to the continued increase of the world population. The twentieth century began with a world population of 1.6 billion and ended with 6 billion inhabitants. The world's population is likely to reach 9.2 billion in 2050 from the current 6.7 billion with most of the growth occurring in developing countries. The growing world population is increasing by about 80 million people per year. At present, about 900 million people in the world are living in poverty. On the other hand, it has been observed that over 70% of agricultural and forest products have not been put to total productivity and have been discarded as waste. Applied mushroom biology not only can convert these huge lignocellulosic biomass wastes into human food but also can produce notable nutriceutical products that have many health benefits. Another significant aspect of applied mushroom biology is using the biota in creating a pollution-free and beneficial environment. The three components of applied mushroom biology are closely associated with three aspects of well-being—food, health, and pollution.

The discipline that is concerned with the principles and practice of mushroom cultivation is known as *mushroom science* (Chang and Miles, 1982). The establishment of principles requires facts arrived at through systematic investigation. The systematic investigation must involve the practical aspects of mushroom cultivation as well as scientific studies. The consistent production of successful mushroom crops necessitates both practical experience and scientific knowledge.

1.3.3 Impact of Applied Mushroom Biology

1.3.3.1 Nongreen Revolution The world population has reached over 6 billion now. It is expected to continue increasing in the twenty-first century.

The amount of food and the level of medical care available to each individual, especially those in less developed countries, will decrease. Environmental pollution and greenhouse gas effects will also become a more serious problem. However, the world has an immense amount of lignocellulosic material resource that, like solar energy, is sustainable. Lignocellulosic material is a kind of biomass which is estimated to amount to 1.09×10^{11} t dry matter on land annually (Chang, 1989), which consists of mainly three components: cellulose, hemicellulose, and lignin. Lignocellulose is a major component of wood and other plant materials. The world's annual yield of cereal straws in 1999 is estimated to be 3570×10^6 t. Since such a large amount of energy is in lignocellulosic biomass (3020 EJ solar energy fixed in biomass per year), it can constitute principal objects for conversion into useful products by man's activities. Note that E is the metric prefix for exa (10^{18}) and joule is the unit of energy.

Although various strategies have been developed to utilize part of the vast quantities of waste lignocellulose generated annually through the activities of agricultural, forestry, and food processing industries, one of the most significant, in terms of producing a higher value product from the waste, is the cultivation of edible mushrooms by solid-state fermentation. More recently, attention has focused on a second area of exploitation following the discovery that many of these mushrooms produce a range of metabolites of intense interest to the pharmaceutical/nutriceutical (e.g., antitumor, immunomodulation agents, and hypocholesterolemic agents), and food (e.g., flavor compounds) industries. Mushrooms, like all other fungi, lack chlorophyll and are nongreen organisms. They cannot convert solar energy through the process of photosynthesis to organic matter as green plants do, but they can produce extensive enzymes that can degrade lignocellulosic materials for their own nutrients for growth and fruiting. Different mushrooms have different lignocellulolytic enzyme profiles (Buswell and Chang, 1994; Buswell et al., 1996b). These are reflected in qualitative variations in the major enzymatic determinants (i.e., cellulases, ligninases) required for substrate bioconversion. For example, *L. edodes*, which is cultivated on highly lignified substrates such as wood or sawdust, produces two extracellular enzymes that have been associated with lignin depolymerization in other fungi (manganese peroxidase and laccase) (Buswell et al., 1995). Conversely, *V. volvacea*, which prefers high-cellulose, low-lignin-containing substrates, produces a family of cellulolytic degrading enzymes, including at least five endoglucanases, five cellobiohydrolases, and two β-glucosidases (Cai et al., 1994, 1998, 1999). *Pleurotus sajor-caju*, the grey oyster mushroom, exhibits both cellulase and ligninase secretions (Buswell et al., 1996a) and therefore is the most adaptable of the three species. It can grow on a wide variety of agricultural waste materials of differing composition in terms of the polysaccharide–lignin ratio. This demonstrates the impressive capacities of mushrooms for *biosynthesis*, which is different from *photosynthesis* by green plants. The species of mushroom fungi not only can convert the agricultural and forestry lignocellulosic wastes through solid fermentation technology into the high-quality protein consumed directly in the form of the mushroom fruiting body but also can convert food processing biomass wastes (e.g., soybean wastes using submerged culture) into fungal protein (Buswell and Chang 1994) or "mycomeat" (Miles and Chang 1988). Soybean waste materials (slurries) are generated in large quantities during the processing of soybean milk and "tofu" (bean curd), which are popular foods in many countries now and are, in some places, discarded without treatment, thereby constituting an environmental pollutant. In addition, mushrooms and their mycelia can provide nutriceutical and pharmaceutical products. As outlined above, by blending the advances in basic biological knowledge with that of practical technology, a mushroom-related industry based on utilization of the lignocellulosic waste materials that are abundantly available in rural and urban areas can have positive global impacts on long-term food nutrition, health, environmental conservation and regeneration, and economic and social change. Therefore, the significant impact of applied mushroom biology on human welfare has been named a "nongreen revolution" (Chang, 1999).

1.3.3.2 Mushroom Bioremediation This component of applied mushroom biology deals mainly with the aspects of benefits to the earth from the activities of mushroom mycelium. Environmental contamination can be ameliorated by the application of mushroom mycelial technologies. For example:

1. The use of bioconversion processes to transform the polluting substances into valuable foodstuffs, for example, the proper treatment and reutilization of spent substrates/composts in order to eliminate the pollution problems (Beyer, 2005; Noble, 2005). One of the most intriguing opportunities offered by mushroom mycelia in the area of bioconversion is the exploitation of their ability to degrade pollutants, many of which are highly carcinogenic, released into the environment as a consequence of human activity.
2. The use of fungi/mushroom mycelia as tools for healing soil, what Stamets (2005) called "mycorestoration," which is the use of fungi/mushrooms to repair or restore the weakened or damaged biosystems of environment. The processes of mycorestoration include the selective use of mushrooms for mycofiltration to filter water, mycoforestry to enact ecoforestry policy, mycoremediation to denature toxic wastes, and mycopesticides to control insect pests. Mycorestoration recognizes the primary role fungi/mushrooms play in determining the balance of biological populations.

1.4 MUSHROOM CULTIVATION

Mushroom cultivation is both a science and an art. The science is developed through research; the art is perfected through curiosity and practical experience. Mushroom growth dynamics involve some technological elements that are in consonance with those exhibited by our common agricultural crop plants. For example, there is a vegetative growth phase, when the mycelia grow profusely, and a reproductive (fruiting) growth phase, when the umbrella-like body that

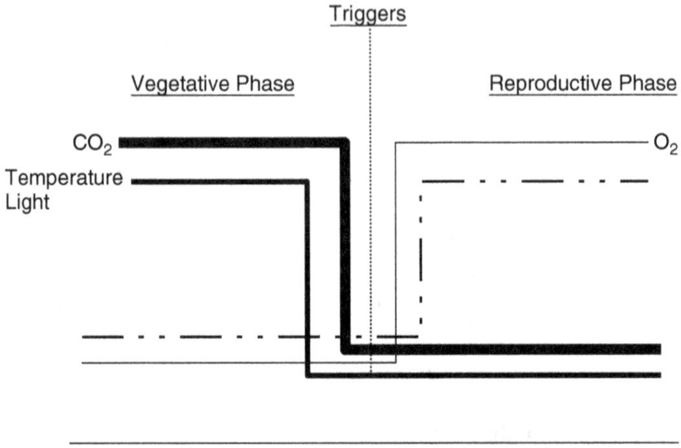

Figure 1.6 Two major phases of mushroom growth and development: vegetative and reproductive. The triggers for the transition from the vegetative phase to the reproductive phase are usually regulated by environmental factors.

we call mushroom develops. In agricultural plants (e.g., sunflowers), when the plants switch from vegetative growth to reproductive growth, retarding tips for further growth (elongation) is an obvious phenomenon of maturity. It is the same principle in mushroom production. After the vegetative (mycelial) phase has reached maturity, what the mushroom farmer needs next is the induction of fruiting. This is the time the mycelial growth tips should be retarded by regulating the environmental factors. These factors, generally called "triggers" or "environmental shocks," such as switching on the light, providing fresh air, and lowering temperature, can trigger fruiting (Figure 1.6).

1.4.1 Major Phases of Mushroom Cultivation

Mushroom farming is a complex business that requires precision. Indeed, it is not as simple as what some people often loosely stipulate. It calls for adherence to precise procedures. The major practical steps/segments of mushroom cultivation are (1) selection of an acceptable mushroom species, (2) secretion of a good-quality fruiting culture, (3) development of robust spawn, (4) preparation of selective substrate/compost, (5) care of mycelial (spawn) running, (6) management of fruiting and mushroom development, and (7) careful harvesting of mushrooms (Chang and Chiu, 1992; Chang 1998). If you ignore one critical step/segment, you are inviting trouble, which could lead to a substantially reduced mushroom crop yield and mushroom marketing value:

1. Before any decision to cultivate a particular mushroom is made, it is important to determine if that species possesses organoleptic qualities

acceptable to the indigenous population or to the international market, if suitable substrates for cultivation are plentiful, and if environmental requirements for growth and fruiting can be met without excessively costly systems of mechanical control.

2. A "fruiting culture" is defined as a culture with the genetic capacity to form fruiting bodies under suitable growth conditions. The stock culture selected should be acceptable in terms of yield, flavor, texture, fruiting time, and so on.
3. A medium through which the mycelium of a fruiting culture has grown and which serves as the inoculum of "seed" for the substrate in mushroom cultivation is called the "mushroom spawn." Failure to achieve a satisfactory harvest may often be traced to unsatisfactory spawn used. Consideration must also be given to the nature of the spawn substrate since this influences rapidity of growth in the spawn medium as well as the rate of mycelial growth and filling of the beds following inoculation.
4. While a sterile substrate free from all competitive microorganisms is the ideal medium for cultivating edible mushrooms, systems involving such strict hygiene are generally too costly and impractical to operate on a large scale. Substrates for cultivating edible mushrooms normally require varying degrees of pretreatment in order to promote growth of the mushroom mycelium to the practical exclusion of other microorganisms. The substrate must be rich in essential nutrients in forms which are readily available to the mushroom and be free of toxic substances that inhibit growth of the spawn. Moisture content, pH, and good gaseous exchange between the substrate and the surrounding environment are important physical factors to consider.
5. Following composting, the substrate is placed in beds where it is generally pasteurized by steam to kill off potential competitive microorganisms. After the compost has cooled, the spawn may be broadcast over the bed surface and then pressed down firmly against the substrate to ensure good contact or inserted 2–2.5 cm deep into the substrate. Spawn running is the phase during which mycelium grows from the spawn and permeates into the substrate. Good mycelial growth is essential for mushroom production.
6. Under suitable environmental conditions, which may differ from those adopted for spawn running, primordial formation occurs and then is followed by the production of fruiting bodies. The appearance of mushrooms normally occurs in rhythmic cycles called "flushes."
7. Harvesting is carried out at different maturation stages depending upon the species and consumer preferences and market value.

1.4.2 Cultivation of Several Selected Mushrooms

The cultivation of edible mushrooms can be divided into two major stages. The first stage involves the preparation of the fruiting culture, stock culture, mother spawn, and planting spawn, while the second stage entails the preparation of the growth substrates for mushroom cultivation. Cultivation conditions for a few selected mushroom species are briefly described in the following sections.

1.4.2.1 Cultivation of Agaricus Composting is prepared in accordance with well-documented commercial procedures (van Griensven, 1988; Chang and Hayes, 1978; Kaul and Dhar, 2007). In phase I of the process (outdoor composting), locally available raw materials are arranged into piles that are periodically turned and watered. The initial breakdown of the raw ingredients by microorganisms takes place in phase I. This phase is usually complete within 9–12 days, when the materials have become pliable, dark brown in color, and capable of holding water. There is normally a strong smell of ammonia. Phase II (indoor fermentation) is pasteurization, when undesirable organisms are removed from the compost. This is carried out in a steaming room where the air temperature is held at 60°C for at least 4 hours. The temperature is then lowered to 50°C for 8–72 hours depending upon the nature of the compost. Carbon dioxide is maintained at 1.5–2% and the ammonia level drops below 10 PPM. Following phase II composting, the substrate is cooled to 30°C for *Agaricus bitorquis* and to 25°C for *A. bisporus* for spawning. Production of phase III or IV composts for growing *Agaricus* mushrooms has been an advanced technological development in recent years in Western countries. The production of phase III compost is phase II compost spawn run in a bulk tunnel and ready for casing when delivered to the grower. If the phase III compost is then cased and spawn developed into the casing layer before dispatching to the growing unit or delivering to growers, it is named as phase IV compost. The successes of bulk phase III and IV depend a lot on the quality of phase I and II processes. Phase II on shelves produces an average of 4.1 crops per year. Since 1999, growers using Phase III production enjoyed an average of 7.1 crops per year. In recent years, phase IV can generate 10–12 crops per year (Dewhurst, 2002; Lemmers, 2003).

1.4.2.2 Cultivation of Lentinula edodes *Lentinula edodes* (xiang gu in Chinese and shiitake in Japanese) was the second most important cultivated edible mushroom, but since 2002 it has become the world number one cultivated mushroom. It can be cultivated either on wood log or on synthetic substrate logs (Quimio et al., 1990; Stamets, 2000; Chang and Miles, 2004).

1. *Biological Nature* *Lentinula edodes* is a heterothallic mushroom. Its sexuality is controlled by two mating factors, A and B, with multiple alleles, and therefore, its life history is a tetrapolar or bifactorial mating system (Chang and Miles, 1984).

Its life cycle starts the germination of basidiospores. After selected mating between two compatibility germinative mycelium, the dikaryon mycelium or fruiting culture is established. From the fruiting culture, the stock culture, mother spawn, and commercially planting spawn can be made. When the spawn is planted on a suitable substrate, under good climatic conditions the fruiting bodies of the mushroom are developed. Then when the mature stage is reached, the spores are discharged and its life cycle is completed.

Lentinula edodes is kind of wood rot fungus. In nature, it grows on dead tree trunks or stumps. In general, the wood for the mushroom growth consists of crude protein 0.38%, fat 4.5%, soluble sugar 0.56%, total nitrogen 0.148%, cellulose 52.7%, lignin 18.09%, and ash 0.56%. Generally speaking, the carbon–nitrogen ratio in substrate should be in the range of 25 : 1–40 : 1 in the vegetative growth stage and from 40 : 1 to 73 : 1 in the reproductive stage. If the nitrogen source is too rich in the reproductive phase, fruiting bodies of the mushroom are usually not formed and developed.

The optimum temperature of spore germination is 22–26°C. The temperature for mycelial growth ranges from 5 to 35°C, but the optimum temperature is 23–25°C. Generally speaking, *L. edodes* belongs to low-temperature mushrooms; the initial and development temperature of fruiting body formation is in the range of 10–20°C and the optimum temperature of fructification for most varieties of the mushroom is about 15°C. Some varieties can fruit in higher temperatures (e.g., 20–23°C). These high-temperature mushrooms usually grow faster and have a bigger and thinner cap (pileus) and a thin and long stalk (stipe). Their fruiting bodies are easily opened and become flat-grade mushrooms, which are considered to be low quality. The optimum pH of the substrate used in making the mushroom bag/log is about 5.0–5.5.

2. *Culture Media and Preparation* The mushroom can grow on a variety of culture media and on different agar formulations, both natural and synthetic, depending on the purpose of the cultivation. Synthetic media are often expensive and time consuming in preparation; hence they are not commonly used for routine purposes.

The potato dextrose agar, or PDA, is the simplest and the most popular medium for growing the mycelium of the mushroom. It is prepared as follows:

(a) *Ingredients*: Diced potato, 200 g; dextrose (or ordinary white cane sugar), 20 g; powdered agar (or agar bars), 20 g; and distilled water (or tap water), 1 L.

(b) *Procedure*: Peeled potatos are washed, weighed, and cut into cubes. They are boiled in a casserole with at least 1 L of water until they become soft (around 15 minutes). The potatos are removed and water is added to the broth to make exactly 1 L. The broth is returned to the casserole, and dextrose and the agar are added. The solution is heated and stirred occasionally until the agar is melted. The hot solution is then poured into clean flat bottles. For pure or stock cultures, the test tubes are filled with at least 10 mL of liquid agar solution. The bottles or test tubes are plugged with cottonwool. When petri dishes are available, these can be used to produce mycelial plugs for inoculation of mother spawn.

Examples of the different formulas for spawn substrates are described below. Mother grain spawn: (i) Wheat/rye grain + 1.5% gypsum or slaked lime. (ii) Cotton seed hull 40%, sawdust 38%, wheat bran 20%, sugar 1%, and gypsum 1%. (iii) Sugar cane bagasse 40%, sawdust 38%, wheat bran 20%, sugar 1%, and gypsum 1%. Planting spawn: A number of materials, mostly agricultural and forest wastes, can be used to prepare mushroom planting spawn. Three of them are given here as examples: sawdust 78%, rice/wheat bran 16%, sugar 1.5%, corn flour 1.7%, ammonium sulfate 0.3%, calcium superphosphate 0.5%, and gypsum 2%; sawdust 64%, wheat bran 15%, spent coffee grounds 20%, and gypsum/lime 1%; and sawdust 78%, sucrose 1%, wheat bran 20%, and calcium carbonate 1%.

The *L. edodes* mushroom is produced on both a cottage and a commercial scale. Some issues associated with the different cultivation styles are summarized below:

1. *Cottage-Scale Cultivation* There are many formulas for the composition of the substrate. The ingredients can be variable from place to place and country to country depending upon the raw materials available and local climatic conditions. In general, after mixing the dry ingredients by hand or with a mechanical mixer, water is added to the mixture so that the final moisture content of the substrate is between 55 and 60%, depending on the capacity of the sawdust to absorb water. The ingredients are then packed into autoclavable polypropylene or high-density polyethylene bags. Although they are more expensive, polypropylene bags are the most popular since polypropylene provides greater clarity than polyethylene. After the bags have been filled (1.5–4 kg wet weight) with the substrate, the end of the bag can be closed either by strings or plugged with a cottonwool stopper. Four formulas in the preparation of the substrate for the cultivation of the mushroom are given here as reference. (i) Sawdust 82%, wheat bran 16%, gypsum 1.4%, potassium phosphate, dibasic 0.2%, and lime 0.4%. (ii) Sawdust 54%, spent coffee grounds 30%, wheat bran 15%, and gypsum 1%. (iii) Sawdust 63%, corncob powder 20%, wheat bran 15%, calcium superphosphate 1%, and gypsum 1%. (iv) Sawdust 76%, wheat bran 18%, corn powder 2%, gypsum 2%, sugar 1.2%, calcium superphosphate 0.5%, and urea 0.3%.
2. *Commercial-Scale Cultivation* In general, the operation can use oak or other hardwood sawdust medium to grow the mushroom. The basic steps are (i) mix the sawdust, supplements, and water; (ii) bag the mixture; (iii) autoclave the bags to 121°C and cool the bags; (iv) inoculate and seal the bags; (v) incubate for 90 days to achieve full colonization of the sawdust mixture, in other words, to allow the mycelium to be established for ready fructification; (vi) fruit the colonized and established sawdust logs/bags/blocks 6 times using a 21-day cycle at 16–18°C; and (vii) harvest, clip steps, grade, box, and cold store for fresh market or harvest, dry, cut steps, grade, and dry again before boxing for dry market.

The major equipment used in production consists of a mixer/conveyor, autoclave, gas boiler, cooling tunnel, laminar flow cabinet, bag sealer, air compressor for humidification, and shelves to incubate.

Incubation can be done in two rooms and in two shipping containers. The two shipping containers can be installed near the fruiting rooms. The temperature during incubation is held between 18 and 25°C.

Fruiting can be done in six rooms so that the blocks/logs can be moved as a unit. With compartmentalization, blocks in each room can be subjected to a cycle of humid cold, humid heat, and dry heat.

1.4.2.3 Cultivation of Pleurotus sajor-caju *Pleurotus sajor-caju* (grey oyster mushroom) is comparable to the high-temperature species in the group of *Pleurotus* (oyster) mushrooms, with high temperatures required for fructification. This mushroom has a promising prospect in tropical/subtropical areas. Its cultivation is easy with relatively less complicated procedures (Chang and Miles, 2004; Kaul and Dhar, 2007):

1. *Biological Nature* The temperature for growth of mycelium is 10–35°C. The optimum growing temperature of the mycelium is 23–28°C. The optimum developmental temperature of the fruiting body is 18–24°C. The optimum pH of the substrate used in making the mushroom bag/bed is 6.8–8.0. The C/N ratio in the substrate is in the range of 30 : 1–60 : 1. A large circulation of air and reasonable light are required for the development of the fruiting bodies.
2. *Spawn Substrate* (i) Wheat grain + 1.5% gypsum or lime. (ii) Cotton seed hull 88%, wheat bran 10%, sugar 1%, and gypsum 1%. (iii) Sawdust 78%, wheat bran 20%, sugar 1%, and gypsum 1%. (vi) Sawdust 58%, spent coffee grounds/spent tea leaves 20%, water hyacinth/cereal straw 20%, sugar 1%, and gypsum 1%.
3. *Cultivation Substrate* (i) Cotton seed hull 95%, gypsum 2%, lime 1%, and calcium superphosphate 2%. (ii) Rice straw 80%, cotton waste 18%, gypsum 1%, and lime 1%. (iii) Water hyacinth 80%, cereal straw 17%, gypsum 2%, and lime 1%.

For demonstration purpose, this mushroom can be nurtured to grow into a tree-like shape (Chang and Li, 1982). The cultivation method, which has been tested to be successful, is as follows: Cotton waste or rice straw mixed with water hyacinth is used as the substrate. Tear large pieces of cotton waste into small parts or cut the straw and water hyacinth into small segments. Add 2% (w/w) lime and mix with sufficient water to get moisture content of about 60–65%. Pile the materials up, cover with plastic sheets, and leave to stand overnight. Load the substrate into small baskets or on shelves for pasteurization or cook the substrate with boiled water for 15 minutes. After cooling to approximately 25°C, mix around 2% (w/w) spawn thoroughly with the substrate and pack into columns of 60-cm-long tubes which have hard plastic [polyvinyl chloride (PVC)] tubing of 100 cm (4 cm in diameter) as central support and with plastic sheets as outside wrapping.

Incubate these columns at around 24–28°C, preferably in the dark. When the mycelium of the mushroom has ramified the entire column of substrate after three to four weeks, remove the plastic wrapping and switch on white light. Watering occasionally is needed to keep the surface from drying. In around three to four days white primordia start to appear over the whole surface. After another two to three days, the *Pleurotus* mushrooms are ready for harvesting. During the cropping period watering is very important if many flushes are required.

1.4.2.4 Cultivation of Volvariella The edible straw mushroom *Volvariella volvacea* is a fungus of the tropics and subtropics and has been traditionally cultivated in rice straw for many yeas in China and South East Asian countries. In 1971, cotton wastes were first introduced as heating material for growing the straw mushroom (Yau and Chang, 1972), and in 1973, cotton wastes had completely replaced the traditional paddy straw to grow the mushroom (Chang, 1974). This was a turning point in the history of straw mushroom cultivation because the cotton waste compost through pasteurization brought the cultivation of the mushroom into an industrial scale first in Hong Kong and then in Taiwan, Thailand, and elsewhere in China. Several techniques are adopted for the cultivation of the mushroom, which thrives in the temperature range of 28–36°C and a relative humidity of 75–85%. Detailed descriptions of the various methods are given by Chang and Quimio (1982), Chang and Miles (2004), Kaul and Dhar (2007), and Quimio et al. (1990). Choice of technologies usually depends on personal preference and the availability of substrates and resources. While the more sophisticated indoor technology is recommended for the industrial-scale production of the mushroom, most of the other technologies are low cost and appropriate for rural area development, especially when production is established at the community level.

1.4.2.5 Cultivation of Agaricus brasiliensis In recent years, *A. brasiliensis*, formerly called *Agaricus blazei* Murill (Wasser et al., 2002), has rapidly become a popular mushroom. It has been proved to be not only a good-tasting and highly nutritious mushroom but also an effective medicinal mushroom, particularly for antitumour active polysaccharides.

Agaricus brasiliensis was a wild mushroom in southeastern Brazil, where it was consumed by the people as a part of their diet. The culture of the mushroom was brought to Japan in 1965 and an attempt to cultivate this mushroom commercially was made in 1978. In 1992, this mushroom was introduced to China for commercial cultivation (Chang and Miles, 2004).

1. *Biological Nature* *Agaricus brasiliensis* belongs to middle-temperature mushrooms. The growth temperature for mycelium ranges from 15 to 35°C and the optimum growth temperature range is 23–27°C. The temperature for fruiting can be from 16 to 30°C and the optimum developmental temperature of fruiting bodies is 18–25°C. The ideal humidity for casing soil is 60–65%. The air humidity in a mushroom house is preferably 60–75% for mycelium growth and 70–85% for fruiting body formation and development. The optimum pH of the compost used in making the mushroom bed is 6.5–6.8. The optimum pH of the casing soil is 7.0. A good circulation of air is required for the development of the fruiting bodies. These conditions are similar to those needed for the cultivation of *A. bisporus*. Under natural conditions, the mushroom can be cultivated for two crops each year. Each crop can harvest three flushes. According to the local climates, the farmer can decide the spawning time in the year in order to have mushrooms for harvest within 50 days after spawning.

2. *Preparation of Mushroom Bed* (Stamets, 2000) *Agaricus brasiliensis* is a kind of mushroom belonging to straw-dung fungi and prefers to grow on substrate rich in cellulose. The waste/by-product agro-industrial materials [e.g., rice straw, wheat straw, bagasse (squeezed residue of sugar cane), cotton seed hull,

corn stalks, sorghum stover, and even wild grasses] can be used as the principal component of the compost for cultivation of the mushroom. It should be noted that these materials have to be air dried first and then mixed with cattle dung, poultry manure, and some chemical fertilizers. The following formulas for making compost are for reference only. They should be modified according to the local available materials and climatic conditions. (i) Rice straw 70%, air-dry cattle dung 15%, cottonseed hull 12.5%, gypsum 1%, calcium superphosphate 1%, and urea 0.5%. (ii) Corn stalks 36%, cottonseed hull 36%, wheat straw 11.5%, dry chicken manure 15%, calcium carbonate 1%, and ammonium sulfate or urea 0.5%. (iii) Rice straw 90.6%, rice bran 2.4%, fowl droppings 3.6%, slaked lime 1.9%, superphosphate 1.2%, and ammonium sulfate/urea 0.3%. (iv) Bagasse 75%, cottonseed hull 13%, fowl droppings 10%, superphosphate 0.5%, and slaked lime 1.5%.

1.4.2.6 Cultivation of Ganoderma lucidum Although the medicinal value of *G. lucidum* has been treasured in China for more than 2000 years, the mushroom was found infrequently in nature. This lack of availability was largely responsible for the mushroom being so highly cherished and expensive. During ancient times in China, any person who picked the mushroom from the natural environment and presented it to a high-ranking official was usually well rewarded (Chang and Miles, 2004).

Artificial cultivation of this valuably mushroom was successfully achieved in the early 1970s and, since 1980 and particularly in China, production of *G. lucidum* has developed rapidly. Currently, the methods most widely adopted for commercial production are the wood log, short wood segment, tree stump, sawdust bag, and bottle procedures (Hsu, 1994; Mizuno et al., 1996; Hung, 1996; Mayzumi et al., 1997; Chang and Buswell, 1999; Stamets, 2000).

Log cultivation methods include the use of natural logs and tree stumps which are inoculated with spawn directly under natural conditions. The third alternative technique involves the use of sterilized short logs about 12 cm in diameter and approximately 15 cm long which allow for good mycelial running. This method provides for a short growing cycle, higher biological efficiency, good-quality fruiting bodies, and, consequently, superior economical benefit. However, this production procedure is more complex and the production costs much higher than the natural log and tree stump methods. For this production procedure, the wood logs should be prepared from broad-leaf trees, preferably from oak. Felling of the trees is usually carried out during the dormant period, which is after defoliation in autumn and prior to the emergence of new leaves the following spring. The optimum moisture content of the log is about 45–55%. The flowchart for the short-log cultivation method is as follows: selection and felling of the tree, sawing/cutting the log into short segments, transfering segments to plastic bags, sterilization, inoculation, spawn running, burial of the log in soil, tending the fruiting bodies during development from the pinhead stage to maturity, harvesting the fruiting bodies, drying the fruiting bodies by electrical driers, and packaging. It should be noted that the prepared logs/segments are usually buried in soil inside a greenhouse or plastic shed. The soil should allow optimum conditions of drainage, air permeability, and water retention, but excessive humidity should be avoided.

Examples of cultivation substrates using plastic bags or bottles as containers include the following (note that these examples are for reference purposes only and can be modified according to the strains selected and the materials available in different localities): (i) sawdust 78%, wheat bran 20%, gypsum 1%, and soybean powder 1%; (ii) bagasse 75%, wheat bran 22%, cane sugar 1%, gypsum 1%, and soybean powder 1%; (iii) cotton seed hull 88%, wheat bran 10%, cane sugar 1%, and gypsum 1%; (iv) sawdust 70%, corn cob powder 14%, wheat bran 14%, gypsum 1%, and cereal straw ash 1%; (v) corn cob powder 78%, wheat/rice bran 20%, gypsum 1%, and straw ash 1%. After sterilization, the plastic bags can be laid horizontally on beds or the ground for fruiting.

1.4.3 Utilization of Mushroom Germplasm

The item of mushroom germplasm is a selective subject only and is not an exhaustive approach to mushroom utilization. It is concerned with the broadening of available mushroom resources in nature in order to conduct successful domesticating and breeding programs, with the aim at developing the cultivation of wild mushrooms and improving all desirable mushroom traits. It cannot be overemphasized that, to fully exploit the opportunities offered by mushrooms which have been properly collected and characterized, it is necessary to ensure a continuous exchange of information between scientists from different disciplines engaged in different areas of mushroom research. Included among the many possible examples that could be quoted are the analytical chemists who analyze the many existing and potentially new growth substrates used for mushroom cultivation in order to certify their suitability from an alimentary standpoint; biochemists who study the fungal enzymes involved in the degradation of the individual components constituting the different substrates; fungal physiologists who focus their attention on the mechanisms underlying carpophore formation; geneticists who are able to comprehend the life cycles of different mushrooms as well as to undertake breeding programs and select strains with desirable characteristics; and growers who transfer to the field scale the knowledge and techniques obtained in the laboratory.

One of the basic requirements for breeding better quality mushrooms in higher yields is the wider availability of a large reserve of phenotypic variation (traits) which can be used for selection purposes by both researchers and the mushroom industry. Since all these phenotypic differences are ultimately under genetic control, mushroom strains with different traits actually possess distinctive gene combinations which can be generated artificially by conventional crossing methods, by protoplast fusion technology, and by transformation with genes cloned using recombinant deoxyribonucleic acid (rDNA) technology. Since the mushrooms themselves are the only source of this genetic material, the genes contained in existing mushroom strains and species represent the total genetic resource, that is, the entire pool of mushroom germplasm. Extinction of a single strain or species would mean the potential loss of many thousands of unique genes that could be used for breeding desirable new strains.

Mushroom germplasm can be preserved by in situ conservation and ex situ preservation. The maintenance of mushrooms in natural preserves as part of a strategy for protecting an ecosystem constitutes in situ conservation. Although this approach is clearly important, it will not be considered here. Mushroom germplasm can also be preserved ex situ as fungal spores or tissue in the form of a culture collection or gene bank. The collection and classification of information pertaining to the morphological, physiological, biochemical, and genetic characteristics of individual mushroom strains can be stored in computer databases called *germplasm databases*. Such databases would provide valuable and readily accessible information for future breeding programs and academic research (Chang et al., 1995). This emerging mushroom germplasm science will address aspects relating to the collection, identification, characterization, utilization and preservation of mushroom germplasm.

Mushrooms have the potential for multipurpose usages ranging from protein enrichment of the human diet, the selective delignification of lignocellulosic materials as part of the recycling process and reinsertion into the food chain and dietary supplements markets, and a contribution to environmental decontamination. The realization of this potential depends upon the availability of mushrooms which possess the characteristics necessary to achieve specified objectives. The pool of available selective mushrooms from nature through the processes of collection, identification, and utilization should be as large as possible in order to ensure that the most appropriate choice of mushroom germplasm can be caught and utilized.

1.5 WORLD MUSHROOM PRODUCTION

The world market for the mushroom industry in 2001 was valued at over U.S.$40 billion (Chang, 2006a). The mushroom industry can be divided into three main categories: edible mushrooms, medicinal mushroom products, and wild mushrooms (Chang, 2006b). International bodies/forums have developed for each of these segments of the mushroom industry that have helped to bring them to the forefront of international attention: (1) International Society of Mushroom Science (ISMS) for edible mushrooms, (2) World Society for Mushroom Biology and Mushroom Products (WSMBMP) for mushroom biology and medicinal mushroom products, and (3) International Workshops on Edible Mycorrhizal Mushrooms for some wild mushrooms. The three international bodies/forums have done much to promote each of their respective fields, not the least of which is bringing scientists together for useful discussions, encouraging research and the dissemination of valuable information. The outlook for many of the known mushroom species is bright. Production of mushrooms worldwide has been steadily increasing, mainly due to contributions from developing countries, such as China, India, Poland, Hungary, and Vietnam. There are also increasing experimentally based evidence to support centuries of observations regarding the nutritional and medicinal benefits of mushrooms. The value of mushrooms has recently been promoted to tremendous levels with medicinal mushrooms trials conducted for human immunodeficiency virus/acquired immunodeficiency syndrome (HIV/AIDS) patients in Africa, generating encouraging results (Chang and Mshigeni, 2000). However, harvests of highly prized edible mycorrhizal mushrooms are continuously decreasing. This has triggered research into devising methods for improved cultivation of wild mushroom. It is hoped that there will be even more research into this area, so that larger quantities of wild mushrooms can be massively harvested through semicultivation methods. Technological development in the mushroom industry in general has seen increasing production capacities, innovations in cultivation technologies, improvements to final mushroom goods, capitalizing the nutritional and medicinal properties of mushrooms, and utilizing the natural qualities of mushrooms for environmental benefits. However, there is always the need to maintain current trends and to continue to seek out new opportunities. The challenge is to recognize opportunities such as increasing consumption capabilities with the increase in world population and to take advantage of this by promoting the consumption of more mushrooms.

Generally, cultivated mushrooms should play a greater role in the endeavor to increase food protein. This is especially true in developing countries, since growth substrates for mushrooms are basically agricultural and industrial discards that are inedible for humans (Chang and Miles, 1984). Biological (bioconversion) efficiency, that is, the yield of fresh weight mushrooms in proportion to the spawning compost in *Agaricus* or to the air-dried substrates in other noncomposting mushrooms, can reach 60–100% for *Agaricus* and 15–100% depending on the cultivation conditions for other species.

The statistics in Table 1.1 illustrate the dramatic increase in the production of farmed mushrooms during the period 1960–2002 (Chang 1999, 2006b; Delcaire, 1978).

TABLE 1.1 World Mushroom Production, 1960–2002

Year	World Production (× 1000 t)
1960	170
1965	301
1970	484
1975	922
1978	1,060
1981	1,257
1983	1,453
1986	2,182
1990	3,763
1994	4,909
1997	6,158
2002	12,250

Whereas in 1997, Asia contributed 74.4% of the total world mushroom tonnage, Europe 16.3%, and North America 7.0%, both Africa and Latin America's shares were less than 1%. This is largely due to lack of know-how, lack of understanding that mushrooms can play vital roles toward enhancing human health when used as dietary food supplements, lack of reliable sources of good-quality mushroom spawn for supporting the efforts of local mushroom growers, lack of venture capital to support mushroom farming entrepreneurs, and absence of systematic government support toward promoting mushroom farming as a valuable nontraditional new food and cash crop (comparable to coffee, tea, cotton, tobacco, etc.).

1.6 MUSHROOM BIOTECHNOLOGY

It has been pointed out that mushroom biotechnology is concerned with mushroom products and encompasses the principles of fermentation technology, mushroom biology/microbiology, and bioprocess. The products have a more generalized or tonic effect, which in some cases may act prophylactically by increasing resistance to disease in humans from the balancing of nutrients in the diet and the enhancing of the immune systems.

1.6.1 Nutritional and Medicinal Value of Mushrooms

The greatest difficulty in feeding humans is to supply a sufficient quantity of the body-building material protein. The other three nutritional categories are the source of energy (carbohydrates and fats); accessory food factors (vitamins); and inorganic compounds which are indispensable to good health. Of course, water, too, is essential.

The moisture content of fresh mushrooms varies within the range 70–95% depending upon the harvest time and environmental conditions, whereas it is about 10–13% in dried mushrooms. The protein content of cultivated species ranges from 1.75 to 5.9% of their fresh weight. It has been estimated that an average value of 3.5–4.0% would be more representative. This means that the protein content of edible mushrooms in general is about twice that of onion (1.4%) and cabbage (1.4%) and 4 and 12 times those of oranges (1.0%) and apples (0.3%), respectively. In comparison, the protein content of common meats is as follows: pork, 9–16%; beef, 12–20%; chicken, 18–20%; fish, 18–20%; and milk, 2.9–3.3%. On a dry-weight basis, mushrooms normally contain 19–35% protein, as compared to 7.3% in rice, 12.7% in wheat, 38.1% in soybean, and 9.4% in corn. Therefore, in terms of the amount of crude protein, mushrooms rank below animal meats but well above most other foods, including milk, which is an animal product. Furthermore, mushroom protein contains all the nine essential amino acids required by humans.

In addition to their good protein, mushrooms are a relatively good source of the following individual nutrients: fat, phosphorus, iron, and vitamins, including thiamine, riboflavin, ascorbic acid, ergosterol, and niacin. They are low in calories, carbohydrates, and calcium. It has also been reported that a total lipid content varying between 0.6 and 3.1% of the dry weight is found in the commonly cultivated mushrooms. At least 72% of the total fatty acids are found to be unsaturated in all the four tested mushrooms (Huang et al., 1985). It should be noted that unsaturated fatty acids are essential and significant for our diet and our health.

In recent years, there has been a trend toward discovering ways of treating mushrooms so as to give them added value. For example, Wermer and Beelman (2002) have reported on growing mushrooms enriched in selenium. By adding sodium selenite to compost over a range of 30–300 PPM, they found that the mushrooms increasingly absorbed selenium according to the amount in the compost, so that it is possible to grow mushrooms containing a desired concentration. Selenium is an essential micronutrient that has generated much recent interest in nutritional and medical research—and more recently within the food industry (Beelman and Royse, 2006). Selenium has numerous physiological functions but is best known as a necessary cofactor for the glutathione peroxidase enzyme system. This system is responsible for removing free radicals from the body, thus reducing oxidative damage.

The desirability of a food product does not necessarily bear any correlation to its nutritional value. Instead, its appearance, taste, and aroma sometimes can stimulate one's appetite (preference). In addition to nutritional value, mushrooms have some unique color, taste, aroma, and texture characteristics which attract their consumption by humans.

The second major attribute of mushrooms, their medicinal properties, has also been drawn to our attention for study, for example, for hypotensive and renal effects (Yip et al., 1987), immunomodulatory and antitumor activities of polysaccharide–protein complex (PSPC) from mycelial cultures (Liu et al., 1995, 1996; Wang et al., 1995a, 1996b, c), immunomodulatory and antitumor activities of lectins from edible mushrooms (Wang et al., 1995b, 1996a, 1997), isolation and characterization of a type I ribosome inactivation protein from *V. volvacea* (Yao et al., 1998), and medicinal effects of *G. lucidum* (Chang and Buswell, 1999; Chang and Miles, 2004). For more detailed coverage of this aspect and comprehensive lists of mushrooms used in dietary supplements and in medicines, the reader is referred to later chapters.

1.6.2 Nutriceuticals and Dietary Supplements

There has been a recent upsurge of interest in mushrooms not only as health vegetables (food) but also as a source of biological active compounds of medicinal value, including use as complementary medicine/dietary supplements for anticancer, antiviral, immunopotentiating, hypocholesterolemic, and hepatoprotective agents. These new compounds, termed *mushroom nutriceuticals* (Chang and Buswell, 1996), are extractable from either the fungal mycelium or fruiting body and represent an important component of the expanding mushroom biotechnology industry.

Of the 14,000–15,000 species of so-called mushrooms in the world, around 400 have known medicinal properties. However, it has been estimated that there are about 1800 species of mushrooms with the potential of medicinal properties. Both these mushrooms and their rootlike structure (called mycelium) produce several medicinal or nutriceutical (general immune-enhancing) compounds, central of which are the polysaccharides (high-molecular-weight strings of sugars), triterpenes, and immunomodulatory proteins. Although virtually all mushrooms and many foods have polysaccharides in their cell walls, certain mushroom species have been found to contain polysaccharides which are particularly effective in retarding the progress of various cancers and other diseases and in alleviating the side effects of chemotherapy and radiation treatment (through cell-level regenerative effects). There are now many studies in Asia, and particularly in China and Japan, documenting life-span increases of cancer patients undergoing conventional cancer treatment plus mushroom extract consumption or injection (Mizuno et al., 1995; Liu, 1999). At the same time, due to the enhancement of the immune systems, it can help people reduce the possibility of being infected by other diseases.

Between 80 and 85% of all medicinal mushroom products are derived from the fruiting bodies, which have been either commercially farmed or collected from the wild, for example, Lentinan, a high-molecular-weight $(1 \rightarrow 3)$-β-D-glucan from *L. edodes* and various products from *G. lucidum*. Only about 15% of all products are based on extracts from mycelia. Notable examples are PSK (trade name Krestin) of a polysaccharide peptide and PSP (polysaccharide-bound peptide) extracted from *Coriolus versicolor*. A smaller percent of mushroom products are obtained from culture filtrates, for example, schizophyllan, a high-molecular-weight $(1 \rightarrow 3),(1 \rightarrow 6)$-$\beta$-D-glucan prepared from *Schizophyllum commune* Fr., and PSPC (a protein-bound polysaccharide complex) from *Tricholoma lobayense* Hein. However, due to increased quality control and year-round production, mycelial products are the wave of the future.

The market value of medicinal mushrooms and their derivative dietary supplements worldwide was about U.S.$1.2 billion in 1991 and about U.S.$3.6 billion in 1994 (Chang, 1996). In 1999, it was estimated to be U.S.$6.0 billion. The market value of *Ganoderma*-based nutriceuticals alone in 1995 was estimated at U.S.$1628.4 million (Chang and Buswell, 1999). The corresponding monetary values were generated by another famous mushroom, *L. edodes*. Ninety-nine percent of all sales of medicinal mushrooms and their derivatives occurred in Asia and Europe with less than 0.1% in North America. The 1999 U.S. market for dietary supplements based mainly on mushrooms was estimated to be U.S.$35 million. However, in recent years, the North American demand is increasing between 20 and 40% annually, depending upon species.

1.7 DEVELOPMENT OF WORLD MUSHROOM INDUSTRY MOVEMENTS

Although mushrooms have been collected from the wild and cultivated artificially for human food and for medicine uses for hundreds and thousands of years, it is

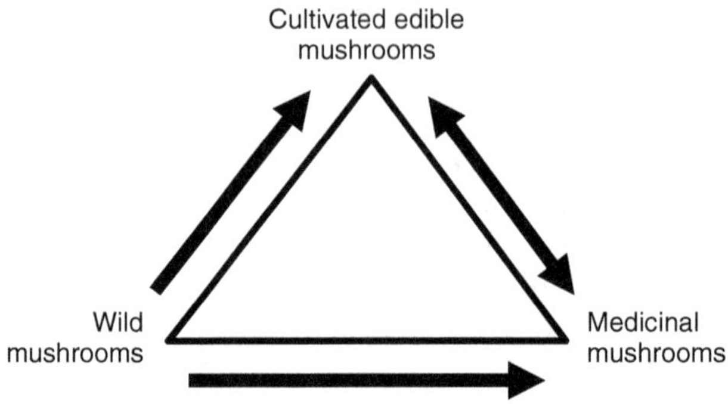

Figure 1.7 Mushroom industry can be considered to be composed of cultivated edible mushrooms, medicinal mushrooms, and wild mushrooms Single arrows indicate that number of both edible and medicinal mushroom species increases from time to time through identification and domestication of unknown and wild mushrooms. Double arrow indicates that most edible species also possess medicinal properties while many medicinal mushrooms can be artificially cultivated.

only recently that the three main segments of the mushroom industry could be identified. These three segments have received international recognition as important interrelated components (Figure 1.7), with each deserving its own special patronage and paths of development: (1) cultivated edible mushrooms (mushroom themselves used directly or indirectly as food), (2) medicinal mushrooms (mushroom derivatives used as nutriceutical therapy/dietary supplements), and (3) wild mushrooms, including edible mycorrhizal, symbiotic, and poisonous mushrooms (collected, up to now, only from the wild). The development of three important international bodies/forums has helped to bring each of these three components of the mushroom industry to the forefront of international attention, showcasing their positive contributions to human welfare (Chang, 2006b).

1.7.1 International Movement for Edible Mushrooms

The movement mainly concerned with mushroom production (mushrooms themselves) was initiated during the First International Conference on Mushroom Science held in Peterborough, the United Kingdom, May 3–11, 1950. Chairman F. C. Atkins with P. J. Bels, E. B. Lambert, and R. L. Edwards were on the organizing committee. The committee members later formed the International Commission on Mushroom Science, which eventually developed into the ISMS.

The Seventeenth International Congress of ISMS will be held May 21–24, 2008, in Cape Town, South Africa. Traditionally, the focus of the ISMS has been on the *A. bisporus* mushroom industry. In recent years, the interests of the ISMS have become more diversified, but *A. bisporus* is still its main concern.

1.7.2 International Movement for Medicinal Mushrooms

The movement mainly concerned with mushroom products (mushroom derivatives) was instituted during the First International Conference on Mushroom Biology and Mushroom Products held in Hong Kong, August 23–26, 1993. Chairman S. T. Chang with J. A. Buswell, V. E. C. Ooi, K. W. K. Liu, and S. W. Chiu were on the organizing committee.

The WSMBMP was launched in January 1994 in response to strong interest expressed at the conference in Hong Kong the previous year. The object of the WSMBMP is to promote the enhancement and application of knowledge related to the basic and applied aspects of mushroom biology and mushroom products (mushroom derivatives possessing medicinal properties from edible, medicinal, and wild mushrooms) through publications, meetings, and other means deemed appropriate. The WSMBMP holds the International Conference for Mushroom Biology and Mushroom Products (ICMBMP) every three years. The sixth one is to be held in Bonn, Germany, in 2008.

The international movement for the medicinal segment of the mushroom industry was given a further boost with the launch of the *International Journal of Medicinal Mushrooms* (IJMM) in 1999 by Solomon P. Wasser as editor-in-chief with Takashi Mizuno, Shu-Ting Chang, and Alexander L. Weis as editors. This then led to the inaugural International Medicinal Mushroom Conference (IMMC) held in Kiev, Ukraine, September 12–14, 2001. It has been agreed that there is an IMMC after an interval of two years. The second IMMC was held in Pattaya, Thailand, July 17–19, 2003, and the third in Port Townsend, Washington, the United States, October 12–17, 2005. IMMC 4 will be in Slovenia in 2007 and IMMC 5 in China in 2009.

1.7.3 International Movement for Wild Mushrooms

The movement, mainly concerned with edible mycorrhizal mushrooms, was born as a pre-Congress activity during the second International Conference on Mycorrhizas in Uppsala, Sweden, in 1999. Two years later, the second International Workshop on Edible Mycorrhizal Mushrooms (IW-EMM) was held in Christchurch, New Zealand, July 3–6, 2001. The third IW-EMM was hosted by the University of Victoria, Canada, August 16–22, 2003, and the fourth was held in Murcia, Spain, November 29–December 2, 2005. The fifth IW-EMM was held in Yunnan, China, in 2007. It should be noted that edible mycorrhizal mushrooms belong to a special group of wild mushrooms which include other symbiotic mushrooms, for example, termite, hallucinogenic, and poisonous mushrooms.

These three international bodies/forums have done much to promote each of their respective fields, not the least of which is bringing together scientists in international forums for useful discussions, encouraging research and the dissemination of valuable information. These three segments of the mushroom-based industry are not for competition but for complementation.

1.8 CONCLUDING REMARKS

As the population of the world is expected to continue increasing in the twenty-first century, so will the amount of food and the level of medical care required by each individual, especially those living in less developed countries. The level of environmental pollution will also become a serious problem. However, the world has an immense amount of lignocellulosic biomass resource which, like solar energy, is sustainable. Currently, the bulk of the lignocellulosic biomass is, to a large extent, considered insignificant or of no commercial value and certainly of no food value, at least in its original form. It should be noted that large amounts of research funds have been set aside to search for increased productivity of the core product, like the oil in the coconut, the cellulose in the tree, the fiber in the sisal, the coffee in the coffee berry, or the grain in the cereal crop. However, little research funding has been reserved for the search for the reuse of many by-products (wastes) from the core products, which are usually considered waste materials. When they are carelessly disposed to the surrounding environment by dumping or burning, these so-called wastes are bound to lead to environmental pollution and consequently to health hazards. It should be emphasized that these lignocellulosic wastes are actually a kind of new natural resource or new raw material. If they could be properly managed and utilized, then eventually economic growth would be promoted. In other words, the by-products in processing the core products can be used/treated as raw materials for the production of second- or third-core products. For example, cereal straw, coffee pulp, spent coffee ground, and sisal waste can be used to grow mushrooms. After harvesting mushrooms, some spent substrates can be used as feeding materials for animals or used for growing earthworms, and afterward, the residues can be used as soil conditioners or crop fertilizers. In the whole exercise, there is no waste produced. This is the concept of zero emissions or total productivity of raw materials (Pauli, 1996). Therefore, the significant impact of applied mushroom biology on human welfare in the twenty-first century could be considered globally as "nongreen revolution."

Since mushrooms, like all other fungi, lack chlorophyll and are nongreen organisms, they cannot convert solar energy through the process of photosynthesis to organic matter, as the green plants do. But they can produce a wide range of enzymes which can degrade lignocellulosic materials for their growth and for fruiting. This serves to demonstrate the magnificent capacities of the mushrooms for biosynthesis, which is different from photosynthesis affected by green plants. Mushrooms not only can become nutritious protein-rich food (through mushroom science) but also can provide nutriceutical and pharmaceutical products (through mushroom biotechnology). In addition, through mushroom bioremediation, the recycling of the by-products (wastes) in the course of each stage of mushroom production using mushroom mycelia can create a pollution-free environment. Therefore, mushrooms, with their great variety of species, can constitute a cost-effective means of supplementing the food nutrition of humankind through the production of edible mushrooms; alleviate the suffering caused by certain kinds of illnesses through medicinal mushrooms and their derivatives as nutriceuticals/dietary supplements (Chang and Buswell, 1996, 2003; Chang and Mshigeni, 2001) and mycomedicinals (Stamets and Yao, 1998); and reduce environmental pollution as well as heal the soils (Stamets, 2005) through mushroom mycelial activities.

The implementation of applied mushroom biology through a well-designed package of multidisciplinary technologies has already had an impact on human welfare at national and regional levels in the twentieth century. It is believed that by blending advances in basic biology with practical technology, mushroom-related industries can have a global and positive impact on long-term food nutrition, health benefits, environmental conservation and regeneration, and economic and social changes.

REFERENCES

Arora, D. (1986). *Mushrooms Demystified.* Berkeley, CA: Ten Speed, p. 959.

Beelman, R. B. and Royse, D. J. (2006). Selenium enrichment of *Pleurotus cornucopiae* (Paulet) Rolland and *Grifola frondosa* (Dicks.:Fr.) S. F. Gray mushrooms. *International Journal of Medicinal Mushrooms, 8*, 77–84.

Beyer, D. (2005). Spent mushroom substrate (SMS) research in the US. *American Medical Group Association Journal, Summer Issue*, 31–32.

Buswell, J. A., Cai, Y. J., and Chang, S. T. (1995). Effect of nutrient nitrogen and manganese on manganese peroxidase and laccase production by *Lentinula edodes. FEMS Microbiology Letters, 128*, 81–88.

Buswell, J. A., Cai, Y. J., and Chang, S. T. (1996a). Ligninolytic enzyme production and secretion in edible mushroom fungi. In *Mushroom Biology and Mushroom Production*, Royse, D. J., editor. University Park: Pan, pp. 113–122.

Buswell, J. A., Cai, Y. J., Chang, S. T., Peberdy, J. F., Fu, S. Y., and Yu, H. S. (1996b). Lignocellulolytic enzyme profiles of edible mushroom fungi. *World Journal of Microbiology and Biotechnology, 12*, 537–542.

Buswell, J. A. and Chang, S. T. (1994). Biomass and extracellular hydrolytic enzyme production by six mushroom species grown on soybean waste. *Biotechnology Letters, 16*, 1317–1322.

Cai, Y. J., Buswell, J. A., and Chang, S. T. (1994). Production of cellulases and hemicellulases by the straw mushroom, *Volvariella volvacea. Mycological Research, 98*, 1019–1024.

Cai, Y. J., Buswell, J. A., and Chang, S. T. (1998). β-Gludosidase components of cellulolytic system of the edible straw mushroom, *Vovlvariella volvacea. Enzyme and Microbial Technology, 22*, 12–129.

Cai, Y. J., Chapman, S. J., Buswell, J. A., and Chang, S. T. (1999). Production and distribution of endoglucanase, cellobiohydrolase, and β-glucosiase components of the cellulolytic system of *Volvariella volvacea*, the edible straw mushroom. *Applied and Environmental Microbiology, 65*, 553–559.

Carluccio, A. (2003). *Complete Mushroom Book.* London: Quadrille Publishing, p. 224.

Chang, S. T. (1974). Production of the straw mushroom (Volvariella volvacea) from cotton wastes. *Mushroom Journal, 21*, 348–353.

Chang, S. T. (1989). Recycling of solid wastes with emphasis on organic wastes. *Green Productivity, March*, 23–16.

Chang, S. T. (1993). Mushroom biology: The impact on mushroom production and mushroom products. In *Mushroom Biology and Mushroom Products.* Chang, S. T., Buswell, J. A., and Chiu, S. W., editors. Hong Kong: Chinese University Press, pp. 3–20.

Chang, S. T. (1996). Mushroom research and development—Equality and mutual benefit. In *Mushroom Biology and Mushroom Products.* Royse, D. J., editor. University Park: Pan, pp. 1–10.

Chang, S. T. (1998). Development of novel agroscience industries based on bioconversion technology. In *Frontiers in Biology: The Challenges of Biodiversity.* Chou, C. H. and Shao, K. T., editors. Taipei: Academia Sinica, pp. 217–222.

Chang, S. T. (1999). Global impact of edible and medicinal mushrooms on human welfare in the 21st century: Nongreen revolution. *International Journal of Medicinal Mushrooms, 1*, 1–7.

Chang, S. T. (2005). The socio-cultural heritage of mushroom science in Southeast Asian Island Communities. *International Journal of Island Affairs (INSYLA), 14*, 43–47.

Chang, S. T. (2006a). The need for scientific validation of culinary–medicinal mushroom products. *International Journal of Medicinal Mushrooms, 8*, 187–195.

Chang, S. T. (2006b). The world mushroom industry: Trends and technological development. *International Journal of Medicinal Mushrooms, 8*, 297–314.

Chang, S. T. and Buswell, J. A. (1996). Mushroom nutriceuticals. *World Journal of Microbiology and Biotechnology, 12*, 473–476

Chang, S. T. and Buswell, J. A. (1999). *Ganoderma lucidum* (Curt.:Fr.) P. Karst. (Aphyllophoromycetidease): A mushrooming medicinal mushroom. *International Journal of Medicinal Mushrooms, 1*, 139–146.

Chang, S. T. and Buswell, J. A. (2003). Medicinal mushrooms—A prominent source of nutriceuticals for the 21st century. *Current Topics in Nutraceutical Research, 1*, 257–280.

Chang, S. T. and Chiu, S. W. (1992). Mushroom production—An economic measure in maintenance of food security. In *Biotechnology: Economic and Social Aspects.* DaSilva, E. J., Ratledge, C., and Sasson, A., editors. New York: Cambridge University Press, pp. 110–141.

Chang, S. T. and Hayes, W. A., editors (1978). *The Biology and Cultivation of Edible Mushrooms.* New York: Academic, p. 819.

Chang, S. T., Kwan, H. S., and Kang, Y. H. (1995). Collection, characterization, and utilization of germ plasm of *Lentinula edodes. Canadian Journal of Botany, 73*, S955–S961.

Chang, S. T. and Li, S. F. (1982). Mushroom culture. In *Advances in Agricultural Microbiology.* Subba Rao, N. S., editors. New Delhi: Oxford and IBH Publishing, pp. 677–691.

Chang, S. T. and Mao, X. L. (1995). *Hong Kong Mushrooms.* Hong Kong: Chinese University Press, p. 470.

Chang, S. T. and Miles, P. G. (1982). Introduction to mushroom science. In *Tropical Mushrooms—Biological Nature and Cultivation Methods.* Chang, S. T. and Quimio, T. H., editors. Hong Kong: Chinese University Press, pp. 3–10.

Chang, S. T. and Miles, P. G. (1984). A new look at cultivated mushrooms. *BioScience, 34*, 358–362.

Chang, S. T. and Miles, P. G. (1989). *Edible Mushrooms and Their Cultivation.* Boca Raton, FL: CRC Press, p. 345.

Chang, S. T. and Miles, P. G. (1992). Mushroom biology—a new discipline. *The Mycologist, 6*, 64–65.

Chang, S. T. and Miles, P. G. (2004). *Mushrooms: Cultivation, Nutritional Value, Medicinal Effect, and Environmental Impact*, 2nd ed. Boca Raton, FL: CRC Press, p. 451.

Chang, S. T. and Mshignei, K. E. (2000). *Ganoderma lucidum*—Paramount among medicinal mushrooms. *Discovery and Innovation, 12*, 97–101.

Chang, S. T. and Mshigeni, K. E. (2001). *Mushrooms and Human Health: With Special Reference to Their Growing Significance as Potent Dietary Supplements.* Windhoek: University of Namibia, p.79.

Chang, S. T. and Quimio, T. H., editors (1982). *Tropical Mushrooms—Biological Nature and Cultivation Methods.* Hong Kong: Chinese University Press, p. 493.

Delcaire, J. R. (1978). Economics of cultivated mushrooms. In *The Biology and Cultivation of Edible Mushrooms.* Chang, S. T. and Hayes, H. A., editors. New York: Academic, pp. 726–793.

Dewhurst, M. (2002). Phase III—The future? *Mushroom Journal, 21*, 17–18.

Fuhrer, B. (2005). *A Field Guide to Australian Fungi.* Melbourne: Bloomings Books, p. 360.

Hall, I. R., Stephenson, S. L., Buchanan, P. K., Wang, Y., and Cole, A. L. J. (2003a). *Edible and Poisonous Mushrooms of the World.* Cambridge: Timber Press, p. 371.

Hall, I. R., Wang, Y. and Amicucci, A. (2003b). Cultivation of edible ectomycorrhizal mushrooms. *Trends in Biotechnology, 21*, 433–438.

Hawksworth, D. L. (2001). The magnitude of fungal diversity: The 1.5 million species estimate revisited. *Mycological Research, 105*, 1422–1432.

Hsu, Z. C. (1994). *New Technology for Cultivation of Ganoderm lucidum* [in Chinese]. Liaoning: Chaoyang Edible Fungi Research Institute, p. 54.

Huang, B. H., Yung, K. H., and Chang, S. T. (1985). The sterol composition of *Volvariella volvacea* and other edible mushrooms. *Mycologia, 77*, 959–963.

Hung, Z. (1996). Artificial cutivation of *Ganoderma lucidum* [in Chinese]. In *Modern Research on Ganoderma lucidum* [in Chinese]. Lin, Z. B., editor. Beijing: Beijing Medical University Press, pp. 61–87.

Kaul, T. N. and Dhar, B. L. (2007). *Biology and Cultivation of Edible Mushrooms.* New Delhi: Westville Publishing House, p. 240.

Lemmers, G. (2003). The merits of bulk Phase III. *Mushroom Journal, 642*, 17–22.

Liu, F., Ooi, V. E. C., and Chang, S. T. (1995). Antitumour components of the culture filtrates from Tricholoma sp. *World Journal of Microbiology and Biotechnology, 11*, 486–490.

Liu, F., Ooi, V. E. C., Liu, W. K., and Chang, S. T. (1996). Immunomodulatory and antitumour activities of polysaccharide-protein complex from the culture filtrates of a local edible mushroom, *Tricholoma lobayense. General Pharmacology, 27*, 621–624.

Liu, G. T. (1999). Recent advances in research of pharmacology and clinical applications of *Ganoderma* P. K Karst. species (Aphyllophoromycetideae) in China. *International Journal of Medicinal Mushrooms, 1*, 63–67.

Mau, J. L., Chang, C. N., Huang, S. J., and Chen, C. C. (2004). Antioxidant properties of methanolic extracts from *Grifola frondosa, Morchella esculenta* and *Termitomyces albuminosus* mycelia. *Food Chemistry, 87,* 111–118.

Mayzumi, F., Okamoto, H., and Mizuno, T. (1997). Cultivation of Reishi (*Ganoderma lucidum*). *Food Reviews International, 13*, 365–382.

Miles, P. G. and Chang, S. T. (1988). "Mycomeat"—A food produced from soybean slurry by fungal mycelium. In *Recent Advances in Biotechnology and Applied Biology.* Chang, S. T., Chan, K. Y, and Woo, N. Y. S., editors. Hong Kong: Chinese University Press, pp. 577–586.

Miles, P. G. and Chang, S. T. (1997). *Mushroom Biology: Concise Basics and Current Developments.* Singapore: World Scientific, p. 194.

Mizuno, T., Naoi, Y., Mayuzumi, F., Ogino, M., and Okamoto, H. (1996). Artificial cultivation of *Ganoderma lucidum* in Japan. In *Ganoderma lucidum.* Mizuno, T., editor. Seoul: IL-Yang Pharmaceutical, p. 298.

Mizuno, T., Saito, H., Nishitoba, T., and Kawagishi, H. (1995). Antitumor-active substances from mushrooms. *Food Reviews International, 11,* 23–61.

Noble, R. (2005). Spent mushroom substrate—An alternative use. *American Medical Group Association Journal, Summer*, 33–35.

Pauli, G. (1996). *Breakthroughs: What Business Can Offer Society.* Surrey: Epsilon, p. 243.

Pegler, D. N. (2002). Useful fungi of the world: The "Poor man's truffles of Arabia" and "Manna of the Israelites". *Myocologist, 16*, 8–9.

Quimio, T. H., Chang, S. T., and Royse, D. J. (1990). Technical guidelines for mushroom growing in the tropics. In *FAO Plant Production and Protection Paper 106.* Rome: Food and Agriculture Organization, p.155.

Shepherd, C. J. and Totterdell, C. J. (1988). *Mushrooms and Toadstools of Australia.* Sydney: Inkata, p.162.

Stamets, P. (2000). *Growing Gourmet and Medicinal Mushrooms.* Berkeley, CA: Ten Speed, p. 339.

Stamets, P. (2005). *Mycelium Running: How Mushroom Can Help Save the World.* Berkeley, CA: Ten Speed, p. 574.

Stamets, P. and Yao, C. C. W. (1998). *Mycomedicinals: An Informational Booklet on Medicinal Mushrooms.* Olympia: MycoMedia, p. 46.

Van der Westhuizen, G. C. A. and Eicker, A. (1994). *Field Guide—Mushrooms of Southern Africa.* Cape Town: Struik Publishers, p. 192.

Van Griensven, L. J. L. D., editor (1988). *The Cultivation of Mushrooms.* Sussex: Darlington Mushroom Lab.

Wang, H. X., Liu, W. K., Ng, T. B., Ooi, V. E. C., and Chang, S. T. (1995a). Immunomodulatory and antitumour activities of a polysaccharide-protein complex from a mycelial culture of *Tricholoma sp.* A local edible mushroom. *Life Science, 57*, 269–281.

Wang, H. X., Ng, T. B., Liu, W. K., Ooi, V. E. C., and Chang, S. T. (1995b). Isolation and characterization of two distinct lectins with anti-proliferative activity from the cultured mycelium of the edible mushroom *Tricholoma mongolicum. International Journal of Peptide and Protein Research, 46*, 508–513.

Wang, H. X., Liu, W. K., Ng., T. B., Ooi, V. E. C., and Chang, S. T. (1996a). The immunomodulatory & antitumour activities of lectins from the mushroom *Tricholoma mongolicum. Immunopharmacology, 31*, 205–211.

Wang, H. X., Ng, T. B., Liu, W. K., Ooi, V. E. C., and Chang, S. T. (1996b). Polysaccharide-peptide complexes from the cultured mycelia of the mushroom *Coriolus versicolor* and their culture medium activate mouse lymphocytes and macrophages. *International Journal of Biochemistry and Cell Biology, 28*, 601–607.

Wang, H. X., Ng, T. B., Liu, W. K., Ooi, V. E. C., and Chang, S. T. (1996c). A polysaccharide-peptide complex from cultured mycelia of the mushroom *Tricholoma mongolicum* with immunoenhancing and antitumour activities. *International Journal of Biochemistry and Cell Biology, 74*, 95–100.

Wang, H. X., Ng, T. B., Ooi, V. E. C., Liu, W. K., and Chang, S. T. (1997). Actions of lectins from the mushroom *Tricholoma mongolicum* on macrophages, splenocytes and life-span in sarcoma-bearing mice. *Anticancer Research, 17*, 416–420.

Wasser, S. P., Didukh, M. Y., Amazonas, M. A., Nevo, E., Stamets, P., and da Éira, A. F. (2002). Is a widely cultivated culinary-medicinal Royal Sun *Agaricus* (the Himematsutake mushroom) indeed *Agaricus blazei* Murrill? *International Journal of Medicinal Mushrooms, 4*, 267–290.

Wasser, S. P. and Weis, A.L. (1999). Medicinal properties of substances occurring in higher basidiomycetes mushrooms: Current perspectives (review). *International Journal of Medicinal Mushrooms, 1*, 31–62.

Wasson, V. P. and Wasson, R. G. (1957). *Mushrooms, Russia and History*. New York: Pantheon Books.

Wermer, A. R. and Beelman, R. B. (2002). Growing high-selenium edible and medicinal button mushrooms (*Agaricus bisporus* (J. Lge) Imbach) as ingredients for functional foods or dietary supplements. *International Journal of Medicinal Mushrooms, 4*, 167–171.

Yao, Q. Z., Yu, M. M., Ooi, L. S. M., Ng, T. B., Chang, S. T., Sun, S. S. M., and Ooi, V. E. C. (1998). Isolation and characterization of a Type I Ribosome-inactivation protein from fruiting bodies of the edible mushroom (*Volvariella volvacea*). *Journal of Agricultural and Food Chemistry, 46*, 788–792.

Yau, C. K. and Chang, S. T. (1972). Cotton waste for indoor cultivation of straw mushroom. *World Crops, 24*, 302–303.

Yip, K. P., Fung, K. P., Chang, S. T., and Tam, S. C. (1987). Purification and mechanism of the hypotensive action of an extract from edible mushroom *Pleurotus sajor-caju*. *Neuroscience Letters Supplement, 28*, S59.

CHAPTER 2

Molecular Analysis and Genomic Studies of Shiitake Mushroom *Lentinula edodes*

Hoi-Shan Kwan and Winnie W. Y. Chum

Department of Biology, The Chinese University of Hong Kong, Hong Kong, China

CONTENTS

2.1 INTRODUCTION

Lentinula edodes (Berk.) Pegler (*L. edodes*) is named xianggu in Chinese and shiitake in Japanese. It is classified in the genus *Lentinula*, the family Tricholomataceae, the order *Agaricales*, and the subphylum Homobasidiomycetes of the phylum Basidiomycota. *Lentinula edodes* is largely cultivated in China, Japan, and other Asian countries and is one of the most popular edible mushrooms in the world because of its taste and nutritional value. It also contains components, such as lentinan, that are well-known for their medicinal utility. Due to its high economic value, many researchers have carried out studies on the strain improvement of *L. edodes*. To achieve this improvement, we have to ascertain

Mushrooms as Functional Foods, Edited by Peter C. K. Cheung

its biological characteristics and understand the molecular mechanisms involved in the growth and development of *L. edodes* at the molecular level. Many molecular methods have been used to type strains of *L. edodes*, to isolate developmentally regulated genes or genes that are expressed in different physiological processes in *L. edodes*, and to carry out more advanced transcriptome analysis. For genetic approaches, the methods have included gene cloning, differential displays with ribonucleic acid (RNA) fingerprinting by arbitrarily primed polymerase chain reaction (RAP-PCR), expressed sequence tags (ESTs), complementary deoxyribonucleic acid (cDNA) representational difference analysis (cDNA-RDA), serial analysis of gene expression (SAGE), cDNA microarray, and the sequencing-by-synthesis approach (454 Life Science). Using these methods, many genes have been reported to participate in the fruiting body development of *L. edodes*. The genes that are differentially expressed at initiation of the fruiting body can be categorized into (1) the initiation—stress response and specific signal transduction; (2) the reconstruction of proteome—protein degradation, modification, and biosynthesis; and (3) the switching of biochemical pathways and structural components. In this chapter, we discuss the findings of these molecular studies, which aim to investigate the growth and developmental processes of *L. edodes*.

2.2 ISOLATION OF GENES

2.2.1 Growth

2.2.1.1 Substrate-Utilizing Genes *Lentinula edodes* generates many extracellular enzymes to degrade extracellular substrates for energy production. It utilizes the cellulose and hemicellulose in wood as its major carbon sources. The enzymes that degrade cellulose and its partial degradation products, *exo*-(1 → 4)-β-D-glucansase (*exo*-cellobiohydrolase) and *endo*-(1 → 4)-β-D-glucansase (*endo*-cellobiohydrolase), have been isolated, as have the enzymes that degrade starch, hemicelluloses, and other water-soluble polysaccharides, (gluco)amylase, hemicellulase, α-L-arabinosidase, β-D-xylosidase, β-D-galactosidase, β-D-mannosidase, and polygalacturonase-pectinase. To degrade the carbohydrate polymers, *L. edodes* has to remove lignin with a lignolytic system. Acid protease is used to degrade proteins. Some water-insoluble fungal cell wall polysaccharides and their partial degradation products, such as laminarinase or *endo*-(1 → 6)-β-D-glucanae (or both), β-D-glucosidase, β-N-acetyl-D-glucosamindase, and chitinase, have been detected by assays. All of these enzymes have been extracted and identified, but only a few genes have been isolated. The molecular studies of the lignocellulytic enzymes that are produced by *L. edodes* are discussed in Section 2.7.4.

Cellulase activity increases during colonization and peaks at the veil break of fruiting body development, which suggests that cellulase is also important for fruiting body development (Ohga and Royse, 2001). Starch is available for fungi growing on plants or plant residues that use enzymes, including glucoamylase

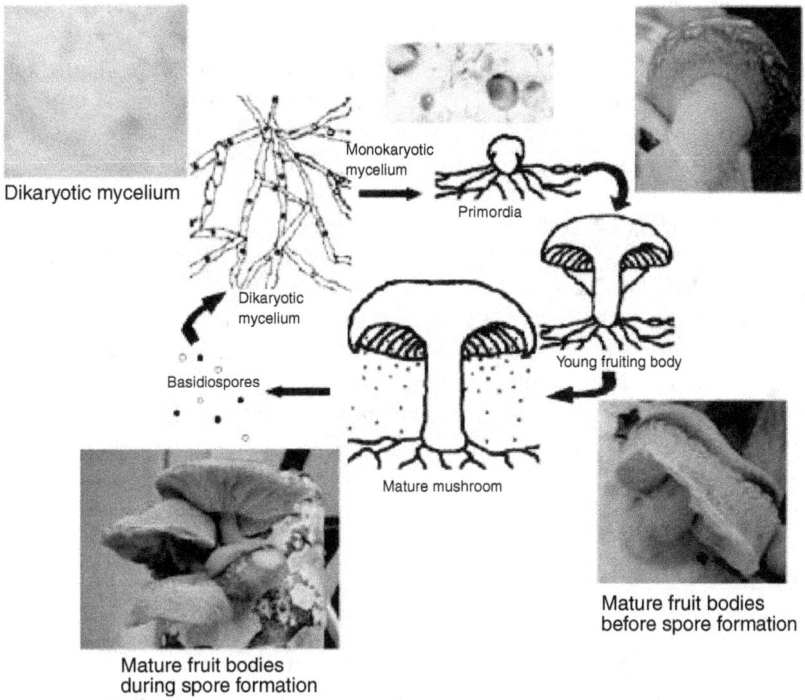

Figure 2.1 Life cycle of *Lentinula edodes* (L-54) grown in sawdust.

(Zhao et al., 2000). Glucoamylase-encoding genes have been cloned from several fungi, for example, *Neurospora crassa* (Stone et al., 1993). The glucoamylase gene (*gla1*) from *L. edodes* has also been cloned to show that its expression is induced by starch and increases during fruiting body formation (Zhao et al., 2000).

2.2.2 Development

Lentinula edodes follows a typical basidiomycete life cycle studied for many years (Figure 2.1) (Carlile and Watkinson, 1997a; Kües, 2000; Kües and Liu, 2000; Miles, 1993; Moore, 1998; Wessels, 1993): Under specific environmental conditions, it produces haploid basidiospores for reproduction. The basidiospores germinate to form monokaryotic mycelium (monokaryon). Two compatible monokaryons mate to form dikaryotic mycelium. Clamp connection occurs to maintain the dikaryotic condition in each hyphal cell after mitosis. When the mycelium has stored enough nutrients and the environmental conditions allow, it can proceed to the fruiting cycle. There are four consecutive stages in a fruiting cycle: induction, pinning, fruiting, and resting. During fruiting, hyphae aggregate to form primordia and then differentiate into specialized mushroom tissues. After growing into fruiting bodies, nuclear fusion occurs at the basidium of the mushroom gills to form karyogamy. Immediately after karyogamy, meiosis proceeds to generate four genetically unique haploid basidiospores. These basidiospores are then dispersed by air currents, and the life cycle begins again (Figure 2.1). The period from dikaryotic mycelium to primordium is critical to fruiting. Lowering the temperature can induce pinning in secondary mycelium. Once primordia start to emerge, mature fruiting bodies can develop.

2.2.2.1 Mating-Type Genes Only two haploid monokaryotic mycelia with different but compatible mating types can fuse to form dikaryon and then primordium (Table 2.1). The mating types are controlled by two main loci: mating types A and B. In related mushrooms, such as *Schizophyllum commune*, mating type A locus is responsible for genes that encode different homeodomains of HD1 and HD2 proteins for transcriptional regulation, whereas mating type B locus is responsible for the genes that regulate pheromone and the pheromone receptor for the signal transduction of mating and fruiting. The mating type genes found in *L. edodes* include a STE3-like pheromone receptor gene (Li et al., 2007). STE3 was reported to be the a-factor receptor gene for mating of *Saccharomyces cerevisiae* (Hagen et al. 1986). Details of the mating systems of *L. edodes* are still unclear and require more molecular studies.

2.2.2.2 Genes Differentially Expressed in Dikaryotic Mycelium In the mycelial stages, the genes responsible for energy production and structural components are expressed at higher levels (Table 2.1, Figure 2.2). The small G-protein RAS regulates these growth processes (Hori et al., 1991; Tanaka et al., 2005). Two transport genes, *LeDep* and *LeStr*, may be associated with the transportation of the signal and sugar, respectively, for signal transduction and energy production (Leung et al., 2000). Apparently, simple but intensive growth to occupy more nutrients and places is most important in the mycelial stages.

2.2.2.3 Genes for Initial Fruiting Bodies/Primordium Formation Induced by certain environmental factors, such as cold shock, the dikaryotic mycelium of *L. edodes* aggregates and differentiates into primordium (Table 2.1, Figure 2.3). The genes for signal transduction and transcriptional regulation, including Gγ, cAMP, *Le.MAPK, Le.DRMIP, Le.nik1, PriA, PriB, LeJun, LeNot1*, and *Le.cdc5*, are highly expressed in primordium. The active signal transduction and transcriptional regulation process are important to initiate and regulate a variety of physiological activities for primordium formation. It has been suggested that the high level of intracellular cyclic adenosine monophosphate (cAMP) is closely related to the onset of fruiting body development and/or primordium formation (Takagi et al., 1988). The *PriA* gene encodes a DNA-binding transcription factor and has higher expression in primordia/immature fruiting bodies than in preprimordial mycelia and mature fruiting bodies (Kajiwara et al., 1992). It may function during fruiting body initiation (Kajiwara et al., 1992). The 565a.a. PRIB protein is a zinc (II) cys6 zinc cluster DNA-binding motif (Endo et al., 1994; Miyazaki et al., 1997) and may control expression of the gene(s) correlated

TABLE 2.1 Genes with Implicated Roles in Growth and Development of *L. edodes*

Biological Processes	Genes	Homologous Protein Product–Protein Function	Developmental Stages[a]	References
		Regulatory Genes		
Signal transduction and transcription regulation	5.8 S rRNA	Ribosomal RNA	Cons.	Kwan et al., 1992b
	Le.recQ	RECQ helicase, basidiospore formation	Cons. (hymenophore)	Katsukawa and Shishido, 2005; Katsukawa et al., 2004
	Le.ras	RAS protein, hymenophore formation	Myc. and MFB (stipe and gill tissue)	Hori et al., 1991; Tanaka et al., 2005
	PriA	PRIA protein, reduced zinc accumulation	Initial fruiting bodies, Pri. formation	Ishizaki and Shishido, 2000; Kajiwara et al., 1992; Leung et al., 2000
	PriB	PRIB protein	Initial fruiting bodies, Pri. formation	Endo et al., 1994; Miyazaki et al., 1997, 2004b
	LeJun	Transcription factor JUN-D	Pri. and immature fruiting body	Leung et al., 2000
	Le.MAPK	Mitogen-activated protein kinase, initiation of fruiting and gill development	Pri. and immature fruiting body	Leung et al., 2000; Szeto et al., 2007
	Le.DRMIP	Developmentally regulated MAPK interacting protein	Pri. and immature fruiting body	Szeto et al., 2007
	pri30080	Gγ protein	Pri.	Miyazaki et al., 2005
	pri1n006	Phosphatidylethanolamine (PE)	Pri.	Miyazaki et al., 2005
	LeNot1	CDC39/NOT1	Pri. and MFB	Leung et al., 2000
	Le.nik1	Histidine kinase	Pri. to MFB (hymenophore)	Szeto et al., 2008

(*continued*)

TABLE 2.1 (*Continued*)

Biological Processes	Genes	Homologous Protein Product–Protein Function	Developmental Stages[a]	References
	Le.cdc5	Le.CDC5	Pri. and MFB	Miyazaki et al., 2004a; Nakazawa et al., 2006
	mfbA	MFBA protein	MFB	Kondoh et al., 1995; Yasuda and Shishido, 1999
	Uck1	UMP-CMP kinase	MFB (gill and pileus)	Kaneko et al., 1998; Miyazaki et al., 2004b
	Le.gal1	Gα protein, basidiospore formation	MFB	Tanaka et al., 2005
	MfbC	MFBC protein	MFB	Miyazaki et al., 2004b
Cell cycle control	*CAP*	Adenylyl-cyclase-associated protein	Cons.	Zhou et al., 1998
	Cip3	14-3-3 protein associated with CAP	Cons.	Zhou et al., 2000
	LeClb	Cyclin B	Pri.	Leung et al., 2000
Mating factor gene	STE3-like pheromone receptor gene	Associated with B-mating type	Myc.	Li et al., 2007
		Substrate-Utilizing Genes		
Lentinan degradation	*Exg1, Exg2*	*exo*-β-1,3-Glucanase	Myc. and MFB	Sakamoto et al., 2005a, 2005b
	Tlg1	Thaumatin-like protein	Postharvest	Sakamoto et al., 2006
Starch utilization	*Gla 1*	Glucoamylase	Pri. and MFB	Zhao et al., 2000

Other Functional Genes				
Nucleotide biosynthesis	*Le.rnr2*	Le.RNR2 (ribonucleotide reductase)	MFB (hymenium and outer region of trama)	Kaneko and Shishido, 2001
Spore formation	*Le.paa*	Le.PR65 (protein phosphatase 2A)	MFB (gill)	Ishizaki et al., 2000
	pri30200	SPS19	Pri.	Miyazaki et al., 2005
	pri30093	Serine/threonine protein kinase	Pri.	Miyazaki et al., 2005
Morphogenesis	*Le.hyd2*	Hydrophobin 2	Myc.	Ng et al., 2000
	Le.hyd 1	Hydrophobin 1	Pri.	Ng et al., 2000; Nishizawa et al., 2002
	mfb10037	Actin	MFB	Miyazaki et al., 2005
	mfb30045	ELN3, stipe elongation		
	Le.cyp1 and 2	Cytochrome P450, stipe elongation	Pri.	Akiyama et al., 2002
	pri30178	Cytochrome P450	Pri.	Miyazaki et al., 2005
	Le.cypfb	Cytochrome P450	MFB (gill)	Hirano et al., 2004
Energy production	*LeGpd-M*	Glycerol-3-phosphate dehydrogenase	Myc. and Pri.	Leung et al., 2000
	pri1n009	Alcohol dehydrogenase	Pri.	Miyazaki et al., 2005
	pri30227	Pyruvate decarboxylase	Pri.	Miyazaki et al., 2005
	pri30119	Citrate synthase	Pri.	Miyazaki et al., 2005
	pri30156	Aconitase	Pri.	Miyazaki et al., 2005
	LeMPPβ	Beta subunit of mitochondrial processing peptidase	MFB	Leung et al., 2000; Zhang et al., 1998
	LeFbp	Fructose-1,6-bisphosphatase	MFB	Leung et al., 2000
	mfb30103	Glucose 6 phosphate	MFB	Miyazaki et al., 2005
Transportation	*LeDep*	Drug efflux pump	Myc.	Leung et al., 2000
	LeStr	Sugar transporter	Myc. and Pri.	Leung et al., 2000
	LePma	H^+ ATPase	Cons.	Leung et al., 2000
	LeApt	α-Adaptin	Cons.	Leung et al., 2000

(*continued*)

TABLE 2.1 (*Continued*)

Biological Processes	Genes	Homologous Protein Product–Protein Function	Developmental Stages[a]	References
Endocytosis	*Le.rab5*	RAB5	Pri. and MFB	Lee et al., 2007
	Le.rab7	RAB7	Pri. and MFB	Lee et al., 2007
	Le.RACK1	RACK1	Pri. and MFB	Lee et al., 2007
Protein degradation	*LeUbi*	Ubiquitin	Pri.	Leung et al., 2000
	pri30170, mfb30076, mfb30106-1	Ubiquitin or polyubiquitin	Pri. and MFB	Miyazaki et al., 2005
	mfb30072	Ubiquitin conjugation enzyme	MFB	Miyazaki et al., 2005
Stress responses	*Le.-FAD1*	Δ9 fatty acid desaturase, fatty acid desaturation	Pri. and MFB	Sakai and Kajiwara, 2003
	Le.-FAD2	Δ12 fatty acid desaturase, fatty acid desaturation	Pri. and MFB	Sakai and Kajiwara, 2003
Flavorful production	*Le1 and 3*	Nuclease	MFB	Kobayashi et al., 2000, 2002
Unknown	mfb30066-1, fgb16	Riboflavin aldehyde–forming enzyme	MFB	Hirano et al., 2004; Miyazaki et al., 2005

[a]Developmental stage in which the gene differentally expressed: Myc., dikaryotic mycelium; Pri., primordium; MFB, mature fruiting body; Cons., constant, genes constantly expressed in all developmental stages.

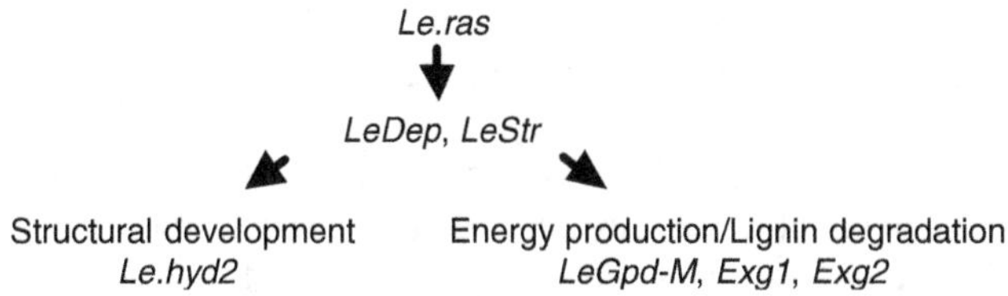

Figure 2.2 Genes that may be related to biological processes in dikaryotic mycelium growth.

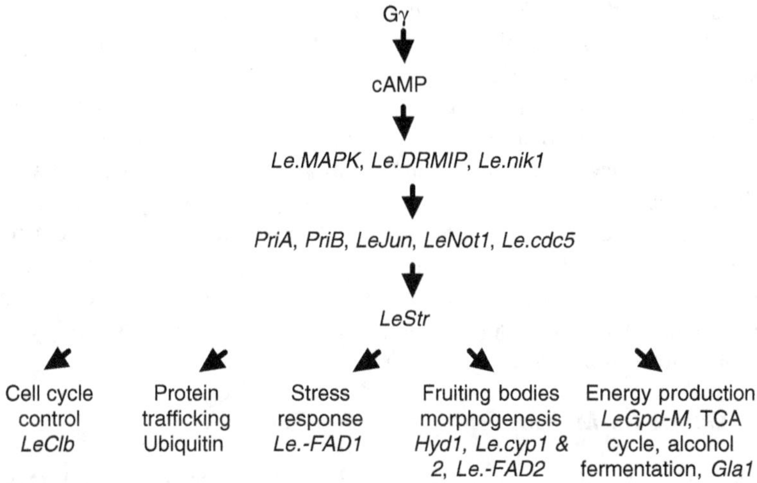

Figure 2.3 Genes that may be related to different biological pathways in primordium development.

to the onset of fruiting body development and/or primordia formation (Endo et al., 1994; Miyazaki et al., 1997, 2004b). The Le.CDC5 protein, encoded by the cDNA homologue to *Schizosaccharomyces pombe* cdc5(+), contains two putative phosphorylation sites of the cAMP-dependent protein kinase (A kinase) in its C-terminus (Miyazaki et al., 2004a). Transcripts of *Le.cdc5* are highly expressed in primordia and immature fruiting bodies, which implies that they may play a role in the beginning and in the early stages of fruiting body development (Miyazaki et al., 2004a). The *Le.MAPK* and *Le.DRMIP* genes are important in the signal transduction pathway, which is discussed in Section 2.2.3.1 (Szeto et al., 2007). The gene for vegetative growth is depressed, whereas the gene for active primordium growth is expressed. *LeNot1* and *LeClb* are highly expressed in primordium to suppress the expression of the genes involved in vegetative growth and to increase cell numbers for the rapid growth of primordium, respectively (Leung et al., 2000). Ubiquitins for the reconstruction of proteome–protein degradation, modification, and biosynthesis are highly expressed in primordium (Leung et al., 2000). Both genes and proteins are regulated for initial fruiting body formation.

Increasing energy demand is necessary for the initial formation of the fruiting bodies. Different metabolic mechanisms, such as the tricarboxylic acid (TCA) cycle, alcohol fermentation, and the utilization of lipid and starch, are active. The gene that encodes the beta subunit of the mitochondrial processing peptidase (β-MPP) expresses higher during the fruiting body formation than during that of the vegetative mycelium (Zhang et al., 1998). Thus, higher mitochondrial activities may be required to meet the energy demands of the rapid growth of the fruiting bodies (Zhang et al., 1998). Details are discussed in Section 2.2.3.2. The expression of the *Le.-FDA1* gene is induced by a lower temperature and fruiting body formation (Sakai and Kajiwara, 2003). Alteration of the fatty acid composition by *Le.-FAD1* may play a role in the stress response for fruiting (Sakai and Kajiwara, 2003). The morphogenesis of fruiting bodies is a more complex mechanism than is mycelial structural development. Different hydrophobins are involved in dikaryotic mycelium and primordium (*Hyd1* and *Le.Hyd2*) (Ng et al., 2000). The *Le.cyp1*, *Le.cyp2*, and *Le.-FAD2* genes for stipe elongation are successively increased in the expression from primordium to mature fruiting bodies (Akiyama et al., 2002; Sakai and Kajiwara, 2003). A high-energy demand is necessary for active signal transduction and transcriptional regulation to regulate the expression and reconstruction of genes and proteins, respectively, for the morphogenesis of complex fruiting bodies.

2.2.2.4 Genes for Mature Fruiting Bodies Formation

Stipe Elongation. Elongationless3 (ELN3) and cytochrome P450 may be involved in the stipe elongation of the fruiting bodies in *L. edodes*. The *eln3* mutation in *C. cinereus* affects stipe elongation during fruiting body formation (Arima et al., 2004). The gene homologue to *eln3* is highly expressed in the mature fruiting bodies of *L. edodes*. Complementary DNAs that were derived from the gill of the fruiting body have been compared with cDNAs from the mycelia by differential screening to identify six fruiting body–specific genes: *fbg03, 08, 13, 14, 16*, and *21* (Hirano et al., 2004). The deduced proteins include cytochrome P450 and the riboflavin aldehyde-forming enzyme (Hirano et al., 2004).

Cytochrome P450 functions in diverse biological pathways among various living organisms. It catalyzes the oxidation of xenobiotic and endogenous natural compounds in different biological pathways. In *L. edodes*, some cytochrome P450 genes are differentially expressed in the primordium to the mature fruiting bodies, especially in stipe rather than pileus and gill tissue (Akiyama et al., 2002; Hirano et al., 2004; Miyazaki et al., 2005).

Gill Development and Basidiospore Formation. Signal transduction for this process may be started from primordium because certain signal transduction genes, such as *Le.ras* and *Le.MAPK*, are differentially expressed in primordium (Table 2.1, Figure 2.4).

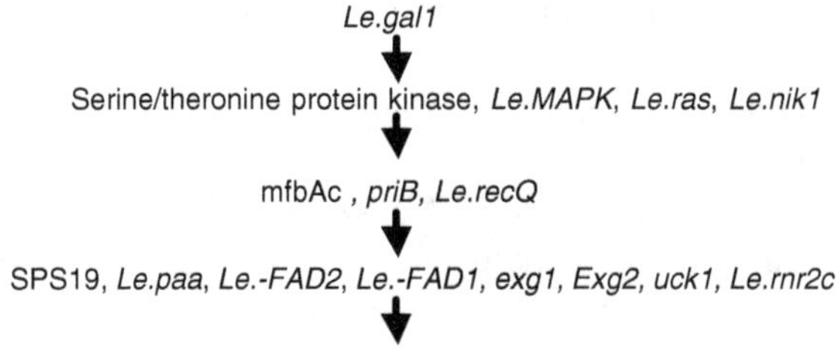

Figure 2.4 Genes that may be related to gill development and basidiospore formation.

During the production of basidiospores, the biosynthesis of nucleic acids, carbohydrates, and lipids is active (Kaneko et al., 1998). Uridine diphosphate (UDP), cytidine diphosphate (CDP), and adenosine diphosphate (ADP) synthesized by the uridine monophosphate–cytidine monophosphate (UMP–CMP) kinase serve as precursors for the synthesis of uridine 5′-triphosphate (UTP) and deoxythymidine triphosphate (dTTP), cytidine triphosphate (CTP) and deoxycytidine triphosphate (dCTP), and adenosine triphosphate (ATP) and deoxyadenosine triphosphate (dATP), and all serve as substrates for RNA and DNA synthesis. UTP and CTP are also involved in the generation of other biosynthesis intermediates, such as UDP-glucose and UDP-galactose in carbohydrate synthesis and CDP-acylglycerol in lipid synthesis (Kaneko et al., 1998).

The mature fruiting body–specific cDNA, mfbAc, encodes a high-molecular-weight cell adhesion protein that contains an Arg-Gly-Asp motif (Kondoh and Shishido, 1995).

The *Le.rnr2*c gene that encodes the protein homologous to the ribonucleotide reductase (RNR) small subunit is highly expressed in the hymenophores of the mature fruiting body of *L. edodes* (Kaneko and Shishido, 2001). In situ RNA–RNA hybridization analysis shows that transcripts of *Le.rnr2* and *uck1* are highly abundant in the hymenium and in other regions of the trama in the hymenophore (Kaneko and Shishido, 2001). The hymenium contains basidia in which two nuclei are fused for meiosis and replication to produce basidiospores (Kaneko and Shishido, 2001). Thus, *Le.rnr2* and *uck1* may play important roles in the nucleotide biosynthesis that is essential for the production of basidiospores (Kaneko and Shishido, 2001). *Le.paa* encoding a regulatory subunit A (PR65) homologue of protein phosphates 2A is actively transcribed during the late stages of fruiting body development (Ishizaki et al., 2000). The *Le.paa* transcript has higher expression in gill tissue than in pileus or stipe; therefore it may play a role in gills in which basidiospores are produced (Ishizaki et al., 2000).

The *Le-FAD1* and *Le-FAD2* genes that encode proteins similar to delta 9 fatty acid desaturases and delta 12 fatty acid desaturases have been cloned and sequenced (Sakai and Kajiwara, 2003, 2005). The transcript level of *Le-FAD2* increases following a shift from 25 to 18°C and also from primordia to mature fruiting bodies, whereas the transcript level of *Le-FAD1* is higher in the primordium and fruiting body than in mycelia cultivated at 18 or 25°C. *Le-FAD1* and *Le-FAD2* transcripts increase during *L. edodes* fruiting, and this correlates with an increase in the unsaturated fatty acid content in total lipids. Therefore, delta 9 and 12 desaturase may be needed in fruiting body formation (Sakai and Kajiwara, 2003, 2005).

By genomic binding site cloning, three target genes of the developmental regulator (PRIB) have been identified in *L. edodes*. These are previously cloned *priB* and *uck1* and a new gene named *mfbC* (Miyazaki et al., 2004b). The product of *mfbC*, MFBC protein, is highly homologous to *Saccharomyces cerevisiae* YJR070 C/Lia1, the protein interacting with a putative translation initiation factor. The *mfbC* transcripts have been found only in mature fruiting bodies, which suggest a need for *mfbC* in the final stage of fruiting body formation (Miyazaki et al., 2004b).

The *Le.recQ* gene homologue that encodes RecQ-type DNA helicase has been isolated (Katsukawa et al., 2004). The expression level of the *Le.recQ* transcripts is similar at all of the developmental stages, and the mycelial growth rate increases with the increase of *Le.recQ* transcripts. This suggests that it is necessary for the good growth of mycelial cells (Katsukawa et al., 2004). As shown by in situ RNA–RNA hybridization, however, the *Le.recQ* transcript level within the hymenophore is higher in the hymenium, subhymenium, and outer region of the trama, which suggests that the *Le.recQ* gene may be involved in basidiospore formation (Katsukawa and Shishido, 2005).

An *exo*-(1 $\rightarrow$ 3)-β-glucanase-encoding gene (*exg1*) is expressed in fruiting bodies but not in vegetative mycelia. The expression is higher in the stipe than in the pileus of young fruiting bodies but is high in the gills of mature fruiting bodies (Sakamoto et al., 2005a). Thus, *exg1* may play a role in fruiting body formation, including the stipe elongation of *L. edodes* (Sakamoto et al., 2005a). The *exg2* gene that encodes exo-(1 $\rightarrow$ 3)-β-glucanase (Sakamoto et al., 2005b), unlike the *exg1* gene, is low in transcription and translation in the gills of mature fruiting bodies but increases after harvesting, which suggests that it is a lentinan-degrading enzyme-encoding gene (Sakamoto et al., 2005b).

Two laccase genes, *lac1* and *lac2*, express higher in the caps of fruiting bodies than in the stipes and primordium (Zhao and Kwan, 1999). Strong laccase activity is present in the caps of fruiting bodies, which indicates that laccase may catalyze the formation of extracellular pigments by oxidation polymerization and is therefore important for fruiting body morphogenesis (Zhao and Kwan, 1999). Another study of the transcriptional regulation of laccase and cellulase genes during the growth and fruiting of *L. edodes*, however, has shown that laccase activity is high during colonization and then declines rapidly during fruiting body formation (Ohga and Royse, 2001). The difference between these studies may be due to the different supplemented sawdust that *L. edodes* was grown on, the different strains of *L. edodes* used, or the different substrates used to determine laccase activity (Ohga and Royse, 2001). No matter what affected the results, laccase activity does change during fruiting body development.

In summary, the molecular studies so far have advanced our understanding of the molecular mechanisms of the fruiting initiation and/or the primordium formation of *L. edodes* (Table 2.1). The results show that during the fruiting initiation and/or the primordium formation the signal is transferred to active transcription, energy production, and protein turnover for fruiting body development. From primordium to young fruiting bodies, the genes expressed are those responsible for morphological changes such as pileus and stipe formation. In addition, some genes may be related to responses to stresses such as temperature change. More studies of the gene expression profiles of dikaryotic mycelium and the mature fruiting bodies and sporulating fruiting bodies of *L. edodes*, however, are required to better understand the growth and developmental processes of *L. edodes*.

2.2.3 Physiological Processes in *Lentinula edodes*

2.2.3.1 Signal Transduction Signal transduction is important in the growth and development of any organism. Several genes that may be involved in signal transduction have been isolated from *L. edodes: Le.ras, Le.Ga, Le.mfbC, Le.recQ, Le.MAPK, Le.DRMIP*, and *Le.nik1*. Mitogen-activated protein kinase (MAPK) is one of the important signal transduction proteins in the cascades that regulates growth and development in many organisms. It is regulated by the phosphorylation cascade of upstream MAPK kinase (MEK) and MEK kinase (MEKK), and MAPK regulates such downstream effectors as transcription factors and growth regulators. The MAPK homologue *Le.MAPK* was identified from the primordium of *L. edodes* (Leung et al., 2000). The expression profiles of *Le.MAPK* and its interacting novel gene that encodes the developmentally regulated MAPK-interacting protein, *Le.DRMIP*, suggest their importance in fruiting body initiation and development (Szeto et al., 2007). Their expressions are highest in the primordial stage, and their transcripts are located in the very young fruiting bodies in which future gill development takes place, which suggests that MAPK and its interacting partner, Le.PRMIP, play some role in cell differentiation and morphogenesis during the gill development of *L. edodes* (Szeto et al., 2007). Fungal histidine kinase plays an essential role in cell wall assembly, virulence, sporulation, hyphal development, fungicide resistance, and osmoregulation (Miller et al., 2002; Nagahashi et al., 1998; Posas et al., 1996; Pott et al., 2000; Yoshimi et al., 2004). The histidine kinase gene in basidiomycetes, namely *Le.nik1*, has been isolated from *L. edodes* by reverse-blot hybridization screening (Szeto et al., 2008). The transcript expression level of *Le.nik1* increases from mycelium to fruiting body, which indicates that *Le.nik1* has a role in the initiation and development of the fruiting body (Szeto et al., 2008). As most of the other signal transduction gene (*Le.ras, Le.Ga, Le.mfbC*, and *Le.recQ*) transcripts are located in the outer region of the hymenophore in *L. edodes* (Hori et al., 1991; Katsukawa and Shishido, 2005; Miyazaki et al., 2004b; Tanaka et al., 2005), *Le.nik1* may play a vital role in trama cell development and in transferring the signal (i.e., basidiospore formation and release) from the middle trama cell to the outer hymenophore (Szeto et al., 2008). *Le.nik1* may regulate important downstream developmental and stress-responding genes (Szeto et al., 2008).

2.2.3.2 Energy Production In the different developmental stages of *L. edodes*, there is the expression of various different genes for different metabolic pathways to produce energy. To meet energy demand during the rapid growth of the fruiting bodies, higher mitochondrial activities may be required. *LeMPP* is highly expressed during the development of the fruiting bodies (Zhang et al., 1998). Surprisingly, both aerobic and anaerobic pathways are used in the primordial stages to produce energy for their active growth and development. The genes that encode enzymes for the TCA cycle, citrate synthase and aconitase, and the genes homologous to the enzymes for alcohol dehydrogenase, alcohol dehydrogenase and pyruvate decarboxylase, are highly expressed in the primordium (Miyazaki et al., 2005). Less energy seems to be needed for mature fruiting bodies, and the genes that encode the enzymes for glycolysis, fructose-1,6-bisphosphatase dehydrogenase (Leung et al., 2000), and glucose-6-phosphate (Miyazaki et al., 2005) are more highly expressed.

2.2.3.3 Structural Proteins in Development Hydrophobins are essential for morphogenesis and pathogenesis in fungi and fruiting body development in mushrooms (Kershaw and Talbot, 1998). Hydrophobin-encoding genes have been isolated from many fungi, such as *S. commune, Agaricus bisporus*, and *Pleurotus ostreatus*. Their expressions are developmentally regulated (Ng et al., 2000). Two hydrophobin-encoding genes (*Le.hyd1* and *Le.hyd2*) have been isolated from *L. edodes* and characterized to be differentially expressed in different developmental stages (Ng et al., 2000). The transcript level of *Le.hyd1* is higher in primordium, and that of *Le.hyd2* is higher in dikaryotic mycelium. The results indicate that these two *L. edodes* hydrophobins have distinct roles in the fruiting body development of *L. edodes*. In mature fruiting bodies, the genes for structural components, including various hydrophobins, are not abundantly expressed. Structural protein production appears to be important from mycelium to primordium but not for mature fruiting bodies.

2.3 MOLECULAR GENETICS

Since the 1930s modern cultivation methods with pure cultured mycelia as inoculum have been developed for mushroom cultivation, and increasing efforts have been made to develop new cultivated strains by mating between strains with desirable characteristics (Hasebe et al., 1991; Terashima et al., 2002a; Tokimoto and Komatsu, 1995). Molecular markers have been used to develop suitable strains for breeding and strain improvement (Table 2.2). They are also used in genetic mapping, identifying and cloning genes, and studying genetic diversity (Table 2.2). PCRs using arbitrary primers (AP-PCRs), restriction fragment length polymorphisms (RFLPs), random-amplified polymorphic DNA (RAPD) markers, amplified fragment length polymorphism (AFLP) analysis, sequence characterized amplified region (SCAR) markers, and inter–simple sequence repeat markers (ISSRs) have been used to generate markers or identify strains with desirable characteristics for mating (Table 2.2).

TABLE 2.2 Approaches Used for Molecular Genetics of *L. edodes*

Molecular Approaches	Investigation	References
Multilocus enzyme electrophoresis	Genotype identification	Royse and May, 1987
Polymerase chain reaction using arbitrary primer (AP-PCR)	Strain typing	Chang et al., 1995; Kwan et al., 1992a
Mitochondrial DNA (mtDNA) restriction fragment length polymorphisms (RFLPs)	Examination of modes of mitochondrial inheritance in sexual crosses and protoplast cell fusions	Fukuda et al., 1995
Random-amplified polymorphic DNA (RAPD)	1. Strain typing	Zhang and Molina, 1995
	2. Construction of genetic linkage map	Kwan and Xu, 2002
DNA fingerprints	Strain characterization by subrepeat regions	Saito et al., 2002
Amplified fragment length polymorphism (AFLP) analysis	Construction of genetic linkage map	Terashima et al., 2002b
	Genetic diversity and strain typing	Terashima et al., 2002a
Sequence-characterized amplified region (SCAR) markers	1. Screen molecular markers linked to mating factors using randomly amplified polymorphic DNA (RAPD)	Tanaka et al., 2004
	2. Strain identification by ISSR	Qin et al., 2006
Inter simple sequence repeat markers (ISSR)	Strain typing in China	Zhang et al., 2007

2.3.1 Generation of Markers

RAPD is based on a PCR with a pair arbitrary primer that has a predominantly G/C composition (60–80%) and a relatively low temperature (Gostimskii et al., 2005; Welsh and McClelland, 1990; Williams et al., 1990). By using RAPD, seven primers have been used to produce polymorphisms in 15 tested strains of *L. edodes*, and it was found that 13 of them had unique DNA fingerprints (Zhang and Molina, 1995).

ISSR is also a PCR-based method, but it uses oligonucleotide primers with repetitive units and the so-called anchor at the 3′ or 5′ end (Gostimskii et al., 2005;

Zietkiewicz et al., 1994). Seventeen Chinese strains of *L. edodes*, including 15 cultivated strains and two wild strains, were clustered into two distinct groups—H (high-temperature) type or B (broad-temperature) type and L (low-temperature) type or M (medium-temperature) type (Zhang et al., 2007). Subrepeat regions within the nuclear rDNA intergenic spacers IGS1 and IGS2 have been amplified to analyze the relationships between 16 commercial cultivars of *L. edodes* (Saito et al., 2002). The DNA fingerprinting from the PCR products of the subrepeat regions (SR1 and SR2) of IGS1 and IGS2 are useful to discriminate among *L. edodes* cultivars (Saito et al., 2002).

SCAR markers are based on PCR using long, specific primers (Gostimskii et al., 2005; Kesseli et al., 1993). A pair of SCAR primers can precisely amplify a single unique fragment from 85 strains of *L. edodes* (Qin et al., 2006).

2.3.2 Typing/Fingerprinting

AP-PCR is a PCR-based method that has been found to be better than rDNA internal transcribed spacer region sequence comparison for *L. edodes* strain typing (Chang et al., 1995; Kwan et al., 1992a). The method provides almost unique DNA profiles for each of the 15 *L. edodes* strains (Kwan et al., 1992a).

AFLP is based on the selective PCR amplification of genomic DNA restriction fragments to produce polymorphic loci (Vos et al., 1995). To identify some major cultivated strains in Japan, six pairs of AFLP primers have been used, and they detected 304 DNA fragments from 13 cultivated strains for wood log cultivation and two strains for sawdust cultivation (Terashima et al., 2002b).

2.3.3 Genetic Mapping

It is estimated that the complete map size of *L. edodes* is about 1200 cM (Kwan and Xu, 2002). More that half of the genome is covered by an RAPD-constructed genetic linkage map that contains 14 linkage groups (Kwan and Xu, 2002).

AFLP markers have also been used to generate a medium-dense genetic linkage map consisting of 11 linkage groups that comprise eight large (over 100-cM) and three small (less then 100-cM) groups, for a total of 1956.7 cM (Terashima et al., 2002b).

2.4 FUNCTIONAL GENOMIC APPROACHES FOR GENE EXPRESSION ANALYSIS

Identifying more differentially expressed genes and analyzing their expression profiles under various developmental stages allow us to gain a better understanding of fruiting body development. The methods used have included RAP-PCR, ESTs, cDNA random sequencing, and SAGE (Table 2.3).

TABLE 2.3 Molecular Approaches Used for Gene Expression Studies of *L. edodes*

Molecular Approaches	Applications	References
RNA fingerprinting by arbitrarily primed PCR (RAP-PCR)	13 cDNA fragments were differentially expressed in primordium	Leung et al., 2000
cDNA representational difference analysis (cDNA-RDA)	105 genes differentially expressed in primordium or mature fruiting body were isolated	Miyazaki et al., 2005
Serial analysis of gene expression (SAGE)	Gene expression profiles of monokaryotic and dikaryotic mycelium, primordium, and fruiting body before and during spore formation	Chum et al., 2006a, 2008
cDNA microarray	Gene expression profiles of monokaryotic mycelium, dikaryotic mycelium, and primordium	Shih, 2003
Suppression subtractive hybridization	Selection of differentially expressed genes in dikaryons against either monokaryotic mycelium parent	Shih, 2003
Expressed sequence tag (EST)	Gene expression profiles of primordium	Unpublished
Yeast two-hybrid system	Isolation of interacting proteins of functional proteins	Lee et al., 2007; Szeto et al., 2007
Sequencing-by-synthesis approach (454 Life Science)	Transcriptome analysis of dikaryotic mycelium and mature fruiting bodies	Chum et al., 2006b; Kwok et al., 2006

2.4.1 Differential Display: RAP-PCR

Some molecular methods, such as AP-PCR and RAP-PCR, have been used to study the genomes of mushrooms and to identify the genes that are differentially expressed during fruiting body development. RAP-PCR has been used for this purpose in *L. edodes*. More than 100 genes have been isolated and sequenced, and 15 have been studied further (Leung et al., 2000). Thirteen RAP fragments were found to be highly homologous to known genes that function in transport across the plasma membrane (drug efflux pump and sugar transporter); cell cycle control (cyclin B); signal transduction and transcriptional regulation (mitogen-activated protein kinase, Cdc39/Not1, *PriA*, and Jun-D); intracellular molecular trafficking (ubiquitin, plasma membrane proton ATPase, and α-adaptin); mitochondrial biogenesis (mitochondrial processing peptidase beta subunit, mitochondrial glycerol-3-phosphate dehydrogenase); and intermediary metabolism (fructose-1,6-bisphosphatase) (Leung et al., 2000).

Six clones identified as fruiting body–specific genes have been isolated by differential screening (Hirano et al., 2004). The deduced proteins include cytochrome P450 and a riboflavin aldehyde-forming enzyme (Hirano et al., 2004).

2.4.2 cDNA Representation Difference Analysis

To study the molecular mechanisms of fruiting in *L. edodes*, cDNA-RDA between vegetatively growing mycelium and two developmental substages, primordium and mature fruiting body, has been used to isolate 105 individual genes (51 in primordium and 54 in the mature fruiting body) (Miyazaki et al., 2005).

2.4.3 SAGE and LongSAGE

SAGE was first proposed by Velculescu et al. (1995). It allows simultaneous comparative and quantitative analysis of the level of transcripts (Yamamoto et al., 2001). SAGE can analyze each transcript without the transcripts, as in microarray hybridization (Vedoy et al., 1999; Velculescu et al., 2000). When compared with differential displays and ESTs, SAGE is more effective and can provide quantitative and comprehensive profiles (Sun et al., 2004; Vedoy et al., 1999; Velculescu et al., 2000). SAGE is based on three main principles as follows:

1. Nine to 14-bp short sequence tags are obtained from a defined region within each transcript that contains sufficient information to identify a transcript uniquely.
2. Ditags (two individual tags ligated randomly) are ligated together to form a concatemer, which is a long DNA sequence that can be cloned for sequencing. Sequencing of the concatemer clones results in the identification of individual tags.
3. SAGE 2000 Software (version 45) (Invitrogen) is used to analyze the expression level of the transcript by counting the number of copies of a particular tag.

LongSAGE is the conventional SAGE method modified by using a different type of the IIS restriction enzyme, *MmeI*, to generate longer SAGE tags (Saha et al., 2002). LongSAGE is based on the acquisition of a sequence tag (15–21 bp) that, theoretically, can be uniquely assigned to a single genomic position (Saha et al., 2002). A LongSAGE tag contains more information for identification and is more reliable in the correct identification of the genomic locus that corresponds to a certain transcript (Wahl et al., 2005).

2.4.3.1 SAGE Profiles: Mycelium to Primordium A total of 6363 tags have been extracted (3278 from dikaryotic mycelium and 3085 from primordium), and 919 tags (293 unique tags) match to an in-house EST database (Chum and Kwan, 2005; Chum et al., 2008). The expression profiles of dikaryotic mycelium and primordium are very different and reveal that the transcriptional expression of a specific set of genes is required for initial fruiting body development. One hundred and thirteen tags are more highly expressed in the dikaryotic mycelium. Some of them match ESTs that encode various putative proteins. These are mycelial hydrophobins, serine-rich proteins, the reduced form of nicotinamide adenine dinucleotide phosphate (NADPH) oxidase I, mitochondrial phosphate carrier, ATP/PDP carrier protein, NADH dehydrogenase I subcomplex, phosphatidylethanolanine-binding protein, and lectin. One hundred and forty-four genes are differentially expressed in primordium. Some of them match ESTs that encode putative proteins, including Gα-binding protein, ADP ribosylation factor 1, hesp-379, methallothionein, cytochrome p450, ATP-synthase, ubiquitin, hydrophobin 1, and riboflavin aldehyde–forming enzyme.

Different structural proteins are abundantly expressed in the two stages. At the primordial stage, the expression of ribosomal proteins is higher than in mycelium, which indicates that protein synthesis is active.

2.4.3.2 SAGE Profiles: Fruiting Bodies Maturation to Sporulation Two LongSAGE libraries were generated to obtain 4000 and 7000 LongSAGE tags from mature fruiting bodies before and during spore formation (Chum et al., 2006b). These LongSAGE libraries have been compared to identify the genes that are more highly expressed in fruiting bodies during sporulation. The expressions of the genes relevant to sporulation were expected to increase, and the products of some genes were stored in basidiospores for use during germination. The LongSAGE tags have also been compared with the SAGE tags obtained in other developmental stages, such as dikaryotic mycelium and primordium, to analyze the transcription profiles of certain interesting genes at different stages (Chum et al., 2006b). Most of the LongSAGE tags, however, do not match the *L. edodes* gene databases that were generated from dikaryotic mycelium and primordium.

2.4.4 cDNA Microarray

DNA microarray analysis is commonly used to investigate the expression of several thousand genes simultaneously, such as screening for interesting genes on a genomic scale. *Lentinula edodes* DNA microarray slides were constructed in-house. About 500 ESTs extracted from primordium and thousands of clones randomly selected from the subtractive library of dikaryotic mycelium were dotted on slides for microarray hybridization. The gene expression of monokaryotic mycelium parents and their dikaryotic mycelium were compared by cDNA microarray hybridization. In addition, 30 strains of *L. edodes* were selected from different regions of mainland China and cultivated in the same conditions. Their mycelial growth rate, fruiting date, number of fruiting bodies, and weight of fruiting bodies were recorded. Eleven of these 30 strains with contrasting characteristics in mycelial growth rates and/or fruiting body amounts were analyzed by cDNA microarray hybridization. The data were analyzed with The Institute for Genome Research (TIGR) MultiExperiment Viewer (MeV), and genes were clustered based on their expression patterns through the experiments (unpublished).

2.4.5 Expressed Sequence Tag

A total of 447 unique ESTs were obtained from the single-pass sequencing of 670 random cDNA clones (unpublished). By BLASTX, 238 ESTs showed moderately to highly significant matches to protein sequences in the databases, and 209 ESTs had weakly significant or no matches. Among the genes with putative identities, 34% are involved in protein synthesis, 21% in metabolism, 7% in RNA synthesis, and 7% in cell defense. Some ESTs are involved in cell signaling (5%), cell structure (5%), or cell division (1%). Dot-blot hybridization was used to analyze the mycelium–primordium expression ratio of 235 ESTs and revealed that 154 ESTs were differentially expressed. Among these, 145 ESTs were more highly expressed in primordium. ESTs are assigned to roles in cell communication, cell defense, RNA and protein synthesis, and metabolism. Thus, these cellular processes may be important in fruiting body initiation. Because the genes involved in different cellular processes show primordial preferential expression patterns, it appears that fruiting body initiation is a complicated process that depends on the integral functions of the genes involved in different cellular roles.

2.4.6 Yeast Two-Hybrid System

Yeast two-hybrid analysis (Fields and Song, 1989) is one of the most powerful tools for the investigation of entire protein interaction networks. It has three main applications:

1. Testing interactions between known proteins
2. Screening libraries for proteins that interact with a known protein
3. Defining the domains and/or amino acids required for an interaction

It has been used to screen proteins that interact with the signal transduction genes: *Le.nik1*, *Le.MAPK*, and *Le.DRMIP* (Szeto et al., 2007), and the endocytosis genes *Le.Rab5* and *Le.RACK1* interact with *Le. Rab7* (Lee et al., 2007).

2.4.7 Sequencing-by-Synthesis Approach (454 Life Science)

In a typical large-scale sequencing project, for example, whole-genome sequencing, it is necessary to clone DNA fragments into vectors to amplify and purify individual templates followed by Sanger sequencing using fluorescent chain-terminating nucleotide analogs and either slab gel or capillary electrophoresis. High-throughput sequencing technologies are being developed to displace the use of vectors and Sanger sequencing as the main generators of sequencing information (Margulies et al., 2005). The cost, complexity, and time required to sequence large amounts of DNA have thus been reduced.

To understand the biological mechanisms of *L. edodes*, it is important to identify more genes. A genome sequence of *L. edodes*, however, is not available, and few genes have been isolated. Most of the isolated genes are related to the initiation of fruiting bodies/primordium formation, and few genes that relate to mature fruiting bodies and sporulating fruiting bodies have been isolated. Large-scale cDNA sequencing can provide more information on the genes that are important in mature fruiting bodies. 454 Life Science has developed a scalable, highly parallel sequencing system with raw throughput that is significantly greater than that of the state-of-the-art capillary electrophoresis instrument. This system uses a novel fiber-optic slide (PicoTitle Plate) of individual wells and is able to sequence 25 million bases at $\geqq$ 99% accuracy in each four-hour run. This is about a 100-fold increase in throughput over the current Sanger sequencing technology (Margulies et al., 2005). To achieve high throughput, an emulsion method for DNA amplification and an instrument for sequencing (Genome Sequencer 20 system) by synthesis using a pyrosequencing protocol optimized for solid support and picoliter-scale volume have been developed. Using this sequencing-by-synthesis approach, more than 5000 and 7000 cDNA contigs from *L. edodes* dikaryotic mycelium and fruiting bodies have been generated.

2.5 TRANSCRIPTIONAL REGULATION

2.5.1 Transcriptional Factors

Many biological processes, such as cell growth, environmental adaptation, and cell development, are regulated at the transcriptional level. The mechanisms of such regulations are conserved among eukaryotes (Struhl, 1995). One of these mechanisms is the specific binding of transcriptional factors (TFs) onto a specific DNA sequence called transcriptional factor binding sites (TFBSs). TFBSs are 5–25 nucleotides located at the 5′ flanking region before the transcription start site. TFs and their corresponding binding sites are important in gene regulation. Current information about TFs and TFBSs in *L. edodes* is very limited. Three transcription factors, *LePriA, PriB*, and *Le-cdc5*, have been identified. Transformants of *L. edodes* that overexpress the *PriA* gene decrease in zinc ion acumination, which indicates that fruiting body development may involve changes in intracellular metal ion concentration (Ishizaki and Shishido, 2000).

2.5.2 Promoter Analysis

Studies of the 2-kb nucleotide sequence, including the S-flanking region of a cell adhesion protein-encoding gene (*mfbA*) that was isolated from *L. edodes*, have revealed that the promoter region contains a TATA box, a GC box, a CAAT box, and a CT-rich sequence element from upstream to downstream (Kondoh and Shishido, 1995). The 3′ noncoding region of the *priB* gene contains several promoter-like motifs—a GC boxlike sequence, two CAAT boxes and two TATA boxlike sequences, and two CT motifs—and has promoter activity in *S. cerevisiae* (Yamazaki et al., 2002). The *PriA* promoter has been introduced into *Coprinopsis cinereus* to drive the expression of the *P. ostreatus* manganese (II) peroxidase gene *mnp* and of the xylanase genes from *Aspergillus oryzae* and *Bacillus subtilis* (Kikuchi et al., 1999, 2004; Ogawa et al., 1998).

2.6 TRANSFORMATION

Cross breeding is the traditional method to improve strains of mushrooms, but it is inefficient. Transgenic breeding is a new way for the genetic improvement of mushrooms to increase their yields and quality. Transformation is an important tool in transgenic breeding and functional genomics studies. Many approaches are introduced through this technique, such as eliminating, destructing, or silencing a gene to reach different targets or needs (Table 2.4). It is powerful in the study of physiology and gene functions. Transformation protocols can be very specific at the species level and even at the strain level. Much effort is required to develop different precautions and protocols for transformation (Table 2.4).

2.6.1 Transformation Methods

Some model fungi have been genetically modified using transformation, including *C. cinereus, P. ostreatus*, and *A. bisporus*. In *L. edodes* studies, transformation has been established for some strains such as dikaryotic strain S-1 (Hirano et al., 2000; Sato et al., 1998). *Lentinula edodes* has been transformed using the restriction enzyme–mediated DNA integration (REMI) method (Hirano et al., 2000; Irie et al., 2003; Sato et al., 1998) and the polyethylene glycol (PEG)–mediated transformation method (Li et al., 2006; Sun et al., 2001), which are described below.

2.6.1.1 PEG-Mediated Transformation PEG-mediated transformation uses PEG as a medium to induce the porous cell membrane for DNA to enter the cell. The PEG-mediated transformation of *L. edodes* protoplast has been used to express vector p301-bG1, which contains a gus gene and a bialaphos resistance gene; both are driven by the glyceraldehyde-3-phosphate dehydrogenase (GPD) gene promoter isolated from *L. edodes* (Sun et al., 2001). The efficiency of PEG-mediated transformation, however, is usually affected by variations in conditions and is too low for most studies (Meyer et al., 2003). A modified PEG-mediated transformation method was used to transfer the *hph* gene into *L. edodes*, resulting in 85–100 transformants per microgram of DNA per 10^7 viable protoplasts (Li et al., 2006).

TABLE 2.4 *L. edodes* **Genes for Transformation**

Genes	Applications	References
PriA gene terminator	Expresses *Streptomyces hygroscopicus* bialaphos resistance gene in *P. ostreatus*	Yanai et al., 1996
	Expresses *E. coli* hygromycin B phosphotransferase *hph* gene in *L. edodes* by REMI	Sato et al., 1998
	Expresses *Bacillus subtilis endo-* β-1, 4-D-xylanase *xyn* in *C. cinereus*	Kikuchi et al., 1999
	Expresses *E. coli* hygromycin B phosphotransferase *hph* gene in *L. edodes* by REMI	Hirano et al., 2000
	Expresses rat cytochrome P450 CYP1A1 in *Coriolus hirsutus*	Orihara et al., 2005
PriA gene promoter	Expresses *P. ostreatus* manganese (II) peroxidase gene *mnp* in *C. cinereus*	Ogawa et al., 1998
	Expresses *Aspergillus oryzae* xylanase XynF1 in *C. cinereus*	Kikuchi et al., 2004
	Expresses laccase gene *lcc1* in *C. cinereus*	Kilaru et al., 2006
GPD promoter and terminator	Express *E. coli* hygromycin B phosphotransferase *hph* gene in *L. edodes* by REMI	Hirano et al., 2000
	Express *E. coli* hygromycin B phosphotransferase *hph* gene in *P. ostreatus* by REMI	Irie et al., 2001
Ras gene promoter	Expresses *S. hygroscopicus* bialaphos resistance gene in *P. ostreatus*	Yanai et al., 1996
	Expresses *P. ostreatus* manganese (II) peroxidase gene *mnp* in *C. cinereus*	Ogawa et al., 1998
	Expresses *E. coli* hygromycin B phosphotransferase *hph* gene in *L. edodes* by REMI	Sato et al., 1998
	Expresses *B. subtilis endo-* β-1, 4-D-xylanase *xyn* in *C. cinereus*	Kikuchi et al., 1999

2.6.1.2 Restriction Enzyme–Mediated Integration REMI transformation uses restriction enzymes to increase efficiency. Different enzymes are added to the cell and enter the nuclear membrane to cleave chromosomal DNA in vivo at their particular restriction sites. The free chromosomal DNA ends that are generated can be ligated to restriction enzyme–linearized plasmid DNA by the host cell enzymes. This was first applied in *S. cerevisiae* to introduce random tagged mutations into the host genome efficiently (Schiestl and Petes, 1991). REMI was used in transformation with pLC1-hph, a recombinant plasmid containing *L. edodes* transcriptional signals, and an *Escherichia coli* hygromycin B phosphotransferase gene into *L. edodes* (Hirano et al., 2000; Sato et al., 1998). An iron sulfur protein (Ip) subunit gene from *L. edodes* was cloned and used to construct a homologous drug-resistant marker, Cbx^R, which was then successfully introduced into *L. edodes* by REMI (Irie et al., 2003).

2.6.1.3 Others

Electroporation. Electroporation, a simple, convenient, and effective technique for transformation, is widely used in many species. The principle of electroporation is the use of an electric field of high intensity that can induce inner membrane permeabilization. This permeabilization remains for a short period after the pulse. It is less often used in mushrooms. Only *Agrocybe aegerita* (Noel and Labarere, 1994) and *Flammulina velutipes* (Kuo et al., 2004) have been transformed using electroporation. There are so far no reports of transformating *L. edodes* with electroporation.

Particle Bombardment. Particle bombardment is the introduction of DNA into intact cells or tissues by using high-velocity microprojectiles via a mechanism that breaches cell walls and membrane. Heavy particles, such as gold, tungsten, and platinum, with the adequate momentum to penetrate into the appropriate tissue are used as carriers. This technique has been successfully applied in different plants, such as tobacco, soybeans, maize, and others (Yin et al., 2004). This method makes it possible to carry DNA directly into cells or tissue. Similar to electroporation, a nonoptimal particle bombardment procedure will decrease cell viability and transformation efficiency. After optimization, its efficiency is very high, and it is convenient for many studies. This method has only been used in transformating *P. ostreatus* (Sunagawa and Magae, 2002) and *Volvariella volvacea* (Guo et al., 2005), but not *L. edodes*.

2.6.2 *Lentinula edodes* Genes Used in Transformation

The GPD gene is a key enzyme of the glycolytic pathway. It is strongly and constitutively expressed in most tissues of *L. edodes* (Hirano et al., 1999). Because only one copy of the GPD gene was detected in *L. edodes*, it was suggested that the GPD promoter is very active and can be a useful component of transformation vectors (Hirano et al., 1999). A plasmid pLG-hph was constructed with an *L. edodes* GPD promoter and terminator and introduced into *L. edodes* by REMI transformation (Hirano et al., 2000). Under the regulation of a GPD promoter, the heterologous gene can be stably expressed in *L. edodes* (Hirano et al., 2000). In addition, GPD expression signals show that the stable and integrative transformation of *P. ostreatus* to hygromycin B resistance is maintained (Irie et al., 2001) and that foreign genes introduced in *L. edodes* by PEG-mediated transformation are effectively expressed (Sun et al., 2001). The *PriA* gene was isolated and found to be highly expressed in the primordia/immature fruiting bodies of *L. edodes*

(Kajiwara et al., 1992). The *PriA* promoter and/or terminator has been used in the transformation of *P. ostreatus* (Yanai et al., 1996), *C. cinereus* (Ogawa et al., 1998), *L. edodes* (Hirano et al., 2000; Sato et al., 1998), and *Coriolus hirsutus* (Orihara et al., 2005). The *Ras* gene is highly and constitutively expressed in *L. edodes* (Hori et al., 1991), and the *Ras* gene promoter has been used in the transformation of *P. ostreatus* (Yanai et al., 1996), *C. cinereus* (Ogawa et al., 1998), and *L. edodes* (Sato et al., 1998). The *Ras* gene promoter has been used with the *PriA* gene promoter to effectively regulate heterologous gene expression in different fungi.

2.7 PROCESS ANALYSIS

2.7.1 Postharvest Studies

Studies of lentinan degradation and fruiting body senescence during the postharvest preservation of *L. edodes* have shown an increase in glucanase activity and isolated three genes, *tlg1, exg1*, and *exg2* (Sakamoto et al., 2005a, 2005b; 2006). The *tlg1* gene encodes a thaumatin-like protein that may be involved in lentinan and cell wall degradation during senescence following harvest and spore diffusion (Sakamoto et al., 2006). The *exg1* gene encodes *exo*-(1 $\rightarrow$ 3)-β-glucanase, and the *exg2* gene encodes *exo*-(1 $\rightarrow$ 3)-β-glucanase.

2.7.2 Stress Responses

2.7.2.1 Studies of Temperature Stress in Mushrooms Although temperature stress is commonly employed in forced fruiting and appears to be a crucial factor in fruiting body induction, it has not been studied at the molecular level in *L. edodes*. Some studies have been carried out, however, in another popular edible mushroom, the phoenix tail mushroom *Pleurotus sajor-caju* (Jeong et al., 2000; Lee et al., 2006).

2.7.2.2 Studies of Molecular Chaperones in Fungi Molecular chaperones are the typical target genes in studies of stress responses in fungi. Several chaperones in *L. edodes* have been studied.

Role of Molecular Chaperones. Molecular chaperones are proteins that mediate appropriate protein folding by binding to unfolded or unassembled proteins and protein translocation from cytosol to target organelles during normal growth. They also protect certain other proteins from misfolding by renaturation under heat and other physiological stresses (Craig et al., 1993). In addition, some molecular chaperones, such as HSP70, have been found to prevent the activation of stress kinase (Gabai et al., 1997). The 70-kilodalton heat-shock proteins (HSP70) are the most well characterized due to their highly conserved amino acid sequences (50–98% similarity) among all species, from bacteria to human (Lindquist and Craig, 1988). The differential expression of the chaperones, *Ssb* and *TCP1*, and the cochaperones, *Mge1* and *Sti1*, during the fruiting body development of *L. edodes* suggests

TABLE 2.5 Lignocellulytic Enzymes Isolated in *L. edodes*

Biopolymer[a]	Enzymes	Genes	References
Cellulose	*endo*-1,4-β-D-Glucanase	*Le.egl*	Kwok et al., 2006
	Cellulase	*Cel7A and 6B*	Lee et al., 2001
Xylan	Xylanase	*xyn11A*	Lee et al., 2005
Lignin	Manganese-dependent peroxidase 1	Mn(II) peroxidase	NCBI accession no. BAE79199
	Laccase	*lac1* and *2*	Zhao and Kwan, 1999

their biological significance to mushroom development (Bian, 2001). Although in most studies the mechanisms are still unclear, the effect of temperature stress on development is still worth studying at the molecular level for insights into the low-temperature-induced forced-fruiting phenomenon.

2.7.3 Lignocellulose Degradation

As it is a white rot fungus, *L. edodes* secretes extracellular enzymes to degrade both lignin and cellulose to leave a light, white, and fibrous residue from wood (Table 2.5). Lignin is a phenylpropanoid polymer that has a complex, heterogeneous structure that is difficult to degrade. Up to 30% of plant material, however, is composed of lignin, which allows plants to have an integrity structure and protects against pests and pathogens. The lignin-degrading enzyme lignin peroxidase has been isolated from the white rot fungus *Phanerochaete chrysosporium*. Lignin is degraded with the presence of hydrogen peroxide, and the haem enzyme releases highly reactive, transient free radicals of oxygen that break the covalent bonds of lignin to release the phenolic compounds that are characteristic of lignin breakdown (Carlile and Watkinson, 1997b; Kirk and Fenn, 1982). To use lignin, *L. edodes* has to generate specific enzymes to degrade lignin. Only a few genes of ligninocellulytic enzymes have been isolated in *L. edodes*. Therefore, the molecular aspects of biodegradation by *L. edodes* are still unclear. Using the sequencing-by-synthesis approach (454 Life Science), *Le.egl* has been isolated from the lignin grown mycelium of *L. edodes*. *Cel7A* and *6B*, *xyn11A*, and *Le.egl* for cellulose and xylan degradation have also been isolated (Kwok et al., 2006; Lee et al., 2001, 2005).. Also, *lac1* and *lac2*, encoding laccase proteins, were isolated from *L. edodes* (Zhao and Kwan, 1999). The gene expressions of *lac1* and *lac2* in different substrates and various developmental stages have been analyzed and shown to be very variable (Zhao and Kwan 1999).

2.7.4 Meiosis

Ribonucleotide reductase small subunit cDNA, *Le.rnr2c*, is most actively transcribed in the hymenophores of the mature fruiting bodies of *L. edodes* (Kaneko and Shishido, 2001), and in situ hybridization analysis has shown this to be the case. This enzyme plays a role in the nucleotide biosynthesis that would be essential both for the production of basidiospores and for the divergence of trama cells into subhymenium cells in the hymenophore (Kaneko and Shishido, 2001).

2.8 CONCLUSION

Our understanding of the growth and development of *L. edodes* is still fragmented. More studies are required to unravel the mechanisms of growth and development in this popular edible mushroom. In the past, molecular genetic studies using RAPD, AP-PCR, and other PCR-based strain-typing methods have been carried out to identify strains with desirable characteristics for breeding and to generate a genetic linkage map of *L. edodes*. More recently, gene cloning and transformation techniques have been developed to identify the functions of the genes and to study the physiological processes of the mushroom. Recently, gene expression profiles under different conditions or developmental stages have been studied by using high-throughput technologies such as cDNA-RDA, SAGE, microarray, and sequencing by synthesis. The transformation efficiency of *L. edodes*, however, is too low to perform functional tests on the novel genes isolated with high-throughput technologies. Current knowledge obtained from the molecular studies of *L. edodes* reveals that gene expression profiles are very different in dikaryotic mycelium to primordium and eventually in mature fruiting bodies. The genes differentially expressed at the initiation of the fruiting body can be categorized into (1) the initiation–stress response and specific signal transduction; (2) the reconstruction of proteome–protein degradation, modification, and biosynthesis; and (3) the switching of biochemical pathways and structural components (Chum et al., 2006a). In particular, certain biological processes, such as postharvest degradation, lignindegradation, and endocytosis, have been studied. Molecular studies of *L. edodes* advance our understanding of its growth and developmental processes and of those of edible mushrooms in general.

REFERENCES

Akiyama, R., Sato, Y., Kajiwara, S., and Shishido, K. (2002). Cloning and expression of cytochrome P450 genes, belonging to a new P450 family, of the basidiomycete *Lentinula edodes*. *Bioscience, Biotechnology and Biochemistry, 66*, 2183–2188.

Arima, T., Yamamoto, M., Hirata, A., Kawano, S., and Kamada, T. (2004). The eln3 gene involved in fruiting body morphogenesis of *Coprinus cinereus* encodes a putative membrane protein with a general glycosyltransferase domain. *Fungal Genetics and Biology, 41*, 805–812.

Bian, X. L. (2001). Isolation and characterization of differentially expressed genes during fruiting body development of xianggu *Lentinula edodes*. Ph.D. thesis. Hong Kong: Chinese University of Hong Kong.

Carlile, M. J. and Watkinson, S. C. (1997a). Life cycle, mating systems and genetic exchange. In *The Fungi*, 4th ed. New York: Academic, pp. 206–220.

Carlile, M. J. and Watkinson, S. C. (1997b). Saprotrophs and ecosystems. In *The Fungi*, 4th ed. New York: Academic, pp. 251–305.

Chang, S. T., Kwan, H. S., and Kang, Y. N. (1995). Collection, characterization, and utilization of germ-plasm of *Lentinula edodes. Canadian Journal of Botony, 73*, S955–S961.

Chum, W. W. and Kwan, H. S. (2005). Modification of LongSAGE for obtaining and cloning long concatemers. *Biotechniques, 39*, 637–638, 640.

Chum, W. W. Y., Kwok, I. S. W., Bian, X. L., Ng, K. T. P., Shih, R. S. M., and Kwan, H. S. (2006a). Differential display, expressed sequence tags and serial analysis of gene expression (SAGE) reveal gene expression profiles of shiitake mushroom *Lentinula edodes* development. In *Eighth International Mycological Congress*. Meyer, W. and Pearce, C., editors. Cairns, Australia: Medimond, pp. 47–51.

Chum, W. W. Y., Kwok, I. S. W., Ng, K. T. P., Bian, X. L., and Kwan, H. S. (2006b). Transcriptome of sporulation in *Lentinula edodes*. In *Eighth International Mycological Congress*. Meyer, W. and Pearce, C., editors. Cairns, Australia: Medimond, pp. 163–166.

Chum, W. W. Y., Ng, K. T. P., Shih, R. S. M., Au, C. H., and Kwan, H. S. (2008). Gene expression studies of the dikaryotic mycelium and primordium of *Lentinula edodes* by serial analysis of gene expression. *Mycological Research*, in press.

Chum, W. W. Y., Shih, R. S. M., Lee, M. T., Au, C. H., and Kwan, H. S. Identification by cDNA microarray analysis of genes with expression related to fruiting body yield in *Lentinula edodes*. Unpublished.

Craig, E. A., Gambill, B. D., and Nelson. R. J. (1993). Heat shock proteins: Molecular chaperones of protein biogenesis. *Microbiological Reviews, 57*, 402–414.

Endo, H., Kajiwara, S., Tsunoka, O., and Shishido, K. (1994). A novel cDNA, priBc, encoding a protein with a Zn(II)2Cys6 zinc cluster DNA-binding motif, derived from the basidiomycete *Lentinus edodes. Gene, 139*, 117–121.

Fields, S. and Song, O. (1989). A novel genetic system to detect protein-protein interactions. *Nature, 340*, 245–246.

Fukuda, M., Harada, Y., Imahori, S., Fukumasa-Nakai, Y., and Hayashi, Y. (1995). Inheritance of mitochondrial DNA in sexual crosses and protoplast cell fusions in *Lentinula edodes. Current Genetics, 27*, 550–554.

Gabai, V. L., Meriin, A. B., Mosser, D. D., Caron, A. W., Rits, S., Shifrin, V. I., and Sherman, M. Y. (1997). Hsp70 prevents activation of stress kinases. A novel pathway of cellular thermotolerance. *Journal of Biological Chemistry, 272*, 18033–18037.

Gostimskii, S. A., Kokaeva, Z. G., and Konovalov, F. A. (2005). Studying plant genome variation using molecular markers. *Genetika, 41*, 480–492.

Guo, L. Q., Lin, J. F., Xiong, S., and Chen, S. C. (2005). Transformation of *Volvariella volvacea* with a thermal hysteresis protein gene by particle bombardment. *Wei Sheng Wu Xue Bao, 45*, 39–43.

Hagen, D. C., McCaffrey, G., and Sprague, G. F. Jr. (1986). Evidence the yeast STE3 gene encodes a receptor for the peptide pheromone a-factor: Gene sequence and implications for the structure of the presumed receptor. *Proceedings of the National Academy of Sciences of the United States of America, 83*, 1418–1422.

Hasebe, K., Murakami, S., and Tsuneda, A. (1991). Cytology and genetics of a sporeless mutant of *Lentinus edodes. Mycologia, 83*, 354–359.

Hirano, T., Sato, T., and Enei, H. (2004). Isolation of genes specifically expressed in the fruit body of the edible basidiomycete *Lentinula edodes*. *Bioscience, Biotechnology and Biochemistry, 68*, 468–472.

Hirano, T., Sato, T., Okawa, K., Kanda, K., Yaegashi, K., and Enei, H. (1999). Isolation and characterization of the glyceraldehyde-3-phosphate dehydrogenase gene of *Lentinus edodes*. *Bioscience, Biotechnology and Biochemistry, 63*, 1223–1227.

Hirano, T., Sato, T., Yaegashi, K., and Enei, H. (2000). Efficient transformation of the edible basidiomycete *Lentinus edodes* with a vector using a glyceraldehyde-3-phosphate dehydrogenase promoter to hygromycin B resistance. *Molecular and General Genetics, 263*, 1047–1052.

Hori, K., Kajiwara, S., Saito, T., Miyazawa, H., Katayose, Y., and Shishido, K. (1991). Cloning, sequence analysis and transcriptional expression of a ras gene of the edible basidiomycete *Lentinus edodes*. *Gene, 105*, 91–96.

Irie, T., Honda, Y., Hirano, T., Sato, T., Enei, H., Watanabe, T., and Kuwahara, M. (2001). Stable transformation of *Pleurotus ostreatus* to hygromycin B resistance using *Lentinus edodes* GPD expression signals. *Applied Microbiology and Biotechnology, 56*, 707–709.

Irie, T., Sato, T., Saito, K., Honda, Y., Watanabe, T., Kuwahara, M., and Enei, H. (2003). Construction of a homologous selectable marker gene for *Lentinula edodes* transformation. *Bioscience, Biotechnology and Biochemistry, 67*, 2006–2009.

Ishizaki, T. and Shishido, K. (2000). Decreased zinc ion accumulation by the basidiomycete *Lentinus edodes* over-expressing *L. edodes* PriA gene. *FEMS Microbiology Letters, 193*, 111–115.

Ishizaki, T., Yokoyama, H., Kajiwara, S., and Shishido, K. (2000). Basidiomycete fungal gene encoding a regulatory subunit a homologue of protein phosphatase 2A. *Bioscience, Biotechnology and Biochemistry, 64*, 438–442.

Jeong, M. J., Park, S. C., Kwon, H. B., and Byun, M. O. (2000). Isolation and characterization of the gene encoding glyceraldehyde-3-phosphate dehydrogenase. *Biochemical and Biophysical Research Communications, 278*, 192–196.

Kajiwara, S., Yamaoka, K., Hori, K., Miyazawa, H., Saito, T., Kanno, T., and Shishido, K. (1992). Isolation and sequence of a developmentally regulated putative novel gene, priA, from the basidiomycete *Lentinus edodes*. *Gene, 114*, 173–178.

Kaneko, S., Miyazaki, Y., Yasuda, T., and Shishido, K. (1998). Cloning, sequence analysis and expression of the basidiomycete *Lentinus edodes* gene uck1, encoding UMP-CMP kinase, the homologue of *Saccharomyces cerevisiae* URA6 gene. *Gene, 211*, 259–266.

Kaneko, S. and Shishido, K. (2001). Cloning and sequence analysis of the basidiomycete *Lentinus edodes* ribonucleotide reductase small subunit cDNA and expression of a corresponding gene in *L. edodes*. *Gene, 262*, 43–50.

Katsukawa, S. and Shishido, K. (2005). Analysis of recQ gene transcript in fruiting bodies of basidiomycetous mushroom *Lentinula edodes*. *Bioscience, Biotechnology and Biochemistry, 69*, 2247–2249.

Katsukawa, S., Yamazaki, T., Kajiwara, S., and Shishido, K. (2004). A recQ gene homolog from the basidiomycetous mushroom *Lentinula edodes*: Sequence analysis and expression. *Bioscience, Biotechnology and Biochemistry, 68*, 2588–2597.

Kershaw, M. J. and Talbot, N. J. (1998). Hydrophobins and repellents: Proteins with fundamental roles in fungal morphogenesis. *Fungal Genetics and Biology, 23*, 18–33.

Kesseli, R. V., Paran, J., and Michelmore, R. W. (1993). Efficient mapping of specifically targeted genomic regions and the tagging of these regions with reliable PCR-based genetic markers. In *Applications of RAPD Technology to Plant Breeding: Selected Papers from the Joint Plant Breeding Symposia Series*. Nienhuis, J., editor. Minneapolis, MN: Crop Science Society of America; American Society for Horticultural Science; American Genetic Association, pp. 31–36.

Kikuchi, M., Kitamoto, N., and Shishido, K. (2004). Secretory production of *Aspergillus oryzae* xylanase XynF1, xynF1 cDNA product, in the basidiomycete *Coprinus cinereus*. *Applied Microbiology and Biotechnology, 63*, 728–733.

Kikuchi, M., Ogawa, K., Yamazaki, T., Kajiwara, S., Sugio, A., Nakamura, S., and Shishido, K. (1999). Secretional expression of a *Bacillus subtilis* xylanase gene in the basidiomycete *Coprinus cinereus*. *FEMS Microbiology Letters, 178*, 277–282.

Kilaru, S., Hoegger, P. J., Majcherczyk, A., Burns, C., Shishido, K., Bailey, A., Foster, G. D., and Kües, U. (2006). Expression of laccase gene lcc1 in *Coprinopsis cinerea* under control of various basidiomycetous promoters. *Applied Microbiology and Biotechnology, 71*, 200–210.

Kirk, T. K. and Fenn, P. (1982). Formation and action of the ligninolytic system in basidiomycetes. In *Fourth Symposium of the British Mycological Society*. Frankland, J. C., Hedger, J. N., and Swift, M. J., editors. Cambridge: Cambridge University Press, pp. 67–90.

Kobayashi, H., Kumagai, F., Itagaki, T., Inokuchi, N., Koyama, T., Iwama, M., Ohgi, K., and Irie, M. (2000). Amino acid sequence of a nuclease (nuclease Le1) from *Lentinus edodes*. *Bioscience, Biotechnology and Biochemistry, 64*, 948–957.

Kobayashi, H., Kumagai, F., Itagaki, T., Koyama, T., Inokuchi, N., Iwama, M., Ohgi, K., and Irie, M. (2002). Amino acid sequence and characterization of a nuclease (nuclease Le3) from *Lentinus edodes*. *Bioscience, Biotechnology and Biochemistry, 66*, 1345–1355.

Kondoh, O., Muto, A., Kajiwara, S., Takagi, J., Saito, Y., and Shishido, K. (1995). A fruiting body-specific cDNA, mfbAc, from the mushroom *Lentinus edodes* encodes a high-molecular-weight cell-adhesion protein containing an Arg-Gly-Asp motif. *Gene, 154*, 31–37.

Kondoh, O. and Shishido, K. (1995). Characterization of the promoter region of a cell-adhesion protein gene derived from the basidiomycete *Lentinus edodes*. *FEMS Microbiology Letters, 130*, 189–192.

Kües, U. (2000). Life history and developmental processes in the basidiomycete *Coprinus cinereus*. *Microbiology and Molecular Biology Reviews, 64*, 316–353.

Kües, U. and Liu, Y. (2000). Fruiting body production in basidiomycetes. *Applied Microbiology and Biotechnology, 54*, 141–152.

Kuo, C. Y., Chou, S. Y., and Huang, C. T. (2004). Cloning of glyceraldehyde-3-phosphate dehydrogenase gene and use of the gpd promoter for transformation in *Flammulina velutipes*. *Applied Microbiology and Biotechnology, 65*, 593–599.

Kwan, H. S., Chiu, S. W., Pang, K. M., and Cheng, S. C. (1992a). Strain typing in *Lentinula edodes* by polymerase chain reaction. *Experimental Mycology, 16*, 163–166.

Kwan, H. S., Pang, K. M., Chiu, S. W., and Cheng, S. C. (1992b). Nucleotide sequence of the 5.8S ribosomal RNA gene of *Lentinula edodes*. *Nucleic Acids Research, 20*, 610.

Kwan, H. S. and Xu, H. L. (2002). Construction of a genetic linkage map of shiitake mushroom *Lentinula edodes* strain L-54. *Journal of Biochemistry and Molecular Biology, 35*, 465–471.

Kwok, I. S. W., Chum, W. W. Y., and Kwan, H. S. (2006). Genes expressed in lignocellulose grown *Lentinula edodes*. In *Eighth International Mycological Congress*. Meyer, W. and Pearce, C., editors. Cairns, Australia: Medimond, pp. 209–211.

Lee, C., Wong, D. W. S. and Robertson, G. H. (2005). Cloning and characterization of the xyn11A gene from *Lentinula edodes*. *Protein Journal, 24*, 21–26.

Lee, C. C., Wong, D. W. S. and Robertson, G. H. (2001). Cloning and characterization of two cellulase genes from *Lentinula edodes*. *FEMS Microbiology Letters, 205*, 355–360.

Lee, J. O., Jeong, M. J., Kwon, T. R., Lee, S. K., Byun, N. O., Chung, M., and Park, S. C. (2006). *Pleurotus sajor-caju* HSP100 complements a thermotolerance defect in hsp104 mutant *Saccharomyces cerevisiae*. *Journal of Biosciences, 31*, 435.

Lee, M. T., Szeto, C. Y., Ng, T. P., and Kwan, H. S. (2007). Endocytosis in the shiitake mushroom *Lentinula edodes* and involvement of GTPase LeRAB7. *Eukaryotic Cell, 6*, 2406–2418.

Leung, G. S. W., Zhang, M., Xie, W. J., and Kwan, H. S. (2000). Identification by RNA fingerprinting of genes differentially expressed during the development of the basidiomycete *Lentinula edodes*. *Molecular and General Genetics, 262*, 977–990.

Li, A. Z., Xu, X. F., Lin, F. X., Cheng, S. M., and Lin, F. C. (2007). Cloning and identification of DNA fragment for the B mating-type factor in *Lentinula edodes* using degenerate PCR. *World Journal of Microbiology and Biotechnology, 23*, 411–415.

Li, G., Li, R., Liu, Q., Wang, Q., Chen, M., and Li, B. (2006). A highly efficient polyethylene glycol-mediated transformation method for mushrooms. *FEMS Microbiology Letters, 256*, 203–208.

Lindquist, S., and Craig, E. A. (1988). The heat-shock proteins. *Annual Review of Genetics, 22*, 631–77.

Margulies, M., et al. (2005). Genome sequencing in microfabricated high-density picolitre reactors. *Nature, 437*, 376–380.

Meyer, V., Mueller, D., Strowig, T., and Stahl, U. (2003). Comparison of different transformation methods for *Aspergillus giganteus*. *Current Genetics, 43*, 371–377.

Miles, P. G. (1993). Biological background from mushroom breeding. In *Genetics and Breeding of Edible Mushrooms*, Chang, S. T., Buswell, J. A., and Miles, P. G., editors. Philadelphia: Gordon & Breach Scientific, pp. 37–64.

Miller, T. K., Renault, S., and Selitrennikoff, C. P. (2002). Molecular dissection of alleles of the osmotic-1 locus of *Neurospora crassa*. *Fungal Genetics and Biology, 35*, 147–155.

Miyazaki, Y., Jojima, T., Ono, T., Yamazaki, T., and Shishido, K. (2004a). A cDNA homologue of *Schizosaccharomyces* pombe cdc5(+) from the mushroom *Lentinula edodes*: Characterization of the cDNA and its expressed product. *Biochimica et Biophysica Acta, 1680*, 93–102.

Miyazaki, Y., Nakamura, M., and Babasaki, K. (2005). Molecular cloning of developmentally specific genes by representational difference analysis during the fruiting body formation in the basidiomycete *Lentinula edodes*. *Fungal Genetics and Biology, 42*, 493–505.

Miyazaki, Y., Sakuragi, Y., Yamazaki, T., and Shishido, K. (2004b). Target genes of the developmental regulator PRIB of the mushroom *Lentinula edodes*. *Bioscience, Biotechnology and Biochemistry, 68*, 1898–1905.

Miyazaki, Y., Tsunoka, O., and Shishido, K. (1997). Determination of the DNA-binding sequences of the Zn(II)2Cys6 zinc-cluster-containing PRIB protein, derived from the basidiomycete *Lentinus edodes* gene. *Journal of Biochemistry, 122*, 1088–1091.

Moore, D. (1998). *Fungal Morphogenesis*. New York: Cambridge University Press.

Nagahashi, S., Mio, T., Ono, N., Yamada-Okabe, T., Arisawa, M., Bussey, H., and Yamada-Okabe, H. (1998). Isolation of CaSLN1 and CaNIK1, the genes for osmosensing histidine kinase homologues, from the pathogenic fungus *Candida albicans*. *Microbiology, 144*, 425–432.

Nakazawa, T., Miyazaki, Y., Kaneko, S., and Shishido, K. (2006). Developmental regulator Le.CDC5 of the mushroom *Lentinula edodes*: Analyses of its amount in each of the stages of fruiting-body formation and its distribution in parts of the fruiting bodies. *FEMS Microbiology Letters, 261*, 60–63.

Ng, W. L., Ng, T. P., and Kwan, H. S. (2000). Cloning and characterization of two hydrophobin genes differentially expressed during fruit body development in *Lentinula edodes*. *FEMS Microbiology Letters, 185*, 139–145.

Nishizawa, H., Miyazaki, Y., Kaneko, S., and Shishido, K. (2002). Distribution of hydrophobin 1 gene transcript in developing fruiting bodies of *Lentinula edodes*. *Bioscience, Biotechnology and Biochemistry, 66*, 1951–1954.

Noel, T. and Labarere, J. (1994). Homologous transformation of the edible basidiomycete *Agrocybe aegerita* with the URA1 gene: Characterization of integrative events and of rearranged free plasmids in transformants. *Current Genetics, 25*, 432–437.

Ogawa, K., Yamazaki, T., Kajiwara, S., Watanabe, A., Asada, Y., and Shishido, K. (1998). Molecular breeding of the basidiomycete *Coprinus cinereus* strains with high lignin-decolorization and -degradation activities using novel heterologous protein expression vectors. *Applied Microbiology and Biotechnology, 49*, 285–289.

Ohga, S. and Royse, D. J. (2001). Transcriptional regulation of laccase and cellulase genes during growth and fruiting of *Lentinula edodes* on supplemented sawdust. *FEMS Microbiology Letters, 201*, 111–115.

Orihara, K., Yamazaki, T., Shinkyo, R., Sakaki, T., Inouye, K., Tsukamoto, A., Sugiura, J., and Shishido, K. (2005). Rat cytochrome P450-mediated transformation of dichlorodibenzo-*p*-dioxins by recombinant white-rot basidiomycete *Coriolus hirsutus*. *Applied Microbiology and Biotechnology, 69*, 22–28.

Posas, F., WurglerMurphy, S. M., Maeda, T., Witten, E. A., Thai, T. C., and Saito, H. (1996). Yeast HOG1 MAP kinase cascade is regulated by a multistep phosphorelay mechanism in the SLN1-YPD1-SSK1" two-component" osmosensor. *Cell, 86*, 865–875.

Pott, G. B., Miller, T. K., Bartlett, J. A., Palas, J. S., and Selitrennikoff, C. P. (2000). The isolation of FOS-1, a gene encoding a putative two-component histidine kinase from *Aspergillus fumigatus*. *Fungal Genetics and Biology, 31*, 55–67.

Qin, L. H., Tan, Q., Chen, M. J., and Pan, Y. J. (2006). Use of intersimple sequence repeats markers to develop strain-specific SCAR markers for *Lentinula edodes*. *FEMS Microbiology Letters, 257*, 112–116.

Royse, D. J. and May, B. (1987). Identification of shiitake genotypes by multilocus enzyme electrophoresis—catalog of lines. *Biochemical Genetics, 25*, 705–716.

Saha, S,. Sparks, A. B., Rago, C., Akmaev, V., Wang, C. J., Vogelstein, B., Kinzler, K. W., and Velculescu, V. E. (2002). Using the transcriptome to annotate the genome. *Nature Biotechnology, 20*, 508–512.

Saito, T., Tanaka, N., and Shinozawa, T. (2002). Characterization of subrepeat regions within rDNA intergenic spacers of the edible basidiomycete *Lentinula edodes*. *Bioscience, Biotechnology and Biochemistry, 66*, 2125–2133.

Sakai, H. and Kajiwara, S. (2003). A stearoyl-CoA-specific Delta 9 fatty acid desaturase from the basidiomycete *Lentinula edodes*. *Bioscience, Biotechnology and Biochemistry, 67*, 2431–2437.

Sakai, H. and Kajiwara, S. (2005). Cloning and functional characterization of a delta 12 fatty acid desaturase gene from the basidiomycete *Lentinula edodes*. *Molecular Genetics and Genomics, 273*, 336–341.

Sakamoto, Y., Irie, T., and Sato, T. (2005a). Isolation and characterization of a fruiting body-specific exo-(1 → 3)-β-glucanase-encoding gene, *exg1*, from *Lentinula edodes*. *Current Genetics, 47*, 244–252.

Sakamoto, Y., Minato, K., Nagai, M., Mizuno, M., and Sato, T. (2005b). Characterization of the *Lentinula edodes* exg2 gene encoding a lentinan-degrading exo-(1 → 3)-β-glucanase. *Current Genetics, 48*, 195–203.

Sakamoto, Y., Watanabe, H., Nagai, M., Nakade, K., Takahashi, M., and Sato, T. (2006). *Lentinula edodes* tlg1 encodes a thaumatin-like protein that is involved in lentinan degradation and fruiting body senescence. *Plant Physiology, 141*, 793–801.

Sato, T., Yaegashi, K., Ishii, S., Hirano, T., Kajiwara, S., Shishido, K., and Enei, I. (1998). Transformation of the edible basidiomycete *Lentinus edodes* by restriction enzyme-mediated integration of plasmid DNA. *Bioscience, Biotechnology and Biochemistry, 62*, 2346–2350.

Schiestl, R. H. and Petes, T. D. (1991). Integration of DNA fragments by illegitimate recombination in *Saccharomyces cerevisiae*. *Proceedings of the National Academy of Sciences of the United States of America, 88*, 7585–7589.

Shih, R. S. M. (2003). Identification and characterization of differentially expressed genes in dikaryons of *Lentinula edodes* by cDNA microarray. M.Phil. thesis. Hong Kong: the Chinese University of Hong Kong.

Stone, P. J., Makoff, A. J., Parish, J. H., and Radford, A. (1993). Cloning and sequence analysis of the glucoamylase gene of *Neurospora crassa*. *Current Genetics, 24*, 205–211.

Struhl, K. (1995). Yeast transcriptional regulatory mechanisms. *Annual Review of Genetics, 29*, 651–674.

Sun, L., Xu, W. H., Cai, H. Q., Hu, Y. L., and Lin, Z. P. (2001). PEG-mediated transformation of *Lentinus edodes*. *Acta Botanica Sinica, 43*, 1089–1092.

Sun, M., Zhou, G., Lee, S., Chen, J., Shi, R. Z., and Wang, S. M. (2004). SAGE is far more sensitive than EST for detecting low-abundance transcripts. *BMC Genomics, 5*, 1.

Sunagawa, M. and Magae, Y. (2002). Transformation of the edible mushroom *Pleurotus ostreatus* by particle bombardment. *FEMS Microbiology Letters, 211*, 143–146.

Szeto, C. Y. Y., Leung, G. S., and Kwan, H. S. (2007). Le.MAPK and its interacting partner, Le.DRMIP, in fruiting body development in *Lentinula edodes*. *Gene, 393*, 87–93.

Szeto, C. Y. Y., Wong, Q. W. L., Leung, G. S., and Kwan, H. S. (2008). Isolation and transcript analysis of two-component histidine kinase gene *Le.nik1* in Shiitake mushroom, *Lentinula edodes*. *Mycological Research, 112*, 108–116.

Takagi, Y., Katayose, Y., and Shishido, K. (1988). Intracellular levels of cyclic AMP and adenylate cyclase activity during mycelial development in fruiting body formation in *Lentinus edodes*. *FEMS Microbiology Letters, 55*, 275–278.

Tanaka, A., Miyazaki, K., Murakami, H., and Shiraishi, S. (2004). Sequence characterized amplified region markers tightly linked to the mating factors of *Lentinula edodes*. *Genome 47*, 156–162.

Tanaka, Y., Kaneko, S., Katsukawa, S., Yamazaki, T., Ohta, A., Miyazaki, Y., and Shishido, K. (2005). Specific distribution in homobasidiomycete hymenophores of the transcripts of Ras protein and G-protein alpha-subunit genes. *FEMS Microbiology Letters, 242*, 169–175.

Terashima, K., Matsumoto, T., Hasebe, K., and Fukumasa-Nakai, Y. (2002a). Genetic diversity and strain-typing in cultivated strains of *Lentinula edodes* (the shii-take mushroom) in Japan by AFLP analysis. *Mycological Research, 106*, 34–39.

Terashima, K., Matsumoto, T. Hayashi, E., and Fukumasa-Nakai, Y. (2002b). A genetic linkage map of *Lentinula edodes* (shiitake) based on AFLP markers. *Mycological Research, 106*, 911–917.

Tokimoto, K. and Komatsu, M. (1995). Selection and breeding of shiitake strains resistant to *Trichoderma* spp. *Revue Canadienne de Botanique, 73*, S962–S966.

Vedoy, C. G., Bengtson, M. H., and Sogayar, M. C. (1999). Hunting for differentially expressed genes. *Brazilian Journal of Medical and Biological Research, 32*, 877–884.

Velculescu, V. E., Vogelstein, B., and Kinzler, K. W. (2000). Analysing uncharted transcriptomes with SAGE. *Trends in Genetics, 16*, 423–425.

Velculescu, V. E., Zhang, L., Vogelstein, B., and Kinzler, K. W. (1995). Serial analysis of gene expression. *Science, 270*, 484–487.

Vos, P., Hogers, R., Bleeker, M., Reijans, M., Vandelee, T., Hornes, M., Frijters, A., Pot, J., Peleman, J., Kuiper, M., and Zabeau, M. (1995). AFLP: A new technique for DNA fingerprinting. *Nucleic Acids Research, 23*, 4407–4414.

Wahl, M. B., Heinzmann, U., and Imai, K. (2005). LongSAGE analysis significantly improves genome annotation: Identifications of novel genes and alternative transcripts in the mouse. *Bioinformatics, 21*, 1393–1400.

Welsh, J. and McClelland, M. (1990). Fingerprinting genomes using PCR with arbitrary primers. *Nucleic Acids Research,18*, 7213–7218.

Wessels, J. G. (1993). Fruiting in the higher fungi. *Advances in Microbial Physiology, 34*, 147–202.

Williams, J. G. K., Kubelik, A. R., Livak, K. J., Rafalski, J. A., and Tingey, S. V. (1990). DNA polymorphisms amplified by arbitrary primers are useful as genetic markers. *Nucleic Acids Research, 18*, 6531–6535.

Yamamoto, M., Wakatsuki, T., Hada, A., and Ryo, A. (2001). Use of serial analysis of gene expression (SAGE) technology. *Journal of Immunological Methods, 250*, 45–66.

Yamazaki, T., Yasuda, T., Miyazaki, Y., Okada, K., Kajiwara, S., and Shishido, K. (2002). A promoter activity in *Saccharomyces cerevisiae* of the 3′-noncoding region of the basidiomycetous mushroom gene. *Journal of General and Applied Microbiology, 48*, 223–231.

Yanai, K., Yonekura, K., Usami, H., Hirayama, M., Kajiwara, S., Yamazaki, M., Shishido, K., and Adachi, T. (1996). The integrative transformation of *Pleurotus ostreatus* using bialaphos resistance as a dominant selectable marker. *Bioscience, Biotechnology and Biochemistry, 60*, 472–475.

Yasuda, T. and Shishido, K. (1999). Analysis of mfbA alleles of the eubasidiomycete *Lentinus edodes*. *Bioscience, Biotechnology and Biochemistry, 63*, 2228–2231.

Yin, Z., Plader, W., and Malepszy, S. (2004). Transgene inheritance in plants. *Journal of Immunological Methods, 45*, 127–144.

Yoshimi, A., Tsuda, M., and Tanaka, C. (2004). Cloning and characterization of the histidine kinase gene Dic1 from *Cochliobolus heterostrophus* that confers dicarboximide resistance and osmotic adaptation. *Molecular Genetics and Genomics, 271*, 228–236.

Zhang, M., Xie, W., Leung, G. S. W., Deane, E. E., and Kwan, H. S. (1998). Cloning and characterization of the gene encoding beta subunit of mitochondrial processing peptidase from the basidiomycete *Lentinula edodes*. *Gene, 206*, 23–27.

Zhang, R. Y., Huang, C. Y., Zheng, S. Y., Zhang, J. X., Ng, T. B., Jiang, R. B., Zuo, X. M., and Wang, H. X. (2007). Strain-typing of *Lentinula edodes* in China with inter simple sequence repeat markers. *Applied Microbiology and Biotechnology, 74*, 140–145.

Zhang, Y. F. and Molina, F. I. (1995). Strain typing of *Lentinula edodes* by random amplified polymorphic DNA assay. *FEMS Microbiology Letters, 131*, 17–20.

Zhao, J., Chen, Y. H., and Kwan, H. S. (2000). Molecular cloning, characterization, and differential expression of a glucoamylase gene from the basidiomycetous fungus *Lentinula edodes*. *Applied and Environmental Microbiology, 66*, 2531–2535.

Zhao, J. and Kwan, H. S. (1999). Characterization, molecular cloning, and differential expression analysis of laccase genes from the edible mushroom *Lentinula edodes*. *Applied and Environmental Microbiology, 65*, 4908–4913.

Zhou, G. L., Miyazaki, Y., Nakagawa, T., Tanaka, K., Shishido, K., Matsuda, H., and Kawamukai, M. (1998). Identification of a CAP (adenylyl-cyclase-associated protein) homologous gene in *Lentinus edodes* and its functional complementation of yeast CAP mutants. *Microbiology, 144*, 1085–1093.

Zhou, G. L., Yamamoto, T., Ozoe, F., Yano, D., Tanaka, K., Matsuda, H., and Kawamukai, M. (2000). Identification of a 14-3-3 protein from *Lentinus edodes* that interacts with CAP (adenylyl cyclase-associated protein), and conservation of this interaction in fission yeast. *Bioscience, Biotechnology and Biochemistry, 64*, 149–159.

Zietkiewicz, E., Rafalski, A., and Labuda, D. (1994). Genome fingerprinting by simple sequence repeat (SSR)-anchored polymerase chain reaction amplification. *Genomics, 20*, 176–183.

CHAPTER 3

Nutritional Value and Health Benefits of Mushrooms

Peter C. K. Cheung

Food and Nutritional Sciences Programme, Department of Biology, The Chinese University of Hong Kong, Hong Kong, China

CONTENTS

3.1 INTRODUCTION

Since the work of Crisan and Sands in the 1970s (Crisan and Sands, 1978), only a few comprehensive reviews have been produced on the nutritional composition of wild and edible mushrooms (Chang and Miles, 2004; de Román et al., 2006). Although the number of cultivated mushroom species (and especially those growing in Asia) has grown recently, there is a lack of knowledge of their chemical composition and nutritional value. This chapter reviews the recent findings on newly developed species of cultivated mushroom and discusses current investigations into the health benefits of edible mushrooms to humans, with emphasis on the prevention of chronic conditions such as cardiovascular disease and diabetes.

Mushrooms as Functional Foods, Edited by Peter C. K. Cheung

3.2 WILD AND CULTIVATED EDIBLE MUSHROOMS

Wild edible mushrooms have been part of the human diet for thousands of years due to their nutritional, organoleptic characteristics and medicinal properties (de Román et al., 2006). An overview of the uses and importance of wild edible mushrooms was published recently in which more than a thousand species of wild mushrooms consumed in 85 countries were reported (Boa, 2004). Among these species, the black truffle (*Tuber melanosporum*), boletes (*Boletus* spp.), and chanterelles (*Cantharellus cibarius*) from Europe and the matsutake (*Tricholoma matsutake*) from China and Japan have the highest economic value.

Fewer than 25 edible mushroom species are widely cultivated and accepted as a food of economic importance (Smith, 1972). In more affluent countries, mushrooms are considered a somewhat expensive type of vegetable and are eaten almost exclusively for their culinary properties and to provide flavor or as a garnish for other foods (Flegg and Maw, 1977). Currently, the possibility of using mushrooms as a source of protein in the human diet in developing countries is being considered (Boras, 1996), as mushrooms are a healthy food that is low in calories and fat but rich in protein, dietary fiber, vitamins, and minerals (Crisan and Sands, 1978). Several studies have been carried out on the chemical composition and nutritional quality of edible mushrooms grown in different parts of the world, including those produced in tropical areas (Aletor, 1995; Sanmee et al., 2003), India (Longvah and Deosthale, 1998), North America (Leichter and Bandoni, 1980), and Europe (Senatore, 1992; Díez and Alvarez, 2001; Manzi et al., 2001; Caglarirmak et al., 2002).

The fruiting bodies of mushrooms are the most common edible form in the human diet (Gray, 1973), but the sclerotia of some mushroom species can also be consumed. The mycelia are less commonly utilized as a human food despite their shorter production time and comparable nutritional value to fruiting bodies (Litchfield, 1967; El-Kattan et al., 1991; Cheung, 1997b).

3.3 PRODUCTION OF CULTIVATED MUSHROOMS

The world production of cultivated mushrooms was estimated to be 6.1 million tons in 1997 and 12.2 million tons in 2002 (Chang, 2006), which represents a doubling in just five years. The world production of *Lentinula edodes* increased from 14.3 to 25.4% (in terms of output, from 180,000 to 1.5 million tons), and that of *Pleurotus* spp. from 2.8 to 14.2% (in terms of output, from 35,000 to 876,000 tons). More than 10 new mushroom species, including *Agaricus blazei, Pleurotus eryngii*, and *Agrocybe aegerita*, have been cultivated in recent years on a small commercial scale, and the potential for expansion is great. However, the output yield of the 10 main cultivated mushrooms makes up 91.6% of the total world production (Table 3.1), with the 6 most important, namely, *Agaricus* (31.8%), *Lentinula* (25.4%), *Pleurotus* (14.2%), *Auricularia* (7.90%), *Flammulina* (4.60%), and *Volvariella* (3.0%), contributing up to 86.9%. In China, *Tremella* is mainly cultivated, whereas in Japan *Hypsizygus marmoreus* and *Grifola frondosa* are the most

TABLE 3.1 World Production of Cultivated Edible Mushrooms in 1997

	Production (Fresh wt × 1000 metric tons)						
Species	China	Japan	Rest of Asia	North America	Latin America	Europe	Africa
Agaricus bisporus	330	NP	68.4	425.3	51.6	990.2	36.0
Lentinula edodes	1397	115.3	47.4	3.60	0.30	0.80	NP
Pleurotus spp.	760	13.3	88.4	1.50	0.20	12.0	0.20
Auricularia spp.	480	NP	5.30	NP	NP	NP	NP
Volvariella volvacea	120	NP	60.8	NP	NP	NP	NP
Flammulina spp.	150	109	25.7	NP	NP	NP	NP
Tremella spp.	130	NP	0.50	NP	NP	NP	NP
Hypsizygus marmoreus	2.10	72.0	0.10	NP	NP	NP	NP
Pholiota nameko	31.0	24.5	NP	NP	NP	NP	NP
Grifola frondosa	2.00	31.0	NP	NP	NP	NP	NP
Hericium erinaceus	0.80	NP	NP	NP	NP	NP	NP
Coprinus comatus	0.50	NP	NP	NP	NP	NP	NP
Others[a]	514.9	2.90	0.80	0.40	NP	0.30	NP
Total	3918.3	368	297.4	430.8	52.1	1003.3	36.2
Percent of total world production	63.6	6.00	4.80	7.00	0.80	16.3	0.60

[a]Mushrooms with emerging commercial potential, e.g., *A. blazei, Lepista nuda, P. eryngii, A. aegerita, Tricholoma giganteum, Auricularia fuscosuccinea, Tremella cinnabarina*, etc.
Note: NP, no production.
Source: From Chang (1999).

common cultivated species. The total mushroom production in China in 1997 was 3.9 million tons, which accounted for 63.6% of the total world output, demonstrating that China has become a leading producer and consumer of cultivated edible mushrooms. *Lentinula edodes* is the leading cultivated mushroom in China, with a production of 1.4 million tons in 1997, which was equal to 35.6% of the total mushroom production in that year.

3.4 NUTRITIONAL COMPOSITION

3.4.1 Conventional Edible Mushrooms

3.4.1.1 Moisture In general, the moisture content of mushrooms ranges from 85 to 95% of their fresh weight. The specific moisture content of some fleshy edible mushrooms is listed as follows: *Grifola frondosa* (86.1%), *Pleurotus ostreatus* (85.2–94.7%), *P. eryngii ferulae* (88.1%), *Pleurotus plumonarius* (87.7%), *L. edodes* (81.8–90%), *Flammulina velutipes* (87.2–89.1%), *Pleurotus cystidious* (86.7%), and *Agaricus bisporus* (92.8–94.8%) (Manzi et al., 1999, 2001; Mau et al., 2001b, c; Yang et al., 2000). However, it must be noted that the moisture content of mushrooms is affected by the time of cropping, watering conditions during cultivation, postharvest period, and temperature and relative humidity during growth (Bano and Rajarathnan, 1988). The moisture content of dry mushrooms is generally less than 10% (w/w), with figures of 9.05% for *Dictypphora indusiata* and 5.30% for *Schizophyllum commune* being previously reported (Longvah and Deosthale, 1998; Mau et al., 2001b, c).

3.4.1.2 Protein and Amino Acids The crude protein content of edible mushrooms is usually high, but varies greatly and is affected by factors such as species and stage of development (Longvah and Deosthale, 1998). The crude protein content [percent dry weight (%DW)] of some common edible mushrooms is as follows: *L. edodes* (15.2–23.0%), *Schizophyllum commune* (16.0–27.0%), *Tricholoma giganteum* (16.1%), *P. ostreatus* (19.9–34.7%), *P. eryngii ferulae* (23.2%), and *Tricholoma terreum* (15.0%) (Longvah and Deosthale, 1998; Manzi et al., 1999; Díez and Alvarez, 2001).

The essential amino acid content (g/100 g protein DW) of mushrooms ranges from 34 to 47%, although one previous report showed that essential amino acid levels can be as high as 61.8% (*Tricholoma portentosum*) and 63.3% (*T. terreum*) (Manzi et al., 1999). The protein of edible mushrooms is comparatively rich in glutamic acid (12.6–24.0%), aspartic acid (9.10–12.1%), and arginine (3.70–13.9%). The essential amino acid profiles of eight mushrooms, together with the Food and Agricultural Organization/World Health Organization (FAO/WHO) requirement patterns (FAO, 1991), are shown in Table 3.2 and reveal that the proteins are deficient in sulfur-containing amino acids, including methionine (1.22–21.6 mg/g protein) and cysteine (15.7–19.1 mg/g protein). However, these edible mushrooms are comparatively rich in threonine (41.2–94.5 mg/g protein) and valine (35.8–88.7 mg/g protein). It has been reported that lysine, leucine, isoleucine, and tryptophan are the limiting amino acids in some edible mushroom proteins (Cheung, 1997b; Manzi et al., 1999; Díez and Alvarez, 2001). The amino acid composition of other wild and cultivated mushrooms was reported previously by Crisan and Sands (1978) and Mdachi et al. (2003).

The free amino acid level in mushrooms is low, ranging from 7.14 to 12.3 mg/g in dry edible mushrooms, with glutamic acid (21.7–23.7%) and alanine (17.7–17.9%) predominating (Manzi et al., 1999). Free amino acids, and especially highly basic amino acids, contribute the main flavor properties of mushrooms (Sugahara et al., 1975; Maga, 1981), but a low level of other free amino acids, such as γ-amino butyric acid (GABA) (10.2–281 mg/100 g DW) and ornithine (88.7–392 mg/100 g DW), has been reported in edible *Pleurotus* mushrooms (Manzi et al., 1999). The profiles of free amino acids in mushrooms are considerably different. Aspartic acid and glutamic acids are monosodium glutamate (MSG)–like components that give the most typical mushroom taste, that is, the umami taste or palatable taste that is the characteristic taste of MSG and 5′-nucleotides (Mau et al., 2001b, c). The content of these MSG-like components in common mushrooms is relatively low, ranging from 22.7 to 47.1 mg/g DW (Tseng and Mau, 1999), 11.2–26.2 mg/g DW in *V. volvacea* (Mau et al., 1997), 10.9–11.9 mg/g DW in *Agrocybe cylindracea* (Mau and Tseng, 1998), and

TABLE 3.2 Essential Amino Acid Profiles (mg/g protein) of Edible Mushrooms

Mushrooms	Ile	Leu	Lys	Met	Cys	Phe	Tyr	The	Val	Trp[a]
Volvariella bombycina	54.1	50.1	54.1	1.22	19.1	60.2	45.8	46.5	35.8	ND
L. ulmarius	58.9	101	46.1	19.1	18.2	51.1	72.1	41.2	56.1	ND
Pleurotus citrinopileatus	35.1	71.2	56.3	25.4	15.8	39.7	32.2	49.2	60.7	ND
T. portentosum	37.2	93.7	86.3	29.6	16.2	43.6	32.1	94.5	77.6	9.60
T. terreum	35.8	81.5	76.3	34.6	17.0	66.1	30.0	90.7	88.7	10.6
L. edodes	33.0	63.8	49.8	21.6	34.0	38.1	26.0	55.5	381	19.2
P. eryngii feruale	41.1	65.6	67.1	16.9	15.7	40.4	34.2	50.4	45.1	12.2
P. ostreatus	44.5	72.8	61.1	20.1	16.8	46.9	40.6	51.6	48.8	40.6
FAO (1991) requirement pattern	28	66	58	25[b]		63[c]		34	35	11

[a]ND, not determined.
[b]Value includes Met and Cys.
[c]Value includes Tyr and Phe.

Source: From FAO, 1991; Cheung, 1997b; Manzi et al., 1999; Díez and Alvarez, 2001; Dabbour and Takruri, 2002a.

3.75–9.06 mg/g DW in *L. edodes* (Lin, 1988). The levels of sweet components, which are mainly soluble sugars (Ala + Gly + Ser + Thr), are also low, ranging from 0.36 to 8.71 mg/g DW. The levels of bitter components (Arg + His + Ile + Leu + Met + Phe + Trp + Val) are high, ranging from 2.37 to 6.46 mg/g DW. However, bitterness from the bitter components can be unequivocally masked by the sweet components (Mau et al., 2001b, c; Yang et al., 2002).

3.4.1.3 Fat Edible mushrooms generally have a low lipid level of less than 10% DW but nevertheless are a source of unsaturated fatty acids, and especially oleic and linoleic acids. In some species, the lipid content may be as low as 2.0% (*L. edodes, S. commune*, and *P. ostreatus*) (Longvah and Deosthale, 1998). The lipid content (%DW) of some other mushrooms are as follows: *Dictyophora indusiata* (2.98%), *T. giganteum* (4.28%), *G. frondosa* (3.10%), *L. edodes* (5.71–6.34%), *Lyophyllum ulmarius* (2.09%), *V. bombycina* (2.75%), and *T. terreum* (6.60%) (Mau et al., 2001b, c; J. H. Yang et al., 2000). The levels of polyunsaturated fatty acids in mushrooms are high, constituting more than 75% of the total fatty acids, of which palmitic (19.2%), oleic (8.3%), and linoleic (68.8–84.0%) acids are the most significant (Cheung, 1997b; Longvah and Deosthale, 1998; Díez and Alvarez, 2001; J. H. Yang et al., 2002). Linolenic acid levels are generally low in mushrooms (Yilmaz et al., 2006), but despite its small quantity, this compound is strongly related to flavor in certain mushrooms, as it is the precursor to 1-octen-3-ol, or the "alcohol" of fungi, and is the principal aromatic compound in most mushrooms (Maga, 1981).

3.4.1.4 Ash and Minerals The ash content in mild edible mushrooms ranges from 6 to 10.9% DW and represents a wide variety of minerals (Zakhary et al.,

1983). Recently, the metal content of wild edible mushroom species in the Mediterranean was reported (Ouzouni et al., 2007). The ash content (%DW) of some edible mushrooms is as follows: *P. ostreatus* (6.90%), *P. eryngii ferulae* (8.60%), *L. edodes* (5.27–5.85%), and *Hericium erinaceus* (9.35%) (Manzi et al., 1999; Mau et al., 2001b, c). Cultivated mushrooms are a good source of minerals, containing macroelements such as calcium, magnesium, sodium, potassium, and phosphorus and microelements such as copper, iron, manganese, and zinc. Some common cultivated edible mushrooms, including *P. ostreatus, L. edodes*, and *A. bisporus*, are rich in potassium (2670–4730 mg/100 g DW) and are a good source of phosphorus (493–1390 mg/100 g DW), magnesium (20–200 mg/100 g DW), zinc (4.70–9.20 mg/100 g DW), and copper (0.52–3.50 mg/100 g DW) but are low in sodium (130–420 mg/100 g DW), calcium (1–25.0 mg/100 g DW), iron (2.80–12.30 mg/100 g DW), and manganese (0.51–2.1 mg/100 g DW) (Zakhary et al., 1983; Verma et al., 1987; Vetter, 1990; Shah et al., 1997).

Mushrooms are known to accumulate heavy metals, but the concentration of these elements is generally assumed to be species dependent, with substrate composition also being an important factor (Kalač and Svoboda, 2000; Demirbaş, 2001; Stijve et al., 2004; Svoboda et al., 2000, 2006). The levels of cadmium (3.60–120 μg/100 g DW) and selenium (3.90–320 μg/100 g DW) vary (Bano and Rajarathnam, 1988; Vetter, 1994a, b, 1995; Mattila et al., 2001), but low levels of lead (2.0–18.0 μg/100 g DW) and mercury (0.09–0.13 mg/100 g DW) have generally been reported (Piepponen et al., 1983; Svoboda et al., 2002). Mushrooms in general and species in the *Boletus* genus in particular are rich in selenium (Cocchi et al., 2006).

3.4.1.5 Vitamins Cultivated mushrooms are a good source of several vitamins, such as riboflavin (vitamin B2), niacin, and folates, with concentrations that vary within the range of 1.8–5.1, 31–65, and 0.30–0.64 mg/100 g DW, respectively, depending on the species. The vitamin B2 content in mushrooms is higher than that generally found in vegetables, and some varieties of *A. bisporus* even have a level of vitamin B2 as high as that found in egg and cheese (Mattila et al., 2001). The vitamin B2 content of other cultivated edible mushrooms is as follows: *P. ostreatus* (2.27–8.97 mg/100 g DW), *A. bisporus* (3.70–5.10 mg/100 g DW), and *L. edodes* (0.90–1.80 mg/100 g DW) (Crisan and Sand, 1978; Bano and Rajaratham, 1986, 1988; Miles and Chang, 1997; Mattila et al., 2001). Cultivated mushrooms are rich in niacin, but again the content varies, from 33.8–109 mg/100 g DW for *P. ostreatus*, 11.9–98.5 mg/100 g DW for *L. edodes*, and 36.2–57.0 mg/100 g DW for *A. bisporus* (Crisan and Sand, 1978; Bano and Rajaratham, 1986, 1988; Miles and Chang, 1997). Mushrooms contain moderately high amounts of folates at concentrations that are of the same magnitude as is generally found in vegetables. Furthermore, the bioavailability of folates is as good as that for folic acids (Clifford et al., 1991), with a content of 640–1412 μg/100 g DW for *P. ostreatus*, 590–933 μg/100 g DW for *A. bisporus*, and 300 μg/100 g DW for *L. edodes* (Bano and Rajaratham, 1986, 1988; Mattila et al., 2001). In addition to riboflavin, niacin, and folates, cultivated mushrooms also contain small amounts of vitamin C and vitamin B1 and traces of vitamins B12 and D2. The variation in the vitamin C content of mushrooms is wide, ranging from 17.0 in *A. bisporus* and 25.0 mg/100 g DW in *L. edodes* to 36.4–144 mg/100 g DW in *P. ostreatus* and 40.4–59.9 mg/100 g DW in *L. edodes*. (Li and Chang, 1985; Beelman and Edwards, 1989; Kurzman, 1997). The vitamin B1 content in mushrooms is quite low, ranging from 0.60 to 0.90 mg/100 g DW but is of the same magnitude as is generally found in vegetables. Traces of vitamin B12 (0.60–0.80 μg/100 g DW) have also been found in cultivated mushrooms (Walker, 1996). Vitamin D is almost absent in cultivated mushrooms, although the level present depends on the cultivation conditions (Mattila et al., 2001). It has been reported that the vitamin D2 content in *A. bisporus* (0.21 μg/100 g DW) cultivated in the dark is lower than that of *L. edodes* (22.0–110 μg/100 g DW) cultivated in natural climatic conditions (Takamura et al., 1991), which is mainly due to the influence of illumination on the conversion of ergocalciferol (provitamin D) to vitamin D (Mattila et al., 2001).

3.4.1.6 Dietary Fiber All fungal cell walls contain a mixture of fibrillar components and amorphous or matrix components. In Basidiomycetes, the main fibrillar components include chitin, which is a straight-chain $(1 \rightarrow 4)$-β-linked polymer of *N*-acetyl-glucosamine, and matrix components, which include various polysaccharides such as $(1 \rightarrow 3)$-α- and $(1 \rightarrow 3)$-β-D-glucans and mannans (Bartnicki-Garcia, 1968). All of these cell wall components are nonstarch polysaccharides that can be classified as dietary fiber [American Association of Cereal Chemists (AACC), 2001].

There is a large variation in the total dietary fiber (TDF) content of edible mushrooms, which depends on their morphological form and species. The fruiting bodies of some mushroom species have a low level of TDF (4.50% DW in *T. giganteum*, 9.26% DW in *D. indusiata*, and 8.74% DW in *P. cystidious*), whereas others have a high TDF level, such as *T. portentosum* (45.0%), *T. terreum* (50.0%), *Auricularia auricula* (49.7%), *Tremella fuciformis* (54.5%), *V. bombycina* (25.7%), *L. ulmarius* (33.2%), and *Pleurotus citrinopileatus* (35.6%) (Cheung, 1997b; Díez and Alvarez, 2001). In general, mushrooms are a good source of dietary fiber, with 100 g of fresh mushrooms providing between 10 and 40.0% of the recommended dietary intake of fiber (Manzi et al., 2001). The TDF in mushrooms is predominantly composed of insoluble dietary fiber (IDF) and a low level of soluble dietary fiber (SDF) of less than 10%. This may be partly due to the differing amounts of chitin found in different mushrooms (Manzi et al., 1999), which in *Pleurotus* varies from 4.70 to 4.90% DW (Yoshida et al., 1986). The β-glucan content ranges from 0.21 to 0.53 g/100 g DW in mushrooms, of which 16.8–46.0% is found in SDF and 53.9–83.2% in IDF (Manzi and Pizzoferrato, 2000). A previous study showed that neutral and amino sugars are the main components in the TDF of *A. bisporus* (84.8%), *Auricularia auricula* (81.4%), *Auricularia fuciformis* (81.4%), *Pleurotus sajor-caju* (77.8%), and *L. edodes* (92.8–92.9%) (Cheung, 1996a, 1997a; Cheung and Lee, 1998), with a small amount of uronic acid (2.10–12.6% of the TDF) and Klason lignin (2.40–12.9%) also being reported (Cheung, 1997a, 1998). Glucose

(43.1–82.8% of the TDF) is the predominant sugar in mushrooms, but the presence of other sugars, such as mannose, xylose, rhamnose, galactose, and uronic acids, in the TDF reflects that the main cell wall polysaccharides in most fungi are hemicelluloses, such as β-glucan, glucuronoxylomannan, pectic substances, and chitin (Cheung, 1996a, 1997a). It is worth noting that mushroom sclerotia, which are the dry compact biomass of fungal hyphae, contain over 80% DW of TDF that is composed mainly of β-glucans (Wong et al., 2003). The structure and function of sclerotial TDF is discussed in more detail in Chapter 4.

3.4.1.7 Carbohydrates The carbohydrate content of edible mushrooms varies with species and ranges from 35 to 70% DW. The carbohydrate content of some edible mushrooms is as follows: *P. indusiata* (67% DW), *G. frondosa* (58.8% DW), *T. giganteum* (70% DW), *L. edodes* (62.3–64.4% DW), *P. cystidious* (63.1% DW), *S. commune* (61.1% DW), and *P. ostreatus* (61.1% DW) (Longvah and Deosthale, 1998; Díez and Alvarez, 2001; Mau et al., 2001b, c). Edible mushrooms are believed to contain a high level of oligosaccharides and only a low level of total soluble sugars (182 mg/100 g DW for *T. portentosun* and 55 mg/100 g DW for *T. terreum*) and glycogen (15.6 mg/100 g DW for *T. portentosun* and 10.6 mg/100 g DW for *T. terreum)* (Díez and Alvarez, 2001). The profiles of soluble sugars differ across species. Arabitol (127 mg/g DW), glucose (4.91–39.4 mg/g DW), mannitol (9.36–50.9 mg/g DW), trehalose (9.71–341 mg/g DW), and inositol (1.43–3.20 mg/g DW) are found in edible mushrooms, with mannitol and trehalose being the two major components in *V. volvocea* and *P. ostreatus*, respectively (Bano and Rajarathnam, 1988; Mau et al., 1997).

3.4.1.8 Energy The energy content of edible mushrooms is generally low, which allows them to be used in low-energy diets. Previous reports have shown that the energy content in *Tricholoma robustus* is 3.02 kcal/g DW, in *Psathyrella antroumbonata* is 2.50 kcal/g DW, in *A. bisporus* is 4.17–4.20 kcal/g DW, in *P. ostreatus* is 4.16–4.23 kcal/g DW, and in the *Boletus* group is 4.20–4.27 kcal/g DW (Aletor, 1995; Manzi et al., 2001). A serving of 100 g of fresh edible mushrooms provides only 1.4–4.4% of the daily energy requirement for a 70-kg adult male who engages in moderate physical activity (LARN, 1996).

3.4.1.9 Other Components Mushrooms contain a very low level of phenolic compounds, with flavonoid and lignan contents that are usually below the limits of detection (Mattila et al., 2001). The content of total 5′-nucleotides is high, ranging from 7.43 to 31.9 mg/g DW. Flavor 5′-nucleotides are found to be 5′-guanosine monophosphate (5′-GMP), 5′-inosine monophosphate (5′-IMP), and 5′-xanthosine monophosphate (5′-XMP) (Chen, 1989). The flavor 5′-nucleotide content in some common mushrooms is particularly high, between 0.62 and 13.6 mg/g DW (Lin, 1988; Mau et al., 2001b, c; Tseng and Mau, 1999). 5′-GMP gives a meaty flavor and is a stronger flavor enhancer than MSG (Litchfield, 1967). The synergistic effects of flavor 5′-nucleotides with MSG-like components can greatly increase the umamic taste of mushrooms (Yamaguchi et al., 1971). The DNA level (0.21–0.26

g/100 g DW) in mushrooms is much lower than the level of RNA (Li and Chang, 1982; Cheung, 1997b). The total nucleic acid content of mushrooms, which ranges from 3.51 to 4.15% (Cheung, 1997b), can be considered safe, as the maximum daily dietary intake of nucleic acid suggested by the Protein Advisory Group of the United Nations System is 4 g (PAG, 1970). The phytic acid (160–360 mg/100 g DW), phytic acid–phosphorus (50–600 mg/100 g DW), and oxalate (80–220 mg/100 g DW) levels of mushrooms are generally not higher than those reported for cowpeas and soybeans (Aletor, 1995). Phytic acids can chelate certain mineral elements, such as calcium, magnesium, iron, and zinc, and render them metabolically unavailable, which interferes with the basic proteins to inhibit certain activities of digestive enzymes (α-amylase, pepsin, and pancreatin) (Huisman, 1991). However, the level of oxalate in edible mushrooms is quite low (80–220 mg/100 g) compared with the values of 2.30–5.80 g/100 g reported earlier for some varieties of guinea corn (Aletor, 1995).

3.5 NEWLY CULTIVATED/NONCONVENTIONAL MUSHROOMS

Some wild-grown edible mushrooms with great nutritional and medicinal potential have been artificially cultivated in mass production by modern agricultural technology. More than 10 new species of mushrooms have been cultivated in recent years on a small scale, and these have a great potential for expansion in the future (Chang, 1999). As the consumption of mushrooms is increasing and the development in cultivation techniques is rapid, new data on the chemical composition and nutritional values of mushrooms are urgently needed (Mattila, et al. 2001).

Several species of edible *Pleurotus* mushrooms and other lesser known edible mushrooms newly developed and cultivated by the Sanming Mycological Institute in Sanming, Fujian, China, were studied to determine their chemical composition and nutritional values (Figure 3.1) (Wong, 2003). Tables 3.3 and 3.4 summarize the nutritional composition of 11 newly cultivated *Pleurotus* mushroom and 12 lesser known cultivated mushroom species, respectively, and Figures 3.2*a*–3.2*z* show pictures of dry specimens of these edible mushrooms. The moisture content of the dried mushrooms was less than 10%. The crude protein content of both groups was reasonably high (14.9–28.8 and 9.40–27.3% DW, respectively), and all of the mushrooms were low in lipids (less than 6.0% DW) and ash (4.55–7.71 and 3.78–9.47% DW, respectively). The TDF of the edible *Pleurotus* mushrooms was 26.7–43.6% DW, and that of the lesser known edible species was 26.7–44.0% DW. The TDF predominantly consisted of IDF (92.6–98.4 and 87.9–98.6% DW for the two groups, respectively) and a small amount of SDF (1.60–7.40 and 1.40–12.1% DW, respectively). Potassium was the major element found in the mushrooms, together with low levels of sodium, magnesium, and calcium in the ash (Wong, 2003). The levels of heavy metal were found to be very low in both groups (Wong, 2003), and all were low in energy (less than 300 kcal/100 g DW) (Wong, 2003).

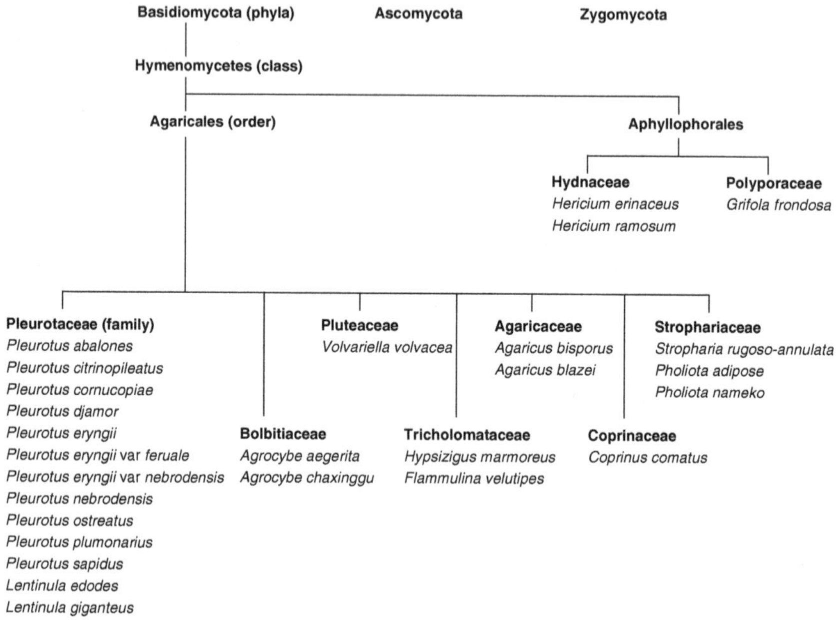

Figure 3.1 The classification of the edible mushrooms being investigated.

3.6 NUTRITIONAL EVALUATION

3.6.1 General Aspects

Mushrooms are rich in high-quality protein, contain a high level of dietary fiber and a high proportion of unsaturated fatty acids, are rich in various vitamins and minerals, and have an acceptably low level of nucleic acids, all of which suggests the suitability of their daily use as a vegetable (Chang, 1999). Edible mushrooms or fungi that are a good and inexpensive source of protein are therefore good candidates for the relief of the problem of chronic protein deficiency. The idea of the development of ribonucleic acid–reduced biomass from fungal hyphae cultivated in submerged fermentation (mycoprotein) originated in the early 1960s in response to severe dietary protein shortages (Litchfield, 1967; PAG, 1970), and today mycoprotein is still being considered as an alternative to meat (Rodger, 2001).

3.6.2 Biological Methods for Nutritional Evaluation

In addition to comparing the amino acid content of a food with the amino acid requirements of humans, extensive evaluation of existing in vitro and in vivo methods for food has been conducted that indicates that the rat balance

TABLE 3.3 Chemical Composition of *Pleurotus* Mushroom

Sample	g/100 g DW						
	Fat %	Protein %	TDF %	IDF %	SDF %	Ash %	Moisture %
Pab	2.27 ± 0.29	16.1 ± 0.07	38.0 ± 0.43	36.6 ± 0.34	1.44 ± 0.09	6.45 ± 0.20	7.33 ± 0.47
Pci	1.52 ± 0.21	27.0 ± 0.32	36.6 ± 0.48	36.0 ± 0.99	0.59 ± 0.76	7.89 ± 0.04	8.33 ± 0.47
Pco	2.01 ± 0.04	17.9 ± 0.31	43.6 ± 0.88	40.7 ± 0.93	2.83 ± 0.08	6.17 ± 0.14	4.97 ± 0.26
Pd	1.46 ± 0.17	31.1 ± 0.20	36.0 ± 1.34	35.3 ± 1.52	0.74 ± 0.22	8.32 ± 0.21	7.33 ± 0.47
Pe	1.93 ± 0.23	19.8 ± 0.80	35.5 ± 0.48	32.9 ± 0.34	2.61 ± 0.16	5.18 ± 0.32	4.00 ± 0.00
Pevf[a]	6.29 ± 0.32	23.9 ± 1.34	37.2 ± 0.13	34.8 ± 0.05	2.39 ± 0.08	5.99 ± 0.28	5.97 ± 0.21
Pevn[b]	2.56 ± 0.74	16.3 ± 0.15	37.7 ± 1.49	36.0 ± 1.42	1.77 ± 0.11	4.92 ± 0.16	7.67 ± 0.12
Pne	1.57 ± 0.11	17.1 ± 0.14	26.7 ± 1.72	24.9 ± 1.67	1.81 ± 0.21	5.92 ± 0.07	8.33 ± 0.47
Po	2.18 ± 0.14	29.6 ± 0.27	42.6 ± 0.56	40.9 ± 0.19	1.67 ± 0.39	7.08 ± 0.90	6.27 ± 0.17
Pp	1.56 ± 0.13	24.0 ± 0.10	30.7 ± 0.45	29.6 ± 0.54	1.08 ± 0.09	6.25 ± 0.17	9.00 ± 0.82
Ps	1.40 ± 0.05	20.1 ± 0.13	37.3 ± 0.85	36.4 ± 1.06	0.87 ± 0.23	6.41 ± 0.19	7.40 ± 0.08

Note: Pleurotus abalones (Pab), *Pleurotus citrinopileatus* (Pci), *Pleurotus cornucopiae* (Pco), *Pleurotus djamor* (Pd), *Pleurotus eryngii* (e), *Pleurotus eryngii* var *ferulae* (Pevf), *Pleurotus eryngii* var *nebrodensis* (Pevn), *Pleurotus nebrodensis* (Pne), *Pleurotus ostreatus* (Po), *Pleurotus plumonarius* (Pp), *Pleurotus sapidus* (Ps). The edible *Pleurotus* mushrooms were cultivated by the Sanming Mycological Institute, Sanming, Fujian, China.

[a]New species from hybridization of *P. eryngii* and *P. feruale*.

[b]New species from hybridization of *P. eryngii* and *P. nebrodensis*.

TABLE 3.4 Chemical Composition of Newly Developed Cultivated Mushrooms

Sample	g/100 g DW						
	Fat %	Protein %	TDF %	IDF %	SDF %	Ash %	Moisture %
Aa	2.97 ± 0.36	25.2 ± 0.14	26.7 ± 1.51	26.2 ± 1.29	0.51 ± 0.22	8.47 ± 0.03	8.00 ± 0.00
Ab	2.44 ± 0.05	30.9 ± 1.74	29.6 ± 3.52	27.6 ± 2.29	1.93 ± 1.24	6.58 ± 0.32	5.17 ± 0.21
Ac	3.07 ± 0.23	21.4 ± 0.18	36.4 ± 1.01	34.9 ± 0.95	1.52 ± 1.53	8.62 ± 0.16	7.63 ± 0.40
Cc	1.95 ± 0.06	10.0 ± 0.30	34.6 ± 5.28	32.8 ± 4.20	1.79 ± 1.13	10.10 ± 0.34	6.23 ± 0.06
Fv	1.91 ± 0.19	17.9 ± 1.79	38.2 ± 2.77	33.8 ± 1.93	4.42 ± 0.99	5.79 ± 0.45	8.33 ± 0.58
Gf	3.92 ± 0.38	17.2 ± 0.60	44.0 ± 1.55	43.1 ± 1.43	0.91 ± 0.13	5.23 ± 0.21	9.33 ± 0.58
GK16[a]	5.56 ± 0.42	23.1 ± 0.61	31.1 ± 1.75	29.6 ± 1.62	1.42 ± 0.12	6.61 ± 0.11	7.27 ± 0.46
He	2.01 ± 0.03	18.8 ± 0.68	34.0 ± 0.98	31.8 ± 0.56	2.12 ± 0.42	4.58 ± 0.19	4.87 ± 0.05
Hm	3.72 ± 0.12	19.9 ± 0.55	32.0 ± 1.35	30.1 ± 1.60	1.89 ± 0.26	7.39 ± 0.13	6.73 ± 0.47
Hr	3.26 ± 0.20	24.7 ± 0.95	26.9 ± 0.84	23.6 ± 0.29	3.25 ± 0.93	9.37 ± 0.14	7.00 ± 0.00
Lg	3.38 ± 0.19	17.9 ± 0.42	34.8 ± 0.53	34.3 ± 0.76	0.50 ± 0.26	4.01 ± 0.39	5.73 ± 0.25
LPK 15[a]	4.58 ± 0.77	21.1 ± 0.65	34.7 ± 0.56	33.5 ± 0.54	1.23 ± 0.07	5.51 ± 0.01	7.97 ± 0.15
Pa	2.61 ± 0.27	28.1 ± 0.67	30.8 ± 0.48	27.7 ± 0.67	3.08 ± 0.20	7.44 ± 0.29	9.67 ± 0.58
Pn	1.84 ± 0.22	16.9 ± 0.04	37.9 ± 0.79	34.8 ± 0.77	3.15 ± 0.11	5.86 ± 0.11	9.00 ± 0.00
Sra	2.10 ± 0.12	29.6 ± 0.03	28.4 ± 0.91	26.2 ± 0.80	2.26 ± 0.31	7.87 ± 0.03	6.00 ± 0.00

Note: Agrocybe aegerita (Aa), *Agaricus blazei* (Ab), *Agrocybe chaxinggu* (Ac), *Coprinus comatus* (Cc), *Flammulina velutipes* (Fv), *Grifola frondosa* (Gf), *Hericium erinaceus* (He), *Hypsizigus marmoreus* (Hm), *Hericium ramosum* (Hr), *Lentinula giganteus* (Lg), *Pholiota adiposa* (Pa), *Pholiota namkeo* (Pn), *Stropharia rugoso-annulata* (Sra).

The other lesser known edible mushrooms were cultivated by the Sanming Mycological Institute, Sanming, Fujian, China.

[a]New species from hybridization of *L. edodes* and *P. feruale*.

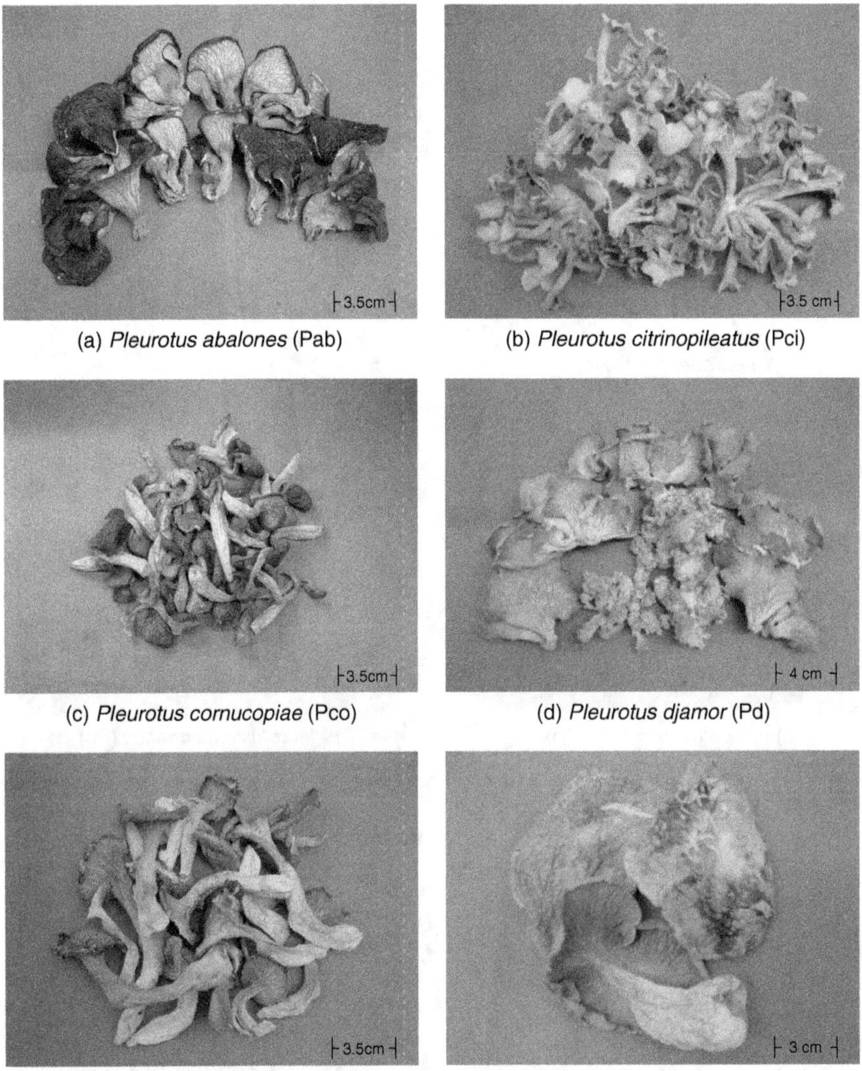

(a) *Pleurotus abalones* (Pab)
(b) *Pleurotus citrinopileatus* (Pci)
(c) *Pleurotus cornucopiae* (Pco)
(d) *Pleurotus djamor* (Pd)
(e) *Pleurotus eryngii* (Pe)
(f) *Pleurotus eryngii var. ferulae* (Pevf)

Figure 3.2 Photos of dried samples of newly developed cultivated mushrooms. (See insert for color representation.)

(g) *Pleurotus eryngii var. nebrodensis* (Pevn)

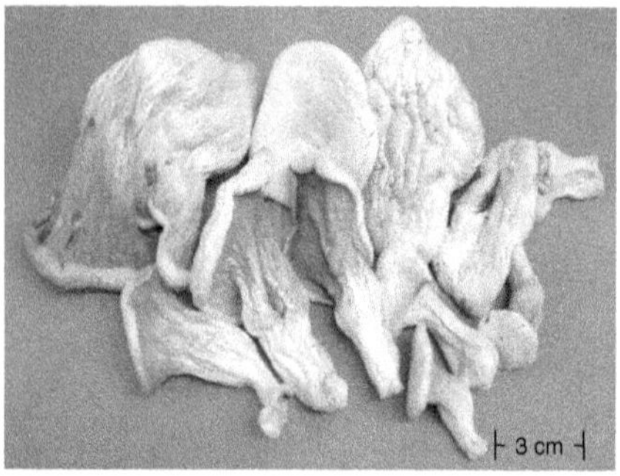

(h) *Pleurotus nebrodensis* (Pne)

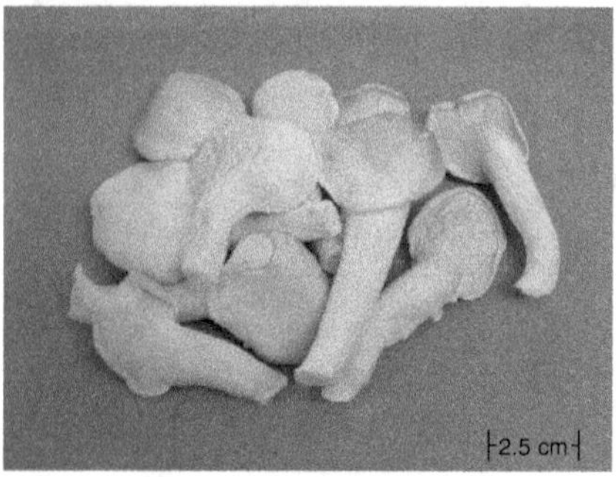

(i) *Pleurotus ostreatus* (Po)

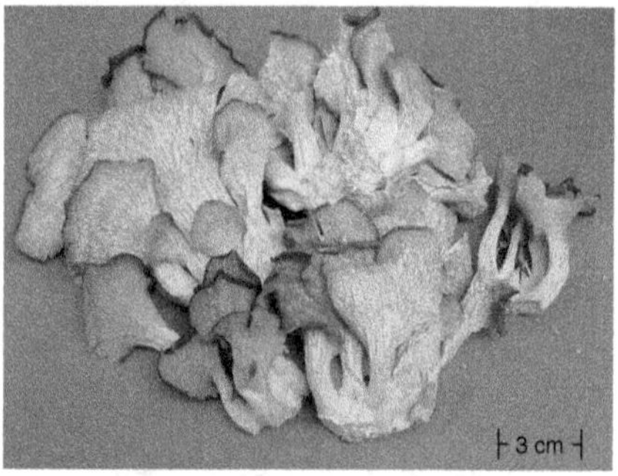

(j) *Pleurotus plumonarius* (Pp)

(k) *Pleurotus sapidus* (Ps)

(l) *Agrocybe aegerita* (Aa)

Figure 3.2 (*Continued*)

(m) *Agaricus blazei* (Ab)

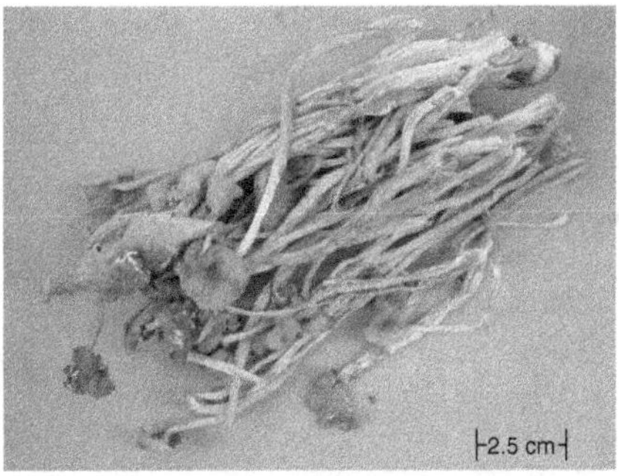

(n) *Agrocybe chaxinggu* (Ac)

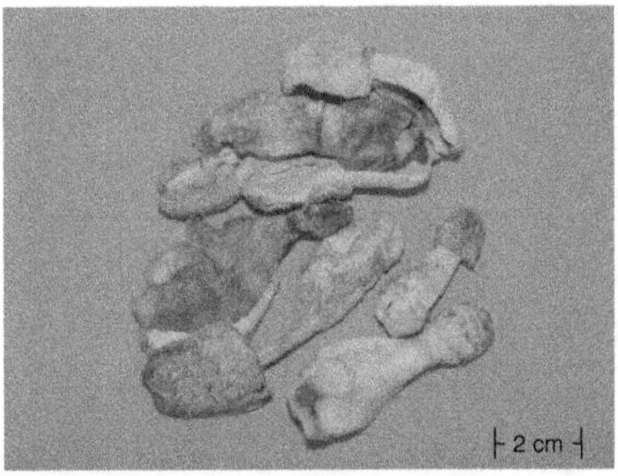

(o) *Coprinus comatus* (Cc)

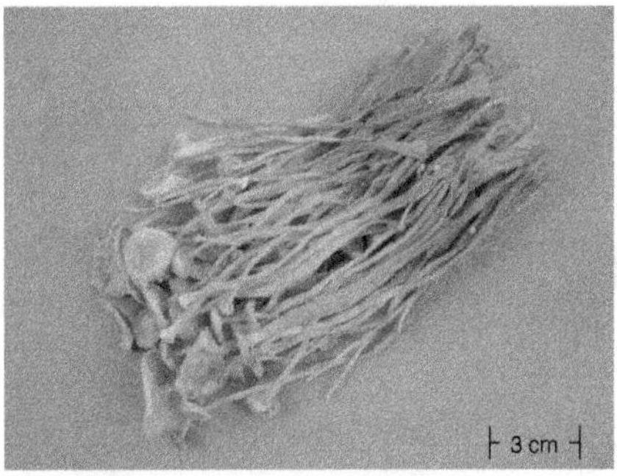

(p) *Flammulina velutipes* (Fv)

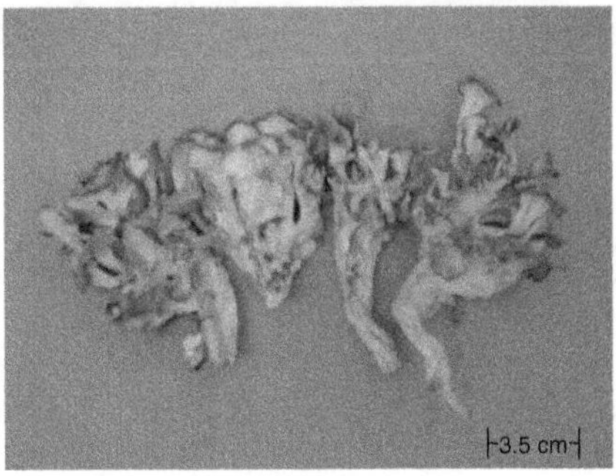

(q) *Grifola frondosa* (Gf)

(r) GK16

Figure 3.2 *(Continued)*

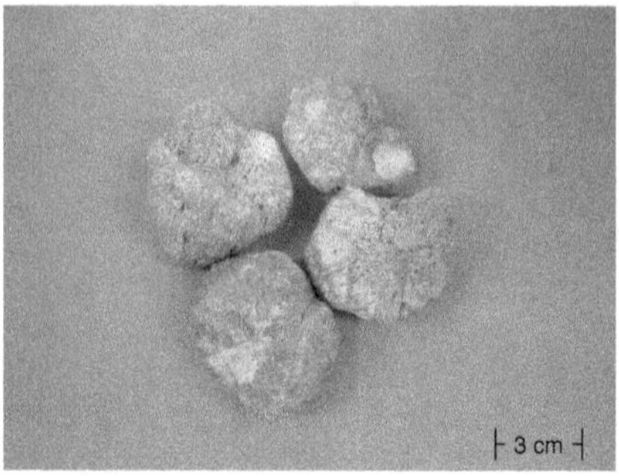

(s) *Hericium erinaceus* (He)

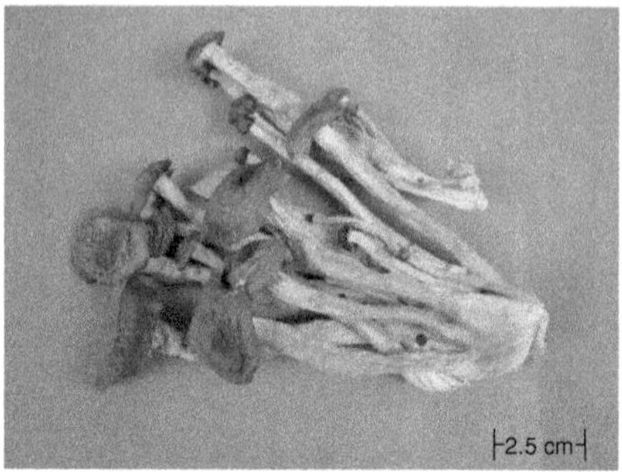

(t) *Hypsizigus marmoreus* (Hm)

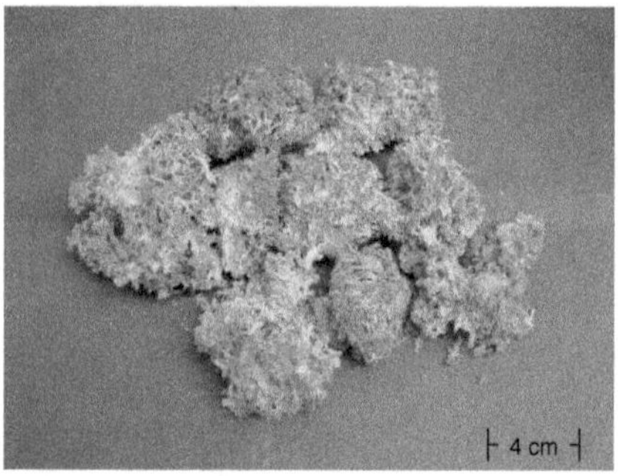

(u) *Hericium ramosum* (Hr)

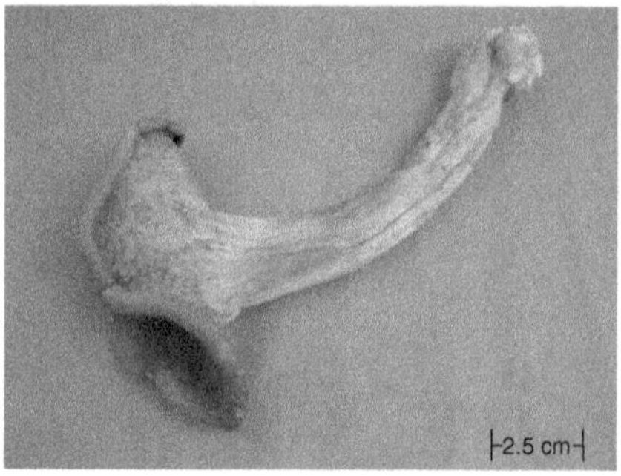

(v) *Lentinula giganteus* (Lg)

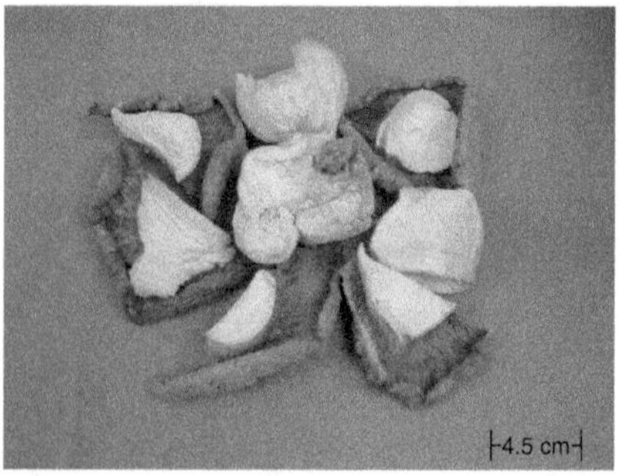

(w) LPK 15

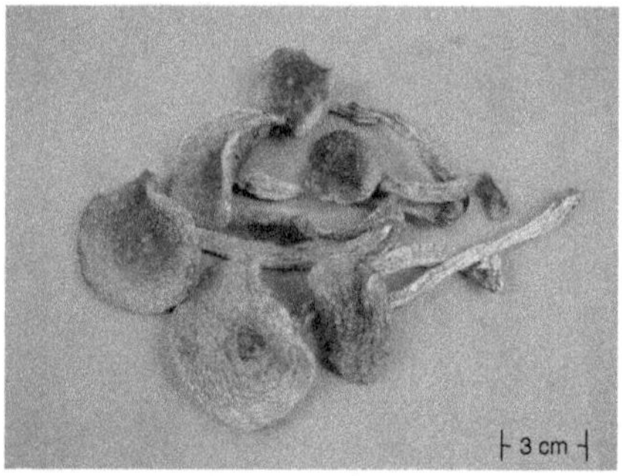

(x) *Pholiota adipose* (Pa)

Figure 3.2 (*Continued*)

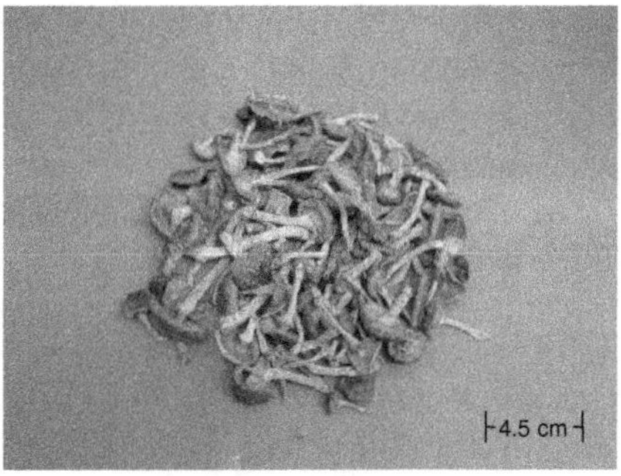

(y) *Pholiota nameko* (Pn)

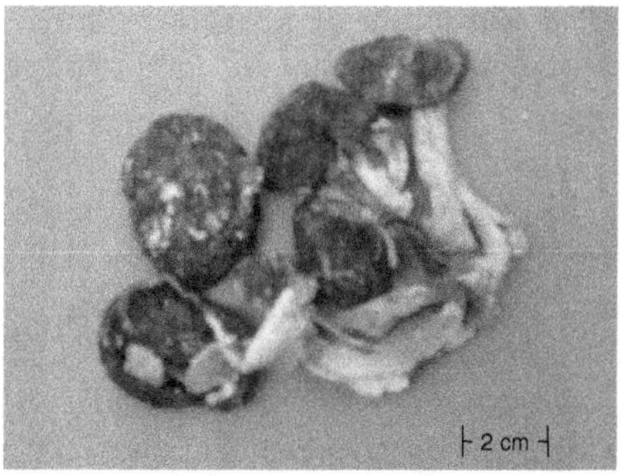

(z) *Stropharia rugoso-annulata* (Sra)

Figure 3.2 (*Continued*)

method is the most practical for predicting protein digestibility in humans (FAO, 1991; Boutrif, 1991). Indeed, the protein efficiency ratio (PER) and net protein ratio (NPR) methods have been used to measure the ability of a protein to support the growth in young and rapidly growing rats in many countries (Boutrif, 1991). However, as the PER method cannot properly account for the protein used for maintenance purposes, the Codex Committee on Vegetable Proteins (CCVP) proposed the protein-digestibility-corrected amino acid score (PDCAAS), which is deemed to be the most suitable routine method for assessing the protein quality of most vegetable protein products and other food products (Codex Alimentarius Commission, 1989). This method has since been extensively used for evaluating the protein quality of various foods sources (Boutrif, 1991; Dabbour and Takruri, 2002b). However, it has also been suggested that the PDCAAS method overestimates the protein quality of certain foods that may contain naturally occurring antinutritional factors and of poorly digestible proteins supplemented with limiting amino acids in rats (Sarwar, 1996).

In vitro methods are relatively rapid, and involve only small amounts of raw materials (Ezquerra et al., 1998). Various tests, such as multienzymatic digestion and measurement of the ensuing pH change and enzymatic pH-stat assays, have been developed and extensively used to measure the in vitro protein digestibility (IVPD) of different protein sources and to determine the protein quality of different foods. Although in vitro enzymatic procedures that involve samples of plant and animal proteins yield only an approximate estimate of protein digestibility, they are still extensively used (Pedersen and Eggum, 1983).

3.6.3 Mushroom Protein Quality

There are only a few publications on the biological quality of protein in edible mushrooms (Crisan and Sands, 1978; Khana and Garcha, 1986; Longvah and Deosthale, 1998).

Pleurotus is one of the main genera of edible mushroom with commercial value, and the biological value of the protein in some of the new strains and cultivated species of this edible mushroom has recently been reported (Wong, 2003; Valencia del Toro et al., 2006).

The IVPD values of *T. portentosum* and *T. terreum* are 72.9 and 73.2%, respectively (Díez and Alvarez, 2001), which are similar to those of legumes but lower than those of protein from animal sources.

The in vivo true protein digestibility (TPD) values for *Terfezia claveryi, P. ostreatus, T. terrum*, and *Agaricus macrosporus* are 61.4, 73.4, 52.6, and 80.5%, respectively (Dabbour and Takruri, 2002a), and similar findings have been observed for *S. commune* (53.2%) and *L. edodes* (76.3%) (Adewusi et al., 1993). These values are again comparable to those of certain legumes, such as *Phaseolus angularis* (70.1%), *Phaseolus calcaratus* (75.8%), *Dolichos lablab* (75.8%) (Wong and Cheung, 1998), *Phaseolus vulgaris* (72.4%) (Carbonaro et al., 2000), and the pinto bean (72.6%) but lower than those of the animal protein sources casein (98.6%), tuna fish (96.6%), cheese (94.4%), nonfat dry milk (NFDM) (87.9%), and rolled oats (98.8%) (McDonough et al., 1990). Based on the FAO (1991) requirement pattern for children, the chemical scores for lysine (the limiting amino acid) in *T. claveryi* and *T. terrum* are 0.71 and 0.67, respectively (Dabbour and Takruri, 2002a), which agrees with previous findings for *T. claveryi* (0.75) (Sawaya et al., 1985). The chemical scores for sulfur-containing amino acids in *P. ostreatus* and *A. macrosporus* are 0.61 and 0.50, respectively (Dabbour and Takruri, 2002a), which are consistent with those reported for *P. ostreaus* and *A. bisporus* (Alofe, 1991; Dannel and Eaker, 1992). Other protein sources, such as legumes, are also found to have a low content of essential sulfur amino acids (methionine and cysteine) (Carbonaro et al., 2000).

A recent study has shown that low intakes of edible mushrooms in rats contribute to small gains in body weight (Dabbour and Takruri, 2002a). Although a meager weight gain (0.11 g/rat/day) of rats fed with *S. commune* was observed, rats fed with *L. edodes* experienced a significantly greater body weight gain (0.97 g/rat/day) (Adewusi et al., 1993).

Nevertheless, the NPR of the mushrooms *T. robustus* (0.80) and *L. edodes* (1.70) (Adewusi et al., 1993) is significantly lower than that of legumes such as *P. angularis* (6.38) and *D. lablab* (6.74) (Wong and Cheung, 1998). It has also been found that the PER and net protein utilization (NPU) values of casein-fed rats (3.55 and 78.5) were significantly higher than those fed a mushroom diet consisting of *T. claveryi* (−1.76 and 32.6), *P. ostreatus* (−0.23 and 38.5), *T. terrum* (−0.98 and 29.1), and *A. macrosporus* (−0.41 and 31.5). A previous report also showed similar PER results in rats fed with dried mushrooms (Adewusi et al., 1993), and similar NPU values for *L. edodes* (45.8) and *S. commune* (23.7) have been reported (Adewusi et al., 1993). The negative PER values indicate that the mushrooms failed to support the growth of rats, although their crude protein contents were high (Dabbour and Takruri, 2002a). The presence of alkaloids and tannins in mushrooms has been reported to cause liver necrosis, and tannins are also known to retard growth by reducing digestion and the absorption of amino acids and minerals (Adewusi et al., 1993). Phenolic compounds that complex with protein decrease amino acid availability (DaDamio and Thompson, 1992). In addition, the unpalatability of a mushroom diet may have had a negative effect on rat growth and on the protein quality of the test food (Sarwar and McDonough, 1990). It has also been suggested that a higher requirement for sulfur-containing amino acids (methionine and cystine), histidine, isoleucine, threonine, and valine in rats leads to the underestimation of the protein quality of mushrooms as a human food (Boutrif, 1991; FAO, 1991). Although previous reports have suggested that mushroom protein is nutritionally incomplete (Bano and Rajarathnam, 1988; Tshinyangu, 1996) and may not be able to support growth as effectively as animal proteins (Dabbour and Takruri, 2002a), a combination of cereal protein, which supplies an adequate level of methionine, and mushroom protein could provide a good balance of amino acids and a good source of dietary protein for human beings.

3.7 HEALTH BENEFITS OF EDIBLE MUSHROOMS

3.7.1 General Aspects

The increasing number of cultivated edible mushrooms being introduced into the market has led to greater attention in the food and nutritional sciences being paid to their potential health benefits to humans. This has resulted in many scientific publications, to the extent that there is now a body of scientific evidence about the specific health effects of mushrooms and their bioactive molecules. In this section, the health benefits of edible mushrooms are explored, with an emphasis on the potential contribution of mushrooms as functional foods and ingredients.

3.7.2 Antioxidants in Mushrooms

3.7.2.1 Bioactive Components and Their Antioxidative Activities

Recently, many studies have found that edible mushrooms possess potent antioxidants. The following sections review the antioxidant properties of mushrooms and give details of the characteristics and biosynthesis of mushroom phenolic antioxidants.

Mushrooms possess many antioxidant properties. Research conducted in Japan that studied the antioxidant activity of the crude ethanol extract of 150 Japanese mushrooms using the peroxide value in the methyl linoleate system showed that many mushrooms, especially those belonging to the *Suillus* genus, had a peroxide value some 80% lower than the control (Kasuga et al., 1993). The same study proposed that there may be an intragenus relationship with antioxidant activity and found that both polar (diethyl ether) and nonpolar (petroleum ether) extracts of oogitake, kugitake, and maitake mushrooms showed high antioxidant activity in assay, which suggests the presence of both polar and nonpolar antioxidants (Kasuga et al., 1993).

It has been reported that polysaccharide extracts isolated from several mushrooms are potent scavengers of hydroxyl and superoxide free radicals but that none of the species tested showed antioxidant activity as measured by the malondialdehyde content of liver microsomes (Liu et al., 1997). Recent research has focused on the determination of the total phenolic content and antioxidant properties of several commercial mushrooms. A study of methanolic extracts from black, red, and snow ear mushrooms found that they had an inhibitory effect on lipid peroxidation, 1,1-diphenyl-2-picrylhydrazyl (DPPH) radical scavenging, and hydroxyl radical scavenging and a strong reducing power and ability to chelate ferrous ions (Mau et al., 2001a). Similar studies on other mushrooms, including *D. indusiata, G. frondosa, H. erinaceus, T. giganteum, F. velutipes, L. edodes, P. cystidiosus*, and *P. ostreatus*, showed that these mushrooms also possess the aforementioned antioxidant properties (J. H. Yang, et al., 2002; Mau et al., 2002). It is therefore likely that most mushrooms possess hydroxyl and DPPH radical scavenging effects, inhibit lipid peroxidation, chelate metals, and have a strong reducing effect.

Methanolic extracts of *D. indusiata, G. frondosa, H. erinaceus*, and *T. giganteum* show polyphenolics to be the major naturally occurring antioxidant components, which have excellent reducing power, scavenging effect, and chelating effect on ferrous ions (Mau et al., 2002). Similar antioxidant properties have also been reported for other edible mushrooms, including *Agrocybe cylindracea* (Tsai et al., 1972) and *H. marmoreus*, both of which belong to the Tricholomataceae family (Lee et al., 2007).

Potent antioxidant activity was found in crude methanol and water extracts of the three common Chinese edible mushrooms *L. edodes* (shiitake mushroom), *Pleurotus tuber-regium*, and *V. volvacea* (straw mushroom). The activity was evaluated by the β-carotene bleaching method, DPPH radical scavenging activity, and erythrocyte hemolysis assay (Cheung, 2001; Cheung et al., 2003). Fractionation of the crude methanol and water extracts of *L. edodes, P. tuber-regium*, and *V. volvacea* further indicated that the dichloromethane and ethyl acetate fractions of these mushrooms have the strongest antioxidant activity, as they showed the lowest median effective concentration (EC_{50}) values (Cheung, 2001).

The antioxidant potential of a lesser known edible mushroom, *A. aegerita*, which belongs to the family Bolbitiaceae, was recently studied (Lo and Cheung, 2005) and was shown to exhibit strong in vitro antioxidant activity as expressed in the inhibition of β-carotene bleaching, DPPH radical scavenging, and the inhibition of erythrocyte hemolysis in crude water and methanol extracts (Lo and Cheung, 2005). In the fractionation of the extracts, the high antioxidant activity was demonstrated by radical scavenging ability and low-density lipoprotein (LDL) oxidation in the ethyl acetate and butanol subfractions to be positively correlated with the total phenolic content (Lo and Cheung, 2005).

A recent report on antioxidant activity and antioxidant compounds in seven wild edible mushrooms (Elmastas et al., 2007) determined the potential antioxidant compounds, including phenolics, α-tocopherol, and β-carotene in methanolic extracts, and their in vitro antioxidant systems, including their reducing power, free-radical scavenging, superoxide anion radical scavenging, total antioxidant activity, and metal-chelating activities (Elmastas et al., 2007).

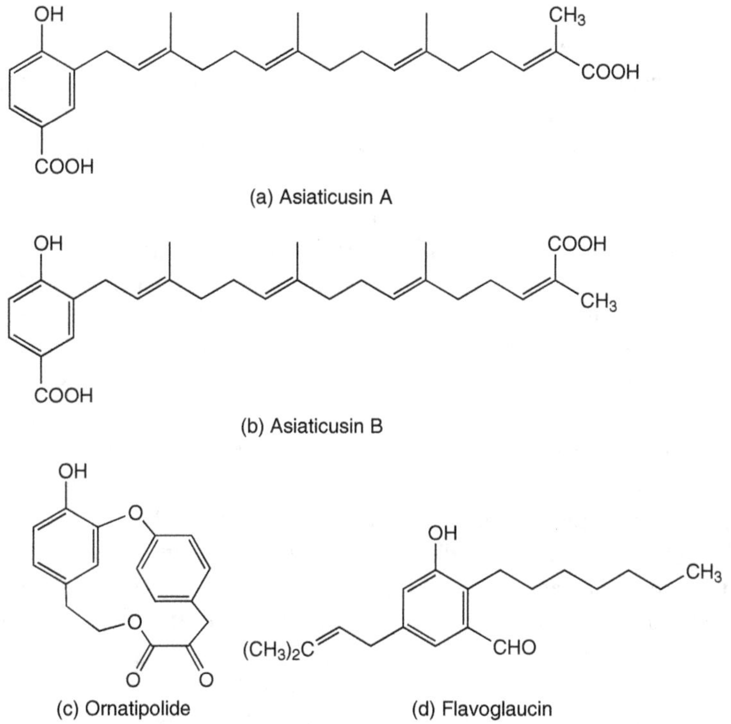

Figure 3.3 The structures of some phenolic compounds isolated from fungi.

Phenolics with antioxidant ability have also been found in other mushroom species. Flavoglaucin, which is a phenolic compound isolated from the mycelial mat of *Eurotium chevalieri*, is an excellent antioxidant in vegetable oil at a concentration of 0.05% (Ishikawa et al., 1984).

The inhibition of lipid peroxidation by mushrooms has also been reported recently, with the finding that *L. edodes* and *V. volvacea* display antioxidant behavior by scavenging the free radicals, such as peroxyl radicals, that are generated in lipid peroxidation (Cheung et al., 2003, 2005). A positive correlation was also found between the total phenolic content in the mushroom extracts and their antioxidative properties. This further confirms that edible mushrooms have a potential as natural antioxidants due to the ability of their phenolics to inhibit lipid oxidation.

3.7.2.2 Characterization of Mushroom Phenolic Antioxidants

Research on the characterization of mushroom antioxidants is scant. In one of the few studies in this area, 750 strains of filamentous fungi from soil were screened for their antioxidant activity (Aoyama et al., 1982). Among them, two strains were studied for their microbial products, during which curvulic acid, procatechuic acid, and citrinin were found to have antioxidant activity, with citrinin appearing to be as effective as α-tocopherol and curvulic acid and procatechuic acid showing a slightly less effective activity than α-tocopherol (Aoyama et al., 1982). However, despite its abundance, citrinin is a known myotoxin and has a limited potential in foods, whereas procatechuic acid is a known antioxidant, and curvulic acid has proved to have a low toxicity at 150 mg/kg body weight in rats (Aoyama et al., 1982).

Figure 3.3*d* shows flavoglaucin, which is a phenolic compound, isolated from the mycelial mat of *E. chevalieri*. The compound was found to possess an excellent antioxidant activity in vegetable oils (Ishikawa et al., 1984), stabilizing lard when used in combination with α-tocopherol at a concentration of 0.04%. However, this fungal antioxidative compound did not show any mutagenic activity on *Salmonella typhimurium* and would thus seem to have limited potential for use in food (Ishikawa et al., 1984). Babitskaya et al. (1996) were probably the earliest investigators to characterize phenolics in the edible mushroom *P. ostreatus*. Using thin-layer chromatography (TLC) and chemical detection methods, they identified the presence of simple phenols and flavones in the mushroom, although they did not carry out the structural identification of compounds exerting an antioxidant effect. Wada et al. (1996) successfully isolated and fully characterized two novel prenylated phenolics from the fruiting body of *Boletinus asiaticus* using both chemical and spectroscopic methods, including infrared (IR) spectrometry, mass spectrometry (MS), and nuclear magnetic resonance (NMR). Two compounds were found that were structural isomers of each other and were named asiaticusin A and asiaticusin B. The molecular formulas of both were found to be $C_{27}H_{36}O_5$, and their structure is as shown in Figures 3.3*a*, *b*. Later, Shibata et al. (1998) isolated and characterized another novel macrolide phenolic compound—ornatipolide—from the fruiting body of another fungus, *Boletus ornatipes*. The structure of this phenolic metabolite is shown in Figure 3.3*c*. In addition to the fruiting body, other structural units of the fungus were also found to contain phenolic compounds. More recently, liquid chromatography–mass spectrometry (LC-MS) analysis identified a polyphenolic compound—epigallocatechin 3-gallate [molecular weight (MW) 458.38]—in the ethyl acetate subfraction of *P. tuber-regium* that showed potential antioxidant activity (Cheung, 2001). In addition, the gas chromatography–mass spectrometry (GC-MS) method has revealed the presence of several phenolic acids, including *trans*-cinnamic acid, hydroxybenzoic acid, protocatechuic acid, and caffeic acid, in *A. bisporus* and *L. edodes* (Mattila et al, 2001).

Other types of phenolic antioxidants, such as sterols, are also present in abundance in mushrooms. Although its antioxidant activity is generally weaker than that of phenolic antioxidants, ergosterol is also found in abundance in mushrooms (Mattila et al., 2002). In some cultivated mushrooms, such as *A. bisporus, P. ostreatus*, and *L. edodes*, over 600 mg of ergosterol can be found in 100 g of mushrooms. Other sterol antioxidants, such as fungisterol, are also present, but in lower quantities (Mattila et al., 2002).

These results indicate that mushrooms can be used as a potential dietary source of phenolic antioxidants to enrich the endogenous antioxidant status of the human body.

3.7.2.3 Biosynthesis of Phenolic Compounds from Mushrooms or Fungi The phenolic compounds in fungi are secondary metabolites derived from intermediates of the shikimic acid pathway, the primary role of which is to provide the essential aromatic amino acids phenylalanine, tyrosine, and tryptophan (Turner, 1971). The intermediates of the shikimic acid pathway are precursors of aromatic compounds, including phenolic compounds. One group of compounds so derived in fungi are simple phenolic compounds, which are usually classified as C_6–C_3, C_6–C_2, and C_6–C_1 depending on the length of the carbon side chain. The C_6–C_3 compounds include cinnamic acid and its derivatives, the C_6–C_2 compounds are phenylacetic acid and its derivatives, and the C_6–C_1 compounds include benzoic acid and its derivatives.

The biosynthesis of the C_6–C_3 compounds in the Basidiomycetes, especially *Lentinula lepideus*, has demonstrated that they possess enzymes such as ammonia-lyases that convert phenylalanine and tyrosine to cinnamic acids (Power et al., 1965). There are three possible pathways for the biosynthesis of C_6–C_1 compounds. In Basidiomycetes the C_6–C_3 compounds are converted into C_6–C_1 compounds. In *Sporobolomyces roseus*, this was found to involve the conversion of C_6–C_3 compounds, such as cinnamic, *p*-coumaric, and caffeic acids, into a C_6–C_1 compound—protocatechuic acid—by β oxidation in the washed cells, probably via benzoic and *p*-hydroxybenzoic acids (Moore et al., 1968). The second pathway is the formation of protocatechuic acid and gallic acid from dehydroshikimic acid. The third pathway involves the stepwise degradation of C_6–C_3 compounds into C_6–C_1 compounds via C_6–C_2 compounds, as identified in *Polyporus tumulosus* (Turner, 1971). The C_6–C_3 compounds also serve as intermediates in the biosynthesis of flavonoids (C_6–C_3–C_6 compounds), which are another type of phenolic antioxidant. This pathway plays a smaller role in fungi, as only two flavonoid compounds have been isolated so far, namely, chlorflavonin and dihydrochalcone, which were isolated from *Aspergillus*

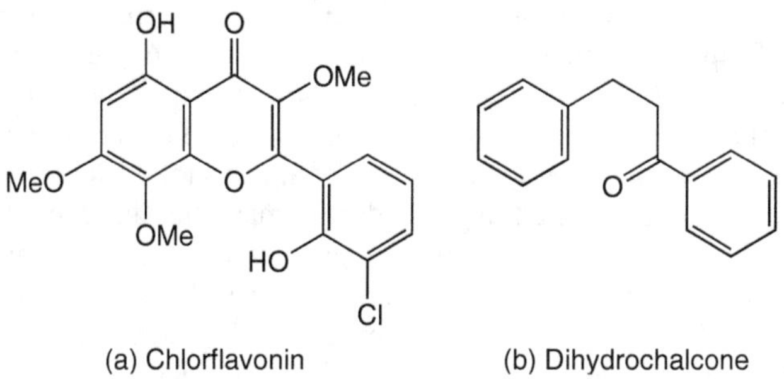

Figure 3.4 Structure of two flavonoids isolated from fungi.

candidus and *Phallus impudicus*, respectively (Bu'Lock, 1967; List and Freund, 1968) (Figure 3.4).

3.7.3 Hypocholesterolemic Effect of Mushrooms

Cardiovascular disease is associated with atherosclerosis, LDL oxidation, and hypercholesterolemia, and thus the regulation of the cholesterol level is important for the prevention and treatment of this disease. Edible mushrooms are an ideal food for the dietetic prevention of atherosclerosis due to their high fiber and low fat content. Indeed, the inclusion of edible mushrooms in a natural hypocholesterolemic and antisclerotic diet is often prescribed in Oriental medicine (Sun et al., 1984). The hypolipidemic effects of some edible mushrooms are summarized in Table 3.5

Initial research on the cholesterol-lowering effects of mushrooms was conducted in Japan in the 1960s (Kaneda and Tokuda, 1966), and it was demonstrated that when rats were fed with a high-fat and high-cholesterol diet supplemented with 5% DW of the fruiting bodies of *L. edodes* (shiitake mushroom) for 10 weeks, the plasma cholesterol levels of the animals decreased significantly (Kaneda and Tokuda, 1966). The adenosine derivative lentinacin or lentysine (currently known as eritadenine) [2(*R*), 3(*R*)-dihydroxy-4-(9-adenyl)-butyric acid] (Figure 3.5) was subsequently isolated and identified to be one of the active hypocholesterolemic components in the shiitake mushroom (Tokita et al., 1972). Eritadenine has also been found to reduce the serum cholesterol level in mice, not only by the inhibition of cholesterol biosynthesis but also by the acceleration of the excretion of ingested cholesterol and its metabolic decomposition (Suzuki and Ohshima, 1976). Various studies have shown that *Lentinula* mushrooms can lower both the blood pressure and the free cholesterol level in plasma and can accelerate the accumulation of lipids in the liver by removing them from circulation (Kabir and Kumura, 1989). Eritadenine affects the metabolism not only of cholesterol but also of phospholipids and fatty acids in rats (Sugiyama et al., 1995; Shimada et al., 2003). The dietary supplementation of eritadenine may therefore decrease phosphatidylcholine biosynthesis by altering the phosphatidylethanolamine concentration (Sugiyama et al., 1995). Similar to soybean protein, eritadenine lowers cholesterol by decreasing the ratio of phosphatidylcholine (PC) to phosphatidylethanolamine (PE) in liver microsomes and altering the composition of PC (Sugiyama and Yamakawa, 1996). Eritadenine can also suppress the metabolism of lipids (linoleic acid) by suppressing Δ 6–desaturase activity (Sugiyama et al., 1997; Shimada et al., 2003). Several other studies on *Lentinula* extracts have shown them to cause a significant decrease in serum cholesterol in young women and people older than 60 years of age in Japan (Hobbs, 1995).

Recently, it has been reported that eritadenine may elicit its effect by the suppression of the hyperhomocysteinic effect of guanidinoacetic acid, which leads to the decreased production of homocysteine and increased cystathionine formation (Fukada et al., 2006). In addition to eritadenine, nucleic acid compounds extracted from *L. edodes* also show an inhibition in platelet agglutination (Wasser and Weis, 1999).

TABLE 3.5 Hypolipidemic Effects of Some Common Edible and Medicinal Mushrooms

Mushrooms	Active Ingredient	Hypolipidemic Effects: ↘ Serum TC	↘ Serum (VLDL-c + LDL-c)	↘ Serum TG	↘ Hepatic TC and TL	↗ Hepatic LDL Receptor mRNA	↗ Fecal Sterol Excretion	References
Agaricus bisporus	Fiber						+	Fukushima et al., 2000
	Eritadenine							Saito et al., 1975
Grifola frondosa	Fiber						+	Fukushima et al., 2001
		+	+	+	+		+	Kubo and Nanba, 1997
Flammulina velutipes	Fiber					+	+	Fukushima et al., 2001
Ganoderma lucidum	—	+						Kabir et al., 1988
Auricularia polytricha	Exobiopolymer	+	+	+				Yang, J. H. et al., 2002
Volvariella volvacea	Fruit body	+	+		+		+	Wasser and Weis, 1999
	Mycelium	+	+		+		+	Cheung, 1996b, 1998
Polyporus confluens	Grifolin, neogrifolin	+						Sugiyama et al., 1992
Auricularia auricula	—	+	+				+	Cheung, 1996b
Tremella fuciformis	—	+	+	+			+	Cheung, 1996b
			+					Wasser and Weis, 1999
Armillariella mellea	—							Wasser and Weis, 1999
Trametes spp.	—	+	+					Wasser and Weis, 1999
Wolfporia cocos	—							Wasser and Weis, 1999
Tremella aurantia	TAP	+						Kiho et al., 2000
Cordyceps sinensis	CSF 30	+		+				Francia et al., 1999
Aspergillus terreus	Mevinolin							Alberts et al., 1989

Note: + , with significant difference between the mushroom groups and the blank control group.

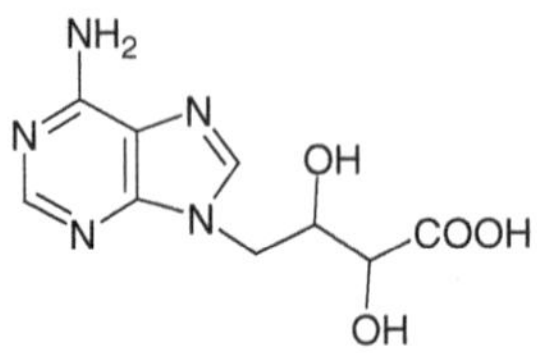

Figure 3.5 Chemical structure of eritadenine.

Lovastatin (mevinolin) and its analogues (Figure 3.6) are powerful inhibitors of 3-hydroxy-3-methyl-glutaryl-CoA (HMG-CoA) reductase and as such are well-known cholesterol-lowering agents (Endo, 1988). They can be isolated from the fruiting bodies of various types of oyster mushroom (*Pleurotus* spp.), including *P. eryngii, P. sapidus, P. ostreatus*, and *P. cornucopiae* (Cimerman and Cimerman, 1995). It has also been found that the addition of 2 and 4% of *P. ostreatus* to a hyperlipidemic diet can prevent the accumulation of cholesterol and triglyceride in both the sera and livers of rats with exogenous, endogenous, or genetically induced hyperlipidemia (Bobek et al., 1991, 1993). A reduction of the serum cholesterol level of up to 80% also resulted from the feeding of the whole mushroom, water, and 30% ethanol extract of *P. ostreatus* to rats. This led to the proposal that the hypolipidemic effect may be attributable to the fiber complex of the mushroom, which limits the reabsorption of cholesterol in the gastrointestinal tract. An undefined substance that influences the metabolism outside of the phase of reabsorption may also contribute to the cholesterol-lowering effect (Bobek et al., 1991, 1993). In another study, dietary fiber extracted from *P. cornucopiae* had a marked antiatherosclerotic effect in vitro (Ryong et al., 1989), and patients with coronary disease showed a decreased atherogenic activity (20–40%) in their sera after the consumption of this mushroom, which confirms that it has a natural cholesterol-lowering agent that is responsible for this hypocholesterolemic effect (Ryong et al., 1989). It has also been found that the addition of 1–5% of oyster mushroom to a hyperlipidemic diet efficiently prevents the accumulation of cholesterol (and especially LDL cholesterol) and triglycerides in both the blood and liver of rats with hyperlipidemia (Bobek et al.,

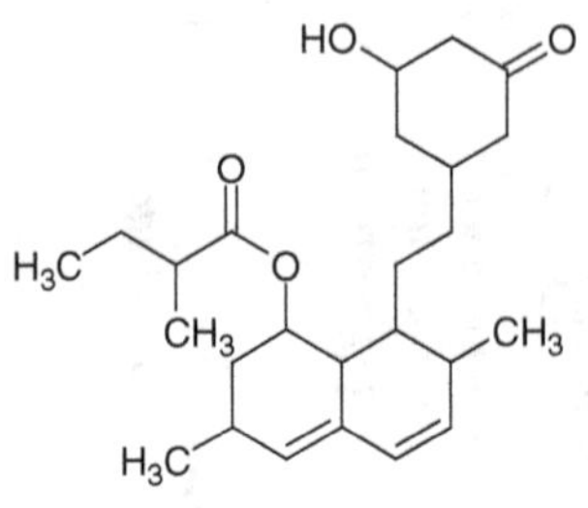

Figure 3.6 Chemical structure of lovastatin.

1998) and also reduces cholesterol biosynthesis by suppressing the activity of hepatic HMG-CoA reductase (Bobek et al., 1995) and accelerated cholesterol catabolism by up-regulating hepatic cholesterol 7α-hydroxylase (Bobek et al., 1994). It has been suggested that the fruiting bodies of oyster mushrooms could be recommended for consumption as a natural cholesterol-lowering agent in the human diet (Cimerman, 1999).

In addition to lovastatin and eritadenine, dietary fiber (nonstarch polysaccharides, mainly β-glucans) has also been suggested to be an important hypocholesterolemic component in mushrooms, and dietary fiber isolated from *Auricularia auricula-judae* (Jew's ear) and *Tremella fuciformis* (white jelly-leaf) can significantly decrease the serum total cholesterol (TC) and LDL cholesterol levels (Cheung, 1996b). Furthermore, exopolysaccharides produced by the submerged fermentation of the mycelium of *V. volvacea* can reduce the levels of serum TC, LDL cholesterol, and liver TC in alimentarially induced hypercholesterolemic rats (Cheung, 1996c). Fibers from *G. frondosa* (maitake mushroom) can greatly increase the fecal sterol excretion, which reduces the total body "sterol pools" (Kubo and Nanba, 1997), and fibers from *F. velutipes* (enokitake mushroom) and *A. bisporus* (button mushroom) can dramatically enhance the hepatic LDL receptor messenger RNA (mRNA), causing the diminution of the serum TC (Fukushima et al., 2000, 2001).

Other mushroom species, such as *A. auricula-judae*, display anticoagulation, antiaggregatory activity in the blood platelets of mice and rats, thus serving to lower their TC, total triglyceride, and lipid levels (Chen, 1989; Sheng and Chen, 1990). The supplementation of 5% DW of *V. volvacea* (straw mushroom) to hamsters fed a hypercholesterolemic diet (0.1% cholesterol and 10% fat) significantly lowered their levels of plasma and hepatic cholesterol and increased the fecal excretion of neutral sterols (Cheung, 1998). Another mushroom, *G. frondosa*, reduced blood pressure in rats without changing the plasma high-density lipoprotein (HDL) level or serum cholesterol level (Mizuno, 1995). It has also been reported that dried *A. aegerita* can significantly reduce the serum TC, triglyceride, atherogenic index, hepatic TC, and total triglyceride levels in rats fed a semisynthetic high-cholesterol diet compared with the control group (Yeung and Cheung, 2002).

The hypocholesterolemic effect of *A. aegerita* has been suggested to be linked with its antioxidant activity (Ng, 2005). Hot water and ethanolic extracts obtained from *A. aegerita* have shown the in vitro inhibition of LDL oxidation, as expressed through thiobarbituric acid reactive substances (Ng, 2005). Apart from having potent antioxidant activity and a high total phenolic content, *A. aegerita* also exhibits an in vivo hypocholesterolemic effect and has potential as a natural source of phenolic antioxidants and a hypocholesterolemic agent.

3.7.4 Hypoglycemic Effect of Mushrooms

An extensive search for traditional plant treatments for diabetes has been conducted (Alarcon-Aguilara et al., 1998) that recognized edible mushrooms as an ideal food for the dietetic prevention of hyperglycemia because of their high dietary fiber and protein and low fat content. Many studies have been conducted on the hypoglycemic activity of whole mushrooms and their fruiting bodies (Horio and Ohtsuru, 2001) and on mushroom bioactive components, including polysaccharides (Kiho et al., 1994a, b, 2000, 2002; Mori et al., 1998; Kurushima et al., 2000) and lectins (Ewart et al., 1975) isolated from the fruiting bodies. Moreover, endo- and exopolymers produced in submerged mycelial cultures have also been found to have a hypoglycemic effect (Kim et al., 1997, 2001). The most common animal models used for the study of the hypoglycemic effects of mushrooms are rats and mice with insulin-dependent diabetes mellitus (IDDM) induced by streptozotocin (STZ) and genetically diabetic mice with non-insulin-dependent diabetes mellitus (NIDDM) (Beattie et al., 1980; Swanston-Flatt et al., 1989; Kiho et al., 2002).

The administration of *G. frondosa* to IDDM STZ diabetic albino Wistar rats at 20% DW in a semipurified diet for 100 days resulted in an increase in insulin excretion and a decrease in the blood glucose level in the animals (Horio and Ohtsuru, 2001). It has also been demonstrated that *G. frondosa* has an antidiabetic effect in NIDDM KK-A^y mice, which is produced by reducing the blood glucose level (Kubo et al., 1994; Kubo and Nanba, 1997). Mushroom polysaccharides, including β-glucan isolated from the fruiting bodies of *A. cylindracea* (Kiho et al., 1994a), *H. erinaceus* (Xue et al., 1989; Mori et al., 1998), and *G. frondosa* (Kurushima et al., 2000), have been investigated for their hypoglycemic effect. The β-glucans isolated from *A. cylindracea* showed remarkable hypoglycemic activity in both normal and STZ-induced diabetic mice when administered intraperitoneally (Kiho et al., 1994a). The $(1 \rightarrow 4)$-linked or $(1 \rightarrow 6)$-linked residues in the $(1 \rightarrow 6)$-β-branched $(1 \rightarrow 3)$-β-D-glucan of the isolated polysaccharides seemed to be necessary for their hypoglycemic effect (Kiho et al., 1994a). The hypoglycemic β-glucans isolated from *H. erinaceus* differed from those of *A. cylindracea* in having a backbone of $(1 \rightarrow 6)$-β-linked-D-glucose with $(1 \rightarrow 6)$-β-linked residue (Mori et al., 1998). The polysaccharide fraction with blood-glucose-depressing effect isolated from *G. frondosa* also has $(1 \rightarrow 6)$-β-linked glucose as the main chain, which is similar to that in *H. erinaceus* but has a $(1 \rightarrow 4)$-α-linked glucopyranosyl residue as a branch chain (X fraction) (Kurushima et al., 2000). This X fraction promotes the responsiveness of the insulin receptors and can lead to some recovery from the NIDDM (Kubo et al., 1994).

Acidic polysaccharides isolated from the fruiting bodies of *A. auricula-judae* (Yuan et al., 1998a, b), *Tremella aurantia* (Kiho et al., 1995, 2002), and *T. fuciformis* (Kiho et al., 1994b) have also been studied for their antidiabetic effects. The antidiabetic acidic polysaccharide isolated from *T. aurantia* has a branched structure with a $(1 \rightarrow 4)$-α-linked-D-mannopyranosyl backbone and side chains of β-D-xylopyranosyl residues, with β-D-glucuropyranosyluronic residues linked to the terminal α-D-mannopyranose (Kiho et al., 2000). *Tremella aurantia* has been shown to exert its hypoglycemic effect in diabetic mouse models of both IDDM and NIDDM following intraperitoneal administration (Kiho et al., 1995). The antidiabetic activity of this mushroom may also be mediated by an increase in the activities of glucokinase, hexokinase, and glucose-6-phosphate dehydrogenase and a decrease in the activity of glucose-6-phosphatase in normal and IDDM diabetic mouse livers after intraperitoneal administration (Kiho et al., 2000). *Tremella aurantia* has also been shown to be effective in lowering the plasma glucose level when administered orally in KK-A^y mice (Kiho et al., 2002). Another acidic polysaccharide from the fruiting bodies of *T. fuciformis* was found to be effective in STZ-induced diabetic mice when administered orally (Kiho et al., 1994b).

A significant reduction in the level of plasma glucose was also observed in STZ-induced diabetic Sprague-Dawley rats fed with exopolymers (mainly polysaccharides) isolated from the submerged mycelial cultures of five types of common edible mushrooms (Kim et al., 2001).

Lectins isolated from mushrooms (*Agaricus campestris* and *A. bisporus*) have been shown to enhance insulin release in isolated Langerhans rat islets (Ahmad et al., 1984). The presence of a non-lectin-type component in *A. campertris* that displays insulin-releasing and insulin-like activity has also been reported (Gray and Flatt, 1998). Guanidine, which is a known hypoglycemic substance related to the biguanide class of oral antidiabetic drugs, has been found in edible mushrooms (Windholz, 1983), but the detailed principles of these active components in mushrooms remain to be elucidated.

3.8 CONCLUSION

The detailed mechanisms of the various health benefits of mushrooms to humans still require intensive investigation, especially given the emergence of new evidence of their health benefits, such as their prebiotic, hypotensive, and hepatoprotective effects. The exploration of newly cultivated mushrooms and their active ingredients with potential therapeutic value therefore remains a challenge.

REFERENCES

Adewusi, S. R. A., Alofe, F. V., Odeyemi, O., Alolabi, O. A., and Okeoz, P. A. (1993). Studies on some edible wild mushrooms from Nigeria: 1. Nutritional, teratogenic and toxic considerations. *Plant Foods for Human Nutrition, 43*, 115–121.

Ahmad, N., Bansal, A. K., and Kidwai, J. R. (1984). Effect of PHA-B fraction of *Agaricus bisporus* lectin on insulin release and $^{45}Ca^{2+}$ uptake by islet of Langerhans in vitro. *Acta Diabetologica, 21*, 63–70.

Alarcon-Aguilara, F. J., Roman-Ramos, R., Perez-Gutierrez, S., Aguilara-Contreras, A., Contreras-Weber, C. C., and Flores-Sanez, J. L. (1998). Study of the antihyperglycemic effect of plants used as antidiabetics. *Journal of Ethnopharmacology, 61*, 101–110.

Alberts, A. W., MacDonald, J. S., Till, A. E., and Tobert, J. A. (1989). Lovastatin. *Cardiovascluar Drug Reviews, 7*, 89–109.

Aletor, V. A. (1995). Compositional studies on edible tropical species of mushrooms. *Food Chemistry, 54*, 265–268.

Alofe, F. V. (1991). Amino acids and trace minerals of three edible wild mushrooms form Nigeria. *Journal of Food Composition and Analysis, 4*, 167–174.

American Association of Cereal Chemists (AACC) (2001). The definition of dietary fiber. *Cereal Foods World, 46*, 112–126.

Aoyama, T., Nakakita, Y., Nakakita, M., and Sakai, H. (1982). Screening for antioxidants of microbial origin. *Agricultural and Biological Chemistry, 46*, 2369–2371.

Babitskaya, V. G., Shcherba, V. V., Oleshko, V. S., and Osadchaya, O. V. (1996). *Pleurotus ostreatus* produces a complex of biologically active compounds. *Applied Biochemistry and Microbiology, 32*, 203–210.

Bano, Z. and Rajarathnam, S. (1986). Vitamin values in *Pleurotus* mushrooms. *Quality of Plant Foods for Human Nutrition, 36*, 11–15.

Bano, Z. and Rajarathnam, S. (1988). *Pleurotus* mushrooms. Part II. Chemical composition, nutritional value, post-harvest physiology, preservation, and role as human food. *CRC Critical Reviews in Food Science and Nutrition, 27*, 87–158.

Bartnicki-Garcia, S. (1968). Cell wall chemistry, morphogenesis and taxonomy of fungi. *Annual Review of Microbiology, 22*, 87–108.

Beattie, G., Lannom, R., Lipsick, J., Kaplan, N. O., and Osler, A. G. (1980). Streptozotocin-induced diabetes in athymic and conventional BALB/c mice. *Diabetes, 29*, 146–150.

Beelman, R. B. and Edwards, C. G. (1989). Variability in the composition and nutritional value of the cultivated mushrooms *Agaricus bisporus*. *Mushroom News, 37*, 20–26.

Boa, E. (2004). *Wild Edible Fungi. A Global Overview of Their Use and Importance to People. Non-Wood Forest Products.* Series No. 17. Rome: Food and Agrigulture Organization.

Bobek, P., Ginter, E., Jurcovicova, M., and Kuniak, L. (1991). Cholesterol-lowering effect of the mushroom *Pleurotus ostreatus* in hereditary hypercholesterolemic rats. *Annals of Nutrition and Metabolism, 35*, 191–195.

Bobek, P., Hromadova, M., and Ozdin, L. (1995). Oyster mushroom (*Pleurotus ostreatus*) reduces the activity of 3-hydroxy-3-methylglutaryl CoA reductase in rat liver microsomes. *Experientia, 51*, 589–591.

Bobek, P., Ondreicka, R., Klvanova, J., and Ozdin, L. (1994). Oyster mushroom (*Pleurotus ostreatus*) decreases serum and liver cholesterol and increase cholesterol 7α-hydroxylase activity and fecal excretion of neutral sterols and bile acids in hypercholesterolemic rats. *Nutrition Research, 14*, 1689–1699.

Bobek, P., Ozdin, L., and Galbavy. S. (1998). Dose- and time-dependent hypocholesterolemic effect of oyster mushroom (*Pleurotus ostreatus*) in rats. *Basic Nutritional Investigation, 14*, 282–286.

Bobek, P., Ozdin, L., and Kuniak, L. (1993). Influence of water and ethanol extracts of the oyster mushroom (*Pleurotus ostreatus*) on serum and liver lipids of the Syrian hamsters. *Die Nahrung, 37*, 571–575.

Boras, M. (1996). Mushroom as a new source for food and its role in food security. *Agricultural Development, 15*, 40–48.

Boutrif, E. (1991). Recent developments in protein quality evaluation. *Food, Nutrition and Agriculture, 1*, 30–40.

Bu'Lock, J. D. (1967). Fungal metabolites with structural function. In *Essays in Biosynthesis and Microbial Development.* London: Wiley, pp. 1–18.

Caglarirmak, N., Unal, K., and Otles, S. (2002). Nutritional value of edible wild mushrooms collected from the Black Sea region of Turkey. *Micologia Aplicada International, 14*, 1–5.

Carbonaro, M., Grant, G., Cappelloni, M., and Pusztai, A. (2000). Perspectives into factors limiting in vivo digestion of legume proteins: Antinutritional compound or storage protein? *Journal of Agricultural and Food Chemistry, 48*, 742–749.

Chang, S. T. (1996). Mushroom nutriceuticals world. *Journal of Microbial Biotechnology, 12*, 473–476.

Chang, S. T. (1999). World production of cultivated edible and medicinal mushrooms in 1997 with emphasis on *Lentinus edodes* (Berk.) sing. in China. *International Journal of Medicinal Mushrooms, 1*, 291–300.

Chang, S. T. (2006). The world mushroom industry: Trends and technological development. *International Journal of Medicinal Mushrooms, 8*, 297–314.

Chang, S. T. and Miles, P. G. (2004). The nutritional attributes of edible mushrooms. In *Mushrooms: Cultivation, Nutritional Value, Medicinal Effect, and Environmental Impact.* Boca Raton, FL: CRC Press, pp. 27–38.

Chen, Q. (1989). Antilipemic effect of polysaccharides from *Auricularia auricula, Tremella fuciformis*, and *Tremella fuciformis* spores. *Zhongguo Yaoke Daxue Xuebao, 20*, 344–347.

Cheung, L. M. (2001). Evaluation of the antioxidant activity and characterization of extracts from three edible Chinese mushrooms. M. Phil. thesis. Hong Kong: Chinese University of Hong Kong.

Cheung, L. M. and Cheung, P. C. K. (2005). Mushroom extracts with antioxidant activity against lipid peroxidation. *Food Chemistry, 89*, 403–409.

Cheung, L. M., Cheung, P. C. K., and Ooi, V. E. C. (2003). Antioxidant activity and total phenolics of edible mushroom extracts. *Food Chemistry, 81*, 249–255.

Cheung, P. C. K. (1996a). Dietary fiber content and composition of some cultivated mushroom fruiting bodies and mycelia. *Journal of Agricultural and Food Chemistry, 44*, 468–471.

Cheung, P. C. K. (1996b). The hypocholesterolemic effect of two edible mushrooms: *Auricularia auricula* (tree-ear) and *Tremella fuciformis* (white jelly-leaf) in hypercholesterolemic rats. *Nutrition Research, 16*, 1721–1725.

Cheung, P. C. K. (1996c). The hypocholesterolemic effect of extracellular polysaccharide from the submerged fermentation of mushroom. *Nutritional Research, 16*, 1953–1957.

Cheung, P. C. K. (1997a). Dietary fiber content and composition of some edible fungi determined by two methods of analysis. *Journal of the Science of Food and Agriculture, 73*, 255–260.

Cheung, P. C. K. (1997b). Chemical evaluation of some lesser-known edible mushroom mycelia produced in submerged culture from soymilk waste. *Food Chemistry, 60*, 61–65.

Cheung, P. C. K. (1998). Plasma and hepatic cholesterol levels and fecal neutral sterol excretion are altered in hamsters fed straw mushroom diets. *Journal of Nutrition, 128*, 1512–1516.

Cheung, P. C. K. and Lee, M. Y. (1998). Comparative chemical analysis of fiber material prepared by enzymatic and chemical methods from two mushrooms (*Pleurotus sajor-caju* and *Pleurotus tuber-regium). Journal of Agricultural and Food Chemistry, 46*, 4854–4857.

Cimerman, N-G. (1999). Medicinal value of the genus *Pleurotus* (Fr.) P. Karst. (Agaricales S. R., basidiomycetes). *International Journal of Medicinal Mushrooms, 1*, 69–80.

Cimerman, N-G. Cimerman, A. (1995). *Pleurotus* fruiting bodies contain the inhibitor of 3-hydroxy-3-methylglutaryl-coenzyme A reductase—Lovastatin. *Experimental Mycology, 19*, 1–6.

Clifford, A. J., Heid, M. K., Peerson, J. M., and Bills, N. D. (1991). Bioavailability of food folates and evaluation of food matrix effects with a rat bioassay. *Journal of Nutrition, 121*, 445–453.

Cocchi, L., Vescovi, L., Petrini, L., and Petrini, O. (2006). Heavy metals in edible mushrooms in Italy. *Food Chemistry, 98*, 277–284.

Codex Alimentarius Commission (1989). Document Alinorm. 89/30. Rome: Food and Agriculture Organization.

Crisan, E. V. and Sands, A. (1978). Nutritional values. In *The Biology and Cultivation of Edible Mushrooms.* Chang, S. T. and Hayes, W. A., editors. New York: Academic, pp. 137–168.

Dabbour, I. and Takruri, H. R. (2002a). Protein quality of four types of edible mushrooms found in Jordan. *Plant Foods for Human Nutrition, 57*, 1–11.

Dabbour, I. and Takruri, H. R. (2002b). Protein digestibility using corrected amino acid score method (PDCAAS) of four types of mushrooms grown in Jordan. *Plant Food for Human Nutrition, 57*, 13–24.

DaDamio, P. A. and Thompson, D. B. (1992). Mushroom (*Agaricus bisporus*), its polyphenol oxidase and polyphenolics affect in vitro iron availability. *Journal of Food Science, 57*, 458–461.

Danner, E. and Eaker, D. (1992). Amino acid and total protein content of the edible mushroom *Cantharellus cibarius (*Fries). *Journal of the Science of Food and Agriculture, 60*, 333–337.

Demirbaş, A. (2001). Concentrations of 21 metals in 18 species of mushrooms growing in the East Black Sea region. *Food Chemistry, 75*, 453–457.

De Román, M., Boa, E., and Woodward, S. (2006). Wild-gathered fungi for health and rural livelihoods. Proceedings for the Nutrition Society, 65, 190–197.

Díez, V. A. and Alvarez, A. (2001). Compositional and nutritional studies on two wild edible mushrooms from northwest Spain. *Food Chemistry, 75*, 417–422.

El-Kattan, M. H., Afify, A. S., and Aly, A. Z. M. (1991). Evaluation of *Pleurotus sajor-caju* fungal pellets as food. *Mushroom Journal of the Tropics, 11*, 13–22.

Elmastas, M., Isildak, O., Turkekul, I., and Temur, N. (2007). Determination of antioxidant activity and antioxidant compounds in wild edible mushrooms. *Journal of Food Composition and Analysis, 20*, 337–345.

Endo, A. (1988). Chemistry, biochemistry, and pharmacology of HMG CoA reductase inhibitors. *Klinische Wochenschrift, 66*, 421–427.

Ewart, R. B. L., Kornfield, S., and Kipnis, D. M. (1975). Effect of lectins on hormone release from isolated rat islets of Langerhans. *Diabetes, 24*, 705–714.

Ezquerra, J. M., García-Carreño, F. L., and Carrillo, O. (1998). In vitro digestibility of dietary protein sources for white shrimp (*Penaeus vannamei*). *Aquaculture, 163*, 123–136.

Flegg, P. B. and Maw, G. (1977). Mushrooms and their possible contribution to world protein needs. *Mushroom Journal, 48*, 395–403.

Food and Agricultural Organization of the United Nations (FAO) (1991). *Protein Quality Evaluation.* Rome: FAO.

Francia, C., Rapior, S., Courtecuisse, R., and Siroux, Y. Y. (1999). Current research findings on the effects of selected mushrooms on cardiovascular diseases. *International Journal of Medicinal Mushrooms, 1*, 169–172.

Fukada, S.-I., Setoue, M., Morita, T., and Sugiyama, K. (2006). Dietary eritadenine suppresses guanidinoacetic acid-induced hyperhomocysteinemia in rats. *Journal of Nutrition, 136*, 2797–2802.

Fukushima, M., Nakano, M., Morii, Y., Ohashi, T., Fujiwara, Y., and Sonoyama, K. (2000). Hepatic LDL receptor mRNA is increased by dietary mushroom *(Agaricus bisporus)* fiber and sugar beet fiber. *Journal of Nutrition, 130*, 2151–2156.

Fukushima, M., Ohashi, T., Fujiwara, Y., Sonoyama, K., and Nakano, M. (2001). Cholesterol-lowering effects of Maitake (*Grifola frondosa*) fiber, Shiitake (*Lentinus edodes*) fiber, and Enokitake (*Flammulina velutipes*) fiber in rats. *Experimental Biology and Medicine, 226*, 758–765.

Gray, W. D. (1973). *The Use of Fungi as Food and in Food Processing, Parts I and II.* London: Butterworths.

Gray, A. M. and Flatt, P. R. (1998). Insulin-releasing and insulin-like activity of *Agaricus campestris* (mushroom). *Journal of Endocrinology, 157*, 259–266.

Hobbs, C. (1995). *Medicinal Mushrooms: An Exploration of Tradition, Healing and Culture.* Santa Cruz, CA: Botanica.

Horio, H. and Ohtsuru, M. (2001). Maitake (*Grifola frondosa*) improve glucose tolerance of experimental diabetic rats. *Journal of Nutritional Science and Vitaminology, 47*, 57–63.

Huisman, J. (1991). Negative effects of antinutritional factors in animal feeds on excretion of nitrogen in the environment and possibilities of reduction digestive physiological effects. *Tijdschrift-voor-Diergeneeskunde, 116*, 391–395

Ishikawa, Y., Marimoto, K., and Hamasaki, T. (1984). Flavoglaucin, a metabolite of *Eurotium chavalieri*, its oxidation and synergism with tocopherol. *Journal of American Oil Chemists' Society, 61*, 1864–1868.

Jordan, P. (1999). *Mushroom Identifier.* New York: Lorenz Books, pp. 29, 52, 59.

Kabir, Y. and Kimura, S. (1989). Dietary mushrooms reduce blood pressure in spontaneously hypertensive rats. *Journal of Nutritional Science and Vitaminology, 35*, 91–94.

Kabir, Y. Kimura, S., and Tamura, T. (1988). Dietary effect of *Ganoderma lucidum* mushroom on blood pressure and lipid levels in spontaneously hypertensive rats (SHR). *Journal of Nutritional Science and Vitaminology, 34*, 433–438.

Kalač, P. and Svoboda, L. (2000). A review of trace element concentrations in edible mushrooms. *Food Chemistry, 69*, 273–281.

Kaneda, T. and Tokuda, S. (1966). Effect of various mushroom preparations on cholesterol levels in rats. *Journal of Nutrition, 90*, 371–376.

Kasuga, A., Aoyagi, Y., and Sugahara, T. (1993). Antioxidative activities of several mushroom extracts. *Journal of Japanese Society of Food Science and Technology, 40*, 56–63.

Khana, P. and Garcha, H. S. (1986). Nucleic acid content and relative nutritive value (RNV) of sporophofore protein of *Pleurotus* spp. *Mushroom Newsletters for the Tropics, 6*, 15–17.

Kiho, T., Kochi, M., Usui, S., Hirano, K., Aizawa, K., and Inakuma, T. (2002). Antidiabetic effect of an acidic polysaccharide (TAP) from *Tremella aurantia* Schw.:Fr. (Heterobasidiomycetes) in genetically diabetic KK-A^y mice. *International Journal of Medicinal Mushrooms, 4*, 291–297.

Kiho, T., Morimoto, H., Kobayashi, T., Usui, S., Ukai, S., Aizawa, K., and Inakuma, T. (2000). Effect of a polysaccharide (TAP) from the fruiting bodies of *Tremella aurantia* on glucose metabolism in mouse liver. *Bioscience, Biotechnology, and Biochemistry, 64*, 417–419.

Kiho, T., Morimoto, H., Sakushima, M., Usui, S., and Ukai, S. (1995). Polysaccharides in fungi. XXXV. Antidiabetic activity of an acidic polysaccharide from the fruiting bodies of *Tremella aurantia. Biological and Pharmaceutical Bulletin, 18*, 1627–1629.

Kiho, T., Sobue, S. and Ukai, S. (1994a). Structural features and hypoglycemic activities of two polysaccharides from a hot-water extract of *Agrocybe cylindracea. Carbohydrate Research, 251*, 81–87.

Kiho, T., Tsujimura, Y., Sakushima, M., Usui, S., and Ukai, S. (1994b). Polysaccharide in fungi. XXXIII. Hypoglycemic activity of an acidic polysaccharide (AC) from *Tremella fuciformis. Yakugaku Zasshi, 114*, 308–315.

Kim, D-H., Yang, B-K., Jeong, S-C., Hur, N-J., Das, S., Yun, J-W., Choi, J-W., Lee, Y-S., and Song, C-H. (2001). A preliminary study on the hypoglycemic effect of the exo-polymers produced by five different medicinal mushrooms. *Journal of Microbiology and Biotechnology, 11*, 167–171.

Kim, M-Y., Park, M-H., and Kim, G-H. (1997). Effects of mushroom protein-bound polysaccharides on the blood glucose levels and energy metabolism in streptozotocin-induced diabetic rats. *Journal of Korean Nutrition, 30*, 743–750.

Kubo, K., Aoki, H., and Nanba, H. (1994). Antidiabetic activity present in the fruit body of *Grifola frondosa* (Maitake). I. *Biological and Pharmaceutical Bulletin, 17*, 1106–1110.

Kubo, K. and Nanba, H. (1997). Anti-hyperliposis effect of Maitake fruit body (*Grifola frondosa*). I. *Biological and Pharmaceutical Bulletin, 20*, 781–785.

Kurushima, H., Kodama, N., and Nanba, H. (2000). Activities of polysaccharides obtained from *Grifola frondosa* on insulin-dependent diabetes mellitus induced by streptozotocin in mice. *Mycoscience, 41*, 473–480.

Kurzman, R. H. (1997). Nutrition from mushrooms: Understanding and reconciling available data. *Mycoscience, 38*, 247–253.

LARN (1996). Livelli di assunzione raccomandati di energia e nutrienti per la popolazione italiana. *Societá Italiana di Nutrizione Umana (SINU)*, Roma.

Lee, Y.-L., Yen, M.-T., and Mau, J.-L. (2007). Antioxidant properties of various extracts from *Hypsizigus marmoreus. Food Chemistry, 104*, 1–9.

Leichter, J. and Bandoni, R. J. (1980). Nutrient content of mushroom grown in British Columbia. *Canadian Institute of Food Science and Technology Journal, 13*, 122–124.

Li, G. S. F. and Chang, S. T. (1982). The nucleic acid content of some edible mushrooms. *European Journal of Applied Microbiology and Biotechnology, 15*, 237–240.

Li, G. S. F. and Chang, S. T. (1985). Determination of vitamin C (ascorbic acid) in some edible mushrooms by differential pulse polarography. *Mushroom Newsletter for the Tropics, 5*, 11–16.

Lin, S. Y. (1988). Studies on the characteristics of taste-active components in shiitake mushroom and the powderization of the concentrate. Master's thesis. Taichung: National Chung Hsing University.

List, P. H. and Freund B. (1968). Odorous substances of *Phallus impudicus* L18. Report on fungi contents. *Planta Medica (Supplement), 23*, 123–132.

Litchfield, J. H. (1967). Submerged culture of mushroom mycelium. In *Microbial Technology*. Peppler, H. J., editor. New York: Reinhold, pp. 107–144.

Liu, F., Ooi, V. E. C., and Chang, S. T. (1997). Free radical scavenging activities of mushroom polysaccharides extracts. *Life Sciences, 60*, 763–771.

Lo, K. M. and Cheung, P. C. K. (2005). Antioxidant activity of extracts from the fruiting bodies of *Agrocybe aegerita* var. alba. *Food Chemistry, 89*, 533–539.

Longvah, T. and Deosthale, Y. G. (1998). Composition and nutritional studies on edible wild mushroom from northeast India. *Food Chemistry, 63*, 331–334.

Maga, J. A. (1981). Mushroom flavor. *Journal of Agricultural and Food Chemistry, 29*, 1–4.

Manzi, P., Aguzzi, A., and Pizzoferrato, L. (2001). Nutritional value of mushrooms widely consumed in Italy. *Food Chemistry, 73*, 321–325.

Manzi, P. and Pizzoferrato, L. (2000). Beta-glucans in edible mushrooms. *Food Chemistry, 68*, 315–318.

Manzi, P., Gambelli, L., Marconi, S., Vivanti, V., and Pizzoferrato, L. (1999). Nutrients in edible mushrooms: An inter-species comparative study. *Food Chemistry, 65*, 477–482.

Mattila, P. H., Piironen, V. I., Uusi-Rauva, E. J., and Koivistoinen, P. E. (2001). Vitamin D contents in edible mushrooms. *Journal of Agricultural and Food Chemistry, 42*, 2449–2453.

Mattila, P., Könkö, K., Euvola, M., Pihlava, J., Astola, J., Vahteristo, L., Hietaniemi, V., Kumpulainan, J., Valtonen, M., and Piironen, V. (2001). Contents of vitamins, mineral elements and some phenolic compound in cultivated mushrooms. *Journal of Agricultural and Food Chemistry, 42*, 2449–2453.

Mattila, P., Lampi, A.-M., Ronkinen, R., Toivo, J., and Piironen, V. (2002). Sterol and vitamin D_2 contents in some wild and cultivated mushrooms. *Food Chemistry, 76*, 293–298.

Mau, J. L., Chao, G. R., and Wu, K. T. (2001a). Antioxidant properties of methanolic extracts from several ear mushrooms. *Journal of Agricultural and Food Chemistry, 49*, 5461–5467.

Mau, J. L., Chyau, C. C., Li, J. Y., and Tseng, Y. H. (1997). Flavor compounds in straw mushrooms *Volvariella volvacea* harvested at different stages of maturity. *Journal of Agricultural and Food Chemistry, 45*, 4726–4729.

Mau, J. L., Lin, H. C., and Chen, C. C. (2001b). Non-volatile taste components of several medicinal mushrooms. *Food Research International, 34*, 521–526.

Mau, J. L., Lin, H. C., Ma, J. T., and Song, S. F. (2001c). Non-volatile taste components of several specialty mushrooms. *Food Chemistry, 73*, 461–466.

Mau, J. L., Lin, H. C., and Song, S. F. (2002). Antioxidant properties of several specialty mushrooms. *Food Research International, 35*, 519–526.

Mau, J. L. and Tseng, Y. H. (1998). Non-volatile taste components of three strains of *Agrocybe cylindracea*. *Journal of Agricultural and Food Chemistry, 46*, 2011–2074.

McDonough, F. E., Steinke, F. H., Sarwar, G., Eggum, B. O., Bressani, R., Huth, P. J., Barbeau, W. E., Mitchell, G. V., and Phillips, J. G. (1990). In vivo rat assay for true protein digestibility: Collaborative study. *Journal of the Association of Official Analytical Chemists, 73*, 801–805.

Mdachi, S. J. M., Nkunya, M. H. H., Nyigo, V. A., and Urasa, I. T. (2003). Amino acid composition of some Tanzanian wild mushrooms. *Food Chemistry, 86*, 179–182.

Miles, P. G. and Chang, S. T. (1997). *Mushroom Biology—Concise Basics and Current Developments.* Singapore: World Scientific, p. 144.

Mizuno, T. (1995). Bioactive biomolecules of mushrooms: Food function and medicinal effect of mushroom fungi. *Food Reviews International, 11*, 7–12.

Moore, K., Rao, P. V., and Towers, G. H. (1968). Degradation of phenylalanine and tyrosine by *Sporobolomyces roseus. Biochemical Journal, 106*, 507–514.

Mori, H., Aizawa, K., Inakuma, T., Ichii, A., Yamauchi, R., and Kato, K. (1998). Structural analysis of the β-D-glucan from the fruit-body of *Hericium erinaceum. Journal of Applied Glycoscience, 45*, 361–365.

Ng, Y. F. (2005). In vitro and in vivo antioxidant activity and hypocholesterolemic effect in extracts of *Agrocybe aegerita*. M.Phil. thesis. Hong Kong: The Chinese University of Hong Kong.

Ouzouni, P. K., Veltsistas, P. G., Paleologos, E. K., and Riganakos, K. A. (2007). Determination of metal content in wild edible mushroom species from regions of Greece. *Journal of Food Composition and Analysis, 20*, 480–486.

Pedersen, B. and Eggum, B. O. (1983). Prediction of protein digestibility by an in vitro enzymatic pH-stat procedure. *Zeitschrift Tierphysiol, 49*, 265–277.

Piepponen, S., Liukkonen-Lilja, H., and Kuusi, T. (1983). The selenium content of edible mushrooms in Finland. *Zeitschrift fur lebensmittel Untersuchung und Forschung, 177*, 257–260.

Power, D. M., Towers, G. H., and Neish, A. C. (1965). Biosynthesis of phenolics acids by certain wood-destroying basidiomycetes. *Canadian Journal of Biochemistry, 43*, 1397–1407.

Protein Advisory Group (PAG) (1970). *Single-Cell Protein Guideline.* Report of FAO/WHO/UNICEF Protein Advisory Group No. 4. New York: United Nations.

Rodger, G. (2001). Mycoprotein—A meat alternative new to the U. S. *Food Technology, 55*, 36–50.

Ryong, L. H., Tertov, V. V., Vasiley, A. W., Tutelyan, V. A., and Orekhov, A. N. (1989). Antiatherogenic and antiatherosclerotic effects of mushroom extracts revealed in human aortic intima cell culture. *Drug Development Research, 17*, 109–117.

Saito, M., Yamashita, T., and Kaneda, T. (1975). Quantitative analysis of eritadenine in Shiitake mushroom and other edible fungi. *Eiyo to Shokuryo, 28*, 503–506.

Sanmee, R., Dell, B., Lumyong, P., Izumori, K., and Lumyong, S. (2003). Nutritive value of popular wild edible mushrooms from Northern Thailand. *Food Chemistry, 82*, 527–532.

Sarwar, G. (1996). *The Protein Digestibility-Corrected Amino Acid Score Method Overestimates Quality of Protein Containing Antinutritional Factors and of Poorly Digestible Proteins Supplemented with Limiting Amino Acids in Rats.* London: American Society for Nutritional Sciences, pp. 748–764.

Sarwar, G. and McDonough, F. (1990). Review of protein quality evaluation methods. *Journal of the Association of Official Analytical Chemists, 73*, 347–350.

Sawaya, W. W., Al-shalhat, A., Al-Sogair, A., and Al-Mohammed, M. (1985). Chemical composition and nutritive value of truffles of Sandi Arabia. *Journal of Food Science, 50*, 450–453.

Shah, H., Khalil, I., and Jabeen, S. (1997). Nutritional composition and protein quality of *Pleurotus* mushroom. *Sarhad Journal of Agriculture, 13*, 621–626.

Sheng, J. and Chen, Q. (1990). Effects of polysaccharides from *Auricularia auricula, Tremella fuciformis, and Tremella fuciformis* spores on experimental thrombin formation. *Zhongguo Yaoke Daxue Xuebae, 21*, 39–42.

Shibata, H., Fukuda, T., Wada, T., Morita, Y., Hashimoto, T., and Asakawa, Y. (1998). Ornatipolide, a novel phenolic metabolite from a basidiomycete, *Boletus ornatipes*. *Bioscience, Biotechnology and Biochemistry, 62*, 1432–1434.

Shimada, Y., Yamakawa, A., Morita, T., and Sugiyama, K. (2003). Effects of dietary eritadenine on the liver microsomal Δ 6-desaturase activity and its mRNA in rats. *Bioscience, Biotechnology and Biochemistry, 67*, 1258–1266.

Senatore, F. (1992). Chemical constituents of some mushrooms. *Journal of the Science of Food and Agriculture, 58*, 499–503.

Smith, J. (1972). Commercial mushrooms production. *Process Biochemistry, 7*, 24–26.

Stijve, T., Goessler, W., and Dupuy, G. (2004). Influence of soil particles on concentrations of aluminium, iron, calcium and other metals in mushrooms. *Deutsche Lebensmittel-Rundschau, 100*, 10–13.

Sugahara, T., Arai, S., Aoyagi, Y., and Kunisaki, N. (1975). Contents of 5′-nucleotides and free amino acids in different varieties of dried shiitake mushroom (*Lentinus edodes* Sing.). *Journal of Japan Society of Nutrition and Food Science, 28*, 477–483.

Sugiyama, K., Akachi, T., and Yamakawa, A. (1995). Hypocholesterolemic action of eritadenine is mediated by a modification of hepatic phospholipid metabolism in rats. *Journal of Nutrition, 125*, 2134–2144.

Sugiyama, K., Kawagishi, H., Tanaka, A., Saeki, S., Yoshida, S., Sakamoto, H., and Ishiguro, Y. (1992). Isolation of plasma cholesterol-lowering components from ningyotake (*Polyporus confluens*) mushroom. *Journal of Nutritional Science and Vitaminology, 38*, 335–342.

Sugiyama, K. and Yamakawa, A. (1996). Dietary eritadenine-induced alteration of molecular species composition of phospholipids in rats. *Lipids, 31*, 399–404.

Sugiyama, K., Yamakawa, A., and Saeki, S. (1997). Correlation of suppressed linoleic acid metabolism with the hypocholesterolemic action of eritadenine in rats. *Lipids, 32*, 859–866.

Sun, M. T., Xiao, J. T., Zhang, S. Q., Liu, Y. J., and Li, S. T. (1984). Therapeutic effect of some foods on hyperlipidermia in man. *Acta Nutritional Sinica, 6*, 127–133.

Suzuki, S. and Ohshima, S. (1976). Influence of shiitake (*Lentinus edodes*) on human serum cholesterol. *Mushroom Science, 9*, 463–470.

Svoboda, L., Havlickova, B., and Kalač, P. (2006). Contents of cadmium, mercury and lead in edible mushrooms growing in a historical silver-mining sea. *Food Chemistry, 96*, 580–585.

Svoboda, L., Kalač, P., Špička, J., and Janoušková, D. (2002). Leaching of cadmium, lead and mercury from fresh and differently preserved edible mushroom, *Xerocomus badius*, during soaking and boiling. *Food Chemistry, 79*, 41–45.

Svoboda, L., Zimmermannová, K., and Kalač, P. (2000). Concentrations of mercury, cadmium, lead and copper in fruiting bodies of edible mushrooms in an emission area of a copper smelter and a mercury smelter. *Science of the Total Environment, 246*, 61–67.

Swanston-Flatt, S. K., Day, C., Flatt, P. R., Gould, B. J., and Bailey, C. J. (1989). Glycemic effects of traditional European plant treatments for diabetes: Studies in normal and streptozotocin diabetic mice. *Diabetes Research, 10*, 69–73.

Takamura, K., Hoshino, H., Sugahara, T., and Amano, H. (1991). Determination of vitamin D_2 in *Shiitake* mushroom by high-performance liquid chromatography. *Journal of Chromatography, 545*, 201–204.

Tokita, F., Shibukawa, N., Yasumoto, T., and Kaneda, T. (1972). Isolation and chemical structure of the plasma-cholesterol reducing substance from Shiitake mushroom. *Mushroom Science, 8*, 783–788.

Tsai, S.-Y., Huang, S.-J., and Mau, J.-L. (2006). Antioxidant properties of hot-water extracts from *Agrocybe cylindracea. Food Chemistry, 98*, 670–677.

Tseng, Y. H. and Mau, J-L. (1999). Contents of sugar, free amino acids and free 5′-nucleotides in mushrooms, *Agaricus bisporus*, during post-harvest storage. *Journal of the Science of Food and Agriculture, 79*, 1519–1523.

Tshinyangu, K. K. (1996). Effect of grass hay substrate on nutritional value of *Pleurotus ostreatus var. Columbus. Nahrung, 40*, 79–83.

Turner, W. B. (1971). Secondary metabolites derived without the intervention of acetate. In *Fungal Metabolites.* London: Academic, pp. 26–61.

Valencia del Toro, G., Vega, R. C., Garin-Aguilar, M. E., and Lara, H. L. (2006). Biological quality of proteins from three strains of *Pleurotus* spp. *Food Chemistry, 94*, 494–497.

Verma, A., Keshervani, G. P., Sharma, Y. K., Sawarkar, N. J., and Singh, P. (1987). Mineral content of edible (dehydrated) mushrooms. *Indian Journal of Nutrition and Dietitians, 24*, 241–245.

Vetter, J. (1990). Mineral element content of edible and poisonous macrofungi. *Acta Alimentaria, 19*, 27–40.

Vetter, J. (1994a). Data on arsenic and cadmium contents of some common mushrooms. *Toxicon, 32*, 11–15.

Vetter, J. (1994b). Mineral elements in the important cultivated mushrooms *Agaricus bisporus* and *Pleurotus ostreatus*. *Food Chemistry, 50*, 277–279.

Vetter, J. (1995). Mineral and amino acid contents of edible cultivated shiitake mushrooms (*Lentinus edodes*). *Zeitschrift fur lebensmittel Untersuchung und Forschung, 201*, 17–19.

Wada, T., Hayashi, Y., and Shibata, H. (1996). Asiaticusin A and B, novel prenylated phenolics from *Boletinus asiaticus* and *B. paluster* (Boletaceae) fungi. *Bioscience, Biotechnology and Biochemistry, 60*, 120–121.

Walker, C. A. (1996). Vitamin B_{12} fortification of *Agaricus bisporus* mushrooms. *Cultivated Mushroom Research Newsletter, 3*, 31–38.

Wasser, S-P. and Weis, A-L. (1999). Medicinal properties of substances occurring in higher basidiomycetes mushrooms: Current perspectives (Review). *International Journal of Medicinal Mushrooms, 1*, 31–62.

Windholz, M. (1983). *The Merck Index*, 10th ed. Rahway, NJ: Merck and Co.

Wong, K. H. and Cheung, P. C. K. (1998). Nutritional assessment of three Chinese indigenous legumes in growing rats. *Nutrition Research, 18*, 1573–1580.

Wong, K. H., Cheung, P. C. K., and Wu, J. Z. (2003). Biochemical and microstructural characteristics of insoluble and soluble dietary fiber prepared from mushroom sclerotia of *Pleurotus tuber-regium, Polyporus rhinoceros* and *Wolfiporia cocos*. *Journal of Agricultural and Food Chemistry, 51*, 7179–7202.

Wong, W. C. (2003). Chemical composition, nutritional values and functional properties of some novel cultivated edible mushrooms. M. Phil. thesis. Hong Kong: The Chinese University of Hong Kong.

Xue, W. J., Yang, W., and Chen, Q. H. (1989). Prevention and treatment of alloxan-induced diabetes in mice by the polysaccharides from *Laminaria japonica* and *Hericium erinaceus*. *Journal of China Pharmaceutical University, 20*, 378–380.

Yang, B-K., Ha, J-Y., Jeong, S-C., Jeon, Y-J., Ra, K-S., Das, S., Yun, J-W., and Song, C-H. (2002). Hypolipidemic effect of an exo-biopolymer produced from submerged mycelial culture of *Auricularia polytricha* in rats. *Biotechnology Letters, 24*, 1319–1325.

Yang, J. H., Lin, H. C., and Mau, J. L. (2000). Non-volatile taste components of several commercial mushrooms. *Food Chemistry, 72*, 465–471.

Yang, J. H., Lin, H. C., and Mau, J. L. (2002). Antioxidant properties of several commercial mushrooms. *Food Chemistry, 77*, 229–235.

Yamaguchi, S., Yoshikawg, T., Ikeda, S., and Ninomiya, T. (1971). Measurement of the relative taste intensity of some α-amino acid and 5′-nucleotides. *Journal of Food Sciences, 36*, 846–849.

Yeung, M. and Cheung, P. C. K. (2002). The hypolipidemic effect of some lesser-known edible and medicinal mushrooms. *Abstracts of the Institute of Food Technologists' Annual Meeting*. Anaheim: Institute of Food Technologists.

Yilmaz, N., Solmaz, M., Turkekul, I., and Elmastas, M. (2006). Fatty acid composition in some wild edible mushrooms growing in the middle Black Sea region of Turkey. *Food Chemistry, 99*, 168–174.

Yoshida, H., Sugahara, T., and Hayashi, Y. (1986). Changes in carbohydrates and organic acids development of mycelium and fruit bodies of Hiratake mushrooms (*Pleurotus ostreatus*). *Nippon Shokuhin Kogyo Gakkaishi, 33*, 519–528.

Yuan, Z., He, P., Cui, J., and Takeuchi, H. (1998a). Hypoglycemic effect of water-soluble polysaccharide from *Auricularia auricula-judae* Quel. on genetically diabetic KK-Ay mice. *Bioscience, Biotechnology and Biochemistry, 62*, 1898–1903.

Yuan, Z., He, P., and Takeuchi, H. (1998b). Ameliorating effects of water-soluble polysaccharides from woody ear (*Auricularia auricula-judae*Quel.) in genetically diabetic KK-A^y mice. *Journal of Nutritional Science and Vitaminology, 44*, 829–840.

Zakhary, J. W., Taiseer, M., Abo-Bakr El-Mhady, A., and EL-Tabey, A. M. S. (1983). Chemical composition of wild mushrooms collected from Alexandria. *Food Chemistry, 11*, 31–41.

CHAPTER 4

Sclerotia: Emerging Functional Food Derived from Mushrooms

Ka-Hing Wong

Department of Biology, The Chinese University of Hong Kong, Hong Kong, China

Peter C. K. Cheung

Food and Nutritional Sciences Programme, Department of Biology, The Chinese University of Hong Kong, Hong Kong, China

CONTENTS

4.1 INTRODUCTION

Unlike plants and animals, filamentous fungi do not form three-dimensional complex tissues via mitotic activity (Elliott, 1994; Moore, 2003). Instead, the complex structures (e.g., strands, rhizomorphs, synnemata, stromata, sclerotia) of fungi develop by the aggregation of the specialized hyphae that result from the differentiations that are governed by tight genetic control (Elliott, 1994; Moore, 2003). These multihyphal structures can be further classified into elongated (e.g., strands) and rounded (e.g., sclerotia) structures. The elongated organs not only

Mushrooms as Functional Foods, Edited by Peter C. K. Cheung

enable the connection of one fungus resource to another (e.g., rhizomorph) but also allow the fungal reproductive organs to reach a location that is favorable for the dispersal of spores (e.g., synnemata). The main functions of the rounded structures (i.e., sclerotia) are to allow the fungus to survive under unfavorable environmental conditions and to provide reserves for the fungus to germinate.

In addition to enzymatic preparation (Wong and Cheung, 2004, 2005a), fractionation (Cheung and Lee, 1999, 2000; Zhang et al., 2003a), and biochemical, microstructural, physicochemical, and functional characterizations (Cheung, 1997; Cheung and Lee, 1998; Wong et al., 2003; Wong and Cheung, 2005a; L. N. Zhang et al., 2001; M. Zhang et al., 2003b), our research team has for the past decade been actively studying the physiological functions (Chau et al., 2007; Cheung and Wong, 2004; Lai and Cheung, 2004; Lai, 2005; Lai et al., 2005; Tao et al., 2006; Wong and Cheung, 2005b; Wong et al., 2005, 2006, 2007; M. Zhang et al., 2001, 2004a, b, 2006a, b) of the nonstarch polysaccharides (NSPs) isolated from three Chinese edible and medicinal mushroom sclerotia, namely *Pleurotus tuber-regium* (Fr.) Sing., *Polyporus rhinocerus* Cooke, and *Wolfiporia cocos* (Schw.) Ryv. et Gilbn. In this chapter, in addition to the concepts, ontogeny, and cultivation of mushroom sclerotia, we share our previous research on the NSPs of the aforementioned sclerotia and their potential to be developed as novel functional foods or nutraceuticals.

4.2 CONCEPTS OF MUSHROOM SCLEROTIA

According to Willetts and Bullock (1992), the sclerotium can be described, from a functional point of view, as a morphologically variable, nutrient-rich, multihyphal aggregated structure that can remain dormant or quiescent when the environment is unfavorable. However, when the conditions improve, the sclerotium can germinate again to produce the fungus. The size of a sclerotium is species dependent and ranges from microscopic (only a few cells, e.g., *Verticillium dahliae*) to enormous (more than 30 cm in diameter, e.g., *Polyporus mylittae*), whereas its shape is usually spherical to oval (Carlisle et al., 2001; Moore, 2003). Sclerotia occur sporadically in the Ascomycotina (e.g., *Claviceps purpurea*; Luttrell, 1980), Basidiomycotina (e.g., *Sclerotium rolfsii*; Chet and Henis, 1975), and Deuteromycotina (e.g., *V. dahliae*). Among these three subdivisions, the most studied sclerotium-forming fungi are those of economic importance, such as the Sclerotiniaceae (e.g., *Monilinia, Botryotinia*, and *Sclerotinia*; Willetts, 1997) and the *Typhula* spp. (Willetts et al., 1990).

4.3 ONTOGENY OF SCLEROTIA

4.3.1 Morphological Aspects

The ontogeny study of sclerotia was reported as early as the nineteenth century by Brefeld (1877) and De Bary (1887). In general, sclerotial ontogeny can be divided into three overlapping stages: (i) initiation (when the hyphae begin to aggregate to form small, discrete initials); (ii) development (when the initials grow to full size, accumulating nutritional reserves from the parent mycelia and droplet excretion); and (iii) maturation (surface delimitation, pigmentation of the peripheral hyphae, conversion of the reserve nutrients into a suitable form for long-term storage, and internal consolidation). These three developmental stages are complicated processes that are accompanied by both morphological and biochemical differentiations under tight genetic control (Carlisle et al., 2001; Chet and Henis, 1975; Cooke, 1970; Townsend and Willetts, 1954).

Although numerous endogenous and exogenous factors are reported to be involved in sclerotial initiation, the sclerotium is basically initiated by the onset of starvation conditions or other circumstances that are unfavorable for continuous mycelial growth (Carlisle et al., 2001; Willetts and Bullock, 1992). The environmental (e.g., light, temperature, pH), mechanical (e.g., when mycelia are cut, torn, grow against the side of the culture vessel), biochemical (e.g., when mycelia grow in contact with staling products, other microorganisms, antibiotics, phenolics, and polyphenoloxidases), nutritional (e.g., C/N ratio, minerals, vitamins), and internal morphogenetic factors that affect sclerotial initiation have been comprehensively reviewed in the literature (Cooke, 1983; Chet and Henis, 1975; Willetts, 1978; Willetts and Bullock, 1992). Thus, these factors will not be discussed here.

There are three main types of sclerotial development—loose, terminal, and lateral—all of which have been discussed in detail by Willetts (1972). When hyphae aggregate to form multihyphal structures such as sclerotia, their repulsion, which occurs in normal hyphal growth, must disappear or even be substituted by attraction for hyphal association. Apart from a few classic studies reviewed by Moore (1984), information on the surface chemistry of hyphae is scarce. Autotrophic agents and the mucilage matrix, which accumulate over the surface of vegetative hyphae, are believed to play an important role in specific cell-to-cell adhesion in the development of multihyphal structures such as sclerotia (Reijnders and Moore, 1985; Willetts, 1972, 1978).

Sclerotial initials develop from one or several knot(s) of aggregated hyphae (e.g., strands in the case of *S. rolfsii*; Townsend and Willetts, 1954) within a mycelial mass. As the sclerotial initials grow and increase in size, the central hyphae exhibit remarkable dichotomous branching, and their cells become swollen and vacuolated. This facilitates their main role of accumulating nutritional reserves (including glycogen, polyphosphate, proteins, and lipids) from the parental mycelia. For the outer hyphae of the developing sclerotia, the cells differentiate to become shortened, followed by wall thickening and melanizing to form a protective layer over the sclerotial surface (rind). The tightly packed structure of the rind together with the deposits between the outer hyphal cells is likely to prevent the apoplastic transfer of the solutes (Carlisle et al., 2001). During sclerotial development, considerable nutrients are absorbed from the substrate, followed by degradation by the various enzymes (e.g., arylesterase and acid phosphatase; Wong and Willetts, 1974), to provide energy and nutrients for the developing sclerotium. Thus, the nutrient status of the substrate directly determines the number and size of the sclerotia produced (Willetts and Wong, 1971). Furthermore, the substratum may be partially enclosed in the sclerotium and surrounded by the melanized rind cells.

Dormant sclerotia may survive for several years, and the pattern of sclerotial germination is species dependent (Willetts and Bullock, 1992). During dormancy, considerable accumulated endogenous reserves, together with the interhyphal materials, are used collectively to sustain the sclerotium during the periods of survival and germination (e.g., *Sclerotinia minor*; Bullock and Willetts, 1996) when conditions are favorable. The germination of sclerotia can be carried out by means of asexual spores (sporogenic), fruiting bodies (carpogenic), mycelia (myceliogenic), or a combination of these. However, only sporogenic and carpogenic germinations are the general characteristics of larger sclerotia, which are produced by airborne fungal pathogens such as mushrooms (Garrett, 1970). Sclerotia germinate as a result of various factors, including exposure to nutrients, appropriate light and temperature conditions (e.g., *P. mylittae*, Carlisle et al., 2001; *S. minor*, Imolehin et al., 1980; *Sclerotinia sclerotiorum*, Huang, 1991), or even stimulation by substances emitted by a host plant (e.g., *Sclerotinia*, Carlisle et al., 2001; *S. minor*, Hall et al., 1982). Obviously, complex and genetic-controlled processes are likely be involved in sclerotial germination (Elliott, 1994; Moore, 2003). As suggested by Carlisle and his co-investigators (2001), the formation of sclerotia, which remain at the site of production, is an effective strategy for the survival of fungi. This is because the area that has been suitable for the growth of the host plants naturally allows the seeds of the host plant to remain, thus allowing the next generation to grow. This then increases the chance of fungal infection in the plants.

4.3.2 Physiological Aspects

4.3.2.1 Translocation As reported by Jennings (1987), during sclerotial development, the insoluble carbohydrates (e.g., glycogen) that accumulate in the mycelium during vegetative growth are converted to soluble forms (e.g., trehalose and mannitol). These chemical changes lead to the development of a turgor gradient between the vegetative mycelium (nutrient base) and the sclerotial initials (nutrient sinks). As a result, the movement of water and nutrients to the developing sclerotia is carried out by a turgor-driven mass flow via a few specialized conducting hyphae (Wilcoxson and Sudia, 1968). As the translocated nutrients are utilized by the developing sclerotia, the turgor gradient is maintained. As suggested by Jennings (1987), the maintenance of metabolic gradients during translocation may even be tightly controlled by evaporation at the surface of the multihyphal structures and by the trehalose in the mycelium, which fine tunes the sugar concentration in both the nutrient base and the sink. In contrast to the early stage of sclerotial development, the nutrient requirement for growth is diminished when the sclerotium is mature because of its large size and compactness. To avoid the accumulation of soluble nutrients, which could affect translocation and other metabolic activities, soluble nutrients are converted to insoluble forms and accumulate as intra- and extracellular reserves, with the greatest deposition in the rind and cortex regions (at the periphery) of the sclerotia (at the end of the translocatory stream) for subsequent use by the fungus, such as in germination under favorable conditions.

4.3.2.2 Exudation Exudation is a common phenomenon during the early stages of sclerotial development (Townsend and Willetts, 1954) and has been extensively reviewed by Colotelo (1978). In early sclerotial development, small droplets begin to form on the surface of the sclerotia, and these coalesce to form a few large droplets when the sclerotia become mature and pigmented. In the case of *Cristulariella* spp., the mature sclerotia are even found to bathe in liquid (Willetts and Bullock, 1992). The droplets may remain on the sclerotia from several days (under evaporation) to several weeks, with some of the constituents being reabsorbed and probably utilized by the sclerotial tissue (Colotelo, 1978). The color of the droplets, which is probably due to the accumulation of oxidized phenolics, varies from clear to pale or dark brown (mature sclerotia only), even on the same sclerotium (Willetts and Bullock, 1982). The composition of the droplets is complex. In addition to carbohydrates (such as trehalose, mannitol, inositol, and glucose) and enzymes (including polyphenoloxidase, peroxidase, glucosidase, and cellulose), amino acids and fatty acids have been reported previously (Cooke, 1969; Jones, 1970). Cooke (1969, 1970) and Jennings (1987) reported that the permeability of the hyphal tip is significantly different from that of the rest of the hypha and exudation from the hyphal tips is probably an active and selective process that assists in dissipating the excessive hydrostatic pressure that is generated during the aforementioned translocation.

4.4 STRUCTURE OF SCLEROTIA

A sclerotium commonly includes a pseudoparenchymatous and melanized "rind" that encases a broad "medulla" of interwoven hyphae. In some sclerotia (e.g., *S. minor*), a narrow layer of close-fitting hyphae, namely the "cortex," is discernible between the rind and the medulla (Willetts and Bullock, 1992). The rind is a continuous layer of tightly packed hyphal tips that become thick walled and pigmented to form an impervious outer surface layer. The medulla constitutes the main part of the sclerotium, the hyphae of which (together with those of the cortex if present) are the main storage area for the intracellular reserves (Carlisle et al., 2001; Moore, 1995; Willetts and Bullock, 1992), whereas the interhyphal space is usually filled with an extracellular matrix (continuous or containing lacunae) (Moore, 1995; Willetts and Bullock, 1992).

4.4.1 Rind

The rind is formed when the hyphal tips at the periphery of the sclerotium become closely packed to form a continuous layer, with the septa lying close to the apices and the terminal cells becoming swollen and rounded (Bullock et al., 1980). In the terminal cells of a young differentiating rind, such organelles as nuclei, mitochondria, small amounts of endoplasmic recticulum (ER), and vacuoles are observed (Willetts and Bullock, 1992). At the early stage of rind development, the cell walls begin to thicken, and many multivesicular bodies that are closely associated with the plasmalemma are observed (*S. minor*; Bullock et al., 1980), which suggests their possible role in the synthesis of the hyphal cell walls of sclerotia (Marchant et al., 1967; Khan and Aldrich, 1973). In mature sclerotia, the cell wall of the rind is the thickest (an eight fold increase in the case of *S. minor*; Bullock et al., 1980). As the rind develops, vacuolation occurs rapidly through an increase in the number of vacuoles, which then coalesce to form one large central vacuole surrounded by a thin layer of cytoplasm. Although mature rind cells are usually devoid of content, large phenol-rich bodies have been observed in the rind cells of the mature sclerotia of Sclerotiniaceae (Kohn and Grenville, 1989a, b) and *Typhula incarrnata* (Willetts et al., 1990). They were probably involved in resistance against antagonistic microorganisms. The factors that control rind differentiation are not clear. In *S. minor*, the sclerotium grows to almost full size before the rind begins to differentiate (Bullock et al., 1980), whereas in *Sclerotinia trifoliorum* and *S. sclerotiorum*, there is considerable enlargement after the rind starts to develop (Cook, 1971). The enlargement of preexisting rind cells and/or the incorporation of new tips into the rind layer from underlying hyphae (e.g., *S. minor*; Bullock et al., 1980) allow the rind to expand to accommodate an increase in the surface area of the sclerotium. These rind cells may also serve as specialized passage cells that allow the sclerotial reserves to pass through the impermeable rind and be utilized for hyphal growth outside of the sclerotium (i.e., for sclerotial germination and secondary sclerotium formation). A characteristic feature of sclerotial rind formation is a change of color from white to buff to dark brown or black, which is caused by an accumulation of melanin (Chet et al., 1967; Chet and Henis, 1968; Jones, 1970). At maturity, most of the sclerotium rind cells have collapsed and are dead; thus, the symplastic transport of water and nutrients across the rind would be unlikely in a normal situation. The barrier role of the pigmented rind is partially supported by the findings of Young and Ashfold (1992), who reported that, as the rind cells differentiate, there is a reduction in the permeability of the sclerotia to the apoplastic tracer, sulforhodamine, which corresponds with the wall thickening and pigmentation of the rind cells.

4.4.2 Cortex

In some sclerotia, as the rind begins to differentiate and the sclerotium grows to almost its full size, a cortex of close-fitting rounded cells becomes discernible. The width of this cortex varies between and even within species (ranging from imperceptible to six cells wide; Townsend and Willetts, 1954; Willetts and Bullock, 1982; Kohn and Grenville, 1989a, b). There is no clear definition of the outer and inner boundaries of the cortex. The outermost cortical cells have some rindlike features, such as pigmentation and vacuolation (although with a thinner cell wall), whereas the innermost cortical cells grade into the medulla (but possess greater density and a rounded shape; Willetts and Bullock, 1992). The laying down of abundant storage bodies within the cortical hyphae is a major feature of cortical development, as the cortex is the region for the accumulation and storage of reserve materials. Similarly to the genus *Sclerotinia*, the cortex in the sclerotia of *T. incarrnata* cannot be distinguished (Willetts et al., 1990).

4.4.3 Medulla

Most of the sclerotium consists of a medulla of prosenchymatous tissue that is formed by the interweaving of the hyphae with a few septa and branches (Willetts et al., 1990; Willetts and Bullock, 1992). The greatest hyphal density is usually in the outer region of the medulla (e.g., *T. incarrnata*; Willetts et al., 1990). The most conspicuous differentiation feature of the medulla is the accumulation of an extracellular matrix in the interhyphal spaces (e.g., *T. incarrnata*; Willetts et al., 1990). In addition to the shrinkage of the extracellular matrix (the air drying of the sclerotia), the presence of lacunae in the extracellular matrix is attributed to the density of the medulla hyphae, which determines whether sufficient materials are produced to fill all of the interhyphal spaces (Willetts and Bullock, 1992). Compared with cortical hyphae, medullary hyphae share similar kinds of storage bodies, the deposition of reserves, and wall thickening.

4.5 CULTIVATION OF MUSHROOM SCLEROTIA

With the enormous improvement in mushroom cultivation technologies, the world production of cultivated edible mushrooms increased more than 12% annually between 1981 and 1997 (to 6.158 million metric tons; Chang, 1999), and their commercial value in 1998 was estimated to be about U.S.$18 billion, which is similar to that of coffee production (Wasser et al., 2000). In China, the production of cultivated edible mushrooms increased drastically from 60,000 metric tons in 1978 to 4.35 million metric tons in 1998 (more that 40 species and accounting for about half of the world's total output), and it is expected to reach more than 6 million metric tons by 2010 (Huang, 1999b, 2000). Thus, China is becoming a significant international edible mushroom producer (Chang, 1999; Huang, 2000). As reported by Huang (1999a), the sclerotium-forming, edible, and/or medicinal mushrooms in China are mainly *Grifola umbelata* (Pers.), *Omphalia lapidenscens* Schröet, *Xylaria nigripes* (Kl.) Sacc., *W. cocos* (Schw.) Ryv. et Gilbn., *P. tuber-regium* (Fr.) Sing., and *P. rhinocerus* Cooke. In addition to *W. cocos, P. tuber-regium* and *P. rhinocerus* are two of the most economically important sclerotium-forming fungi to gain popularity in China recently (Huang, 2000). Although the main source for edible and medicinal use at present is still wild sclerotia, their natural habitats have been gradually destroyed in recent years due to insufficient protection from the rapid development of agricultural and urban areas. For the conservation and exploration of mushroom sclerotia as a functional food, research on the cultivation of mushroom sclerotia under well-controlled artificial conditions needs to be comprehensively conducted, so that their scale and efficiency of production can be improved and their commercialization facilitated.

4.5.1 Sclerotia of *Pleurotus tuber-regium* (Fries) Singer

Pleurotus tuber-regium is an edible and medicinal mushroom from the Basidiomycotina, which are mainly distributed in tropical and subtropical regions such as China, Australia, and Africa (especially Nigeria) (Oso, 1975, 1977; Singer, 1961; Zadrazil, 1996; Zoberi, 1973). *Pleurotus tuber-regium* is the only *Pleurotus* species in which the fruiting bodies arise from a sclerotium (Isikhuemhen and Nerud, 1999). The sclerotium of *P. tuber-regium* is large and subterranean and is spherical to oval in shape (about 10–25 cm in diameter toward the end of the growing season, which is from April to September), and new fruiting bodies (that are light brown in color, up to 10 cm in size, and depressed in the center) are formed on it in the subsequent growing season (Oso, 1977; Zoberi, 1972, 1973). The rind of the sclerotia is shiny and dark brown, whereas the internal structure is hard, powdery, and white (Oso, 1977).

Although it is quite expensive, the sclerotium of *P. tuber-regium* is popularly consumed in Nigeria (Okhuoya and Etugo, 1993; Oso, 1977) and considered to be a delicacy (Okhuoya and Okogbo, 1990). The sclerotium is chopped or ground into powder before being put into soup, *egusi*, or melon seed ball preparations (Akobundu and Eluchie, 1992; Nwokolo, 1987; Oso, 1977; Zadrazil, 1996; Zoberi, 1973). In addition to using *P. tuber-regium* sclerotial powder as a tablet disintegrant (Iwuagwu and Onyekweli, 2002), its successful incorporation into pork sausage has also been reported by Akobundu and Eluchie (1992). In addition, the sclerotia of *P. tuber-regium* can be kept for years without losing their nutritional quality as a foodstuff or even their ability to produce fruiting bodies (Nwokolo, 1987; Zadrazil, 1996). Furthermore, the sclerotium of *P. tuber-regium* is also used for medicinal purposes by medical practitioners in Nigeria (Oso, 1977; Zoberi, 1972, 1973) to cure headaches, stomach ailments, colds, constipation, fever, asthma, smallpox, nervous disorders, and high blood pressure (Oso, 1977; Fasidi and Olorunmaiye, 1994; Zadrazil, 1996). In China, *P. tuber-regium* is commonly known as the "tiger milk mushroom" and is mainly found in the southern region, particularly in Yunnan Province (Deng et al., 2000). Recently, because of the nutraceutical benefits of its various polysaccharide fractions, such as antitumor and immunopotentiating effects (Cheung and Lee, 2000; M. Zhang et al., 2001), the consumption of *P. tuber-regium* sclerotia is growing in popularity and economic importance (Huang, 2000).

Pleurotus tuber-regium is a classic, wood-rotting fungus that can utilize a wide range of broad-leaf and needle-leaf trees. The successful cultivation of *P. tuber-regium* sclerotia has been reported by numerous studies (Fasidi and Ekuere, 1993; Jiang et al., 2000; Okhuoya and Okogbo, 1990). In China, in addition to the logs of broad-leaf trees, *P. tuber-regium* sclerotia have been cultivated using substrate bags. The two general formulations of compost used were (1) saw dust (78%), wheat bran (20%), white sugar (1%), $CaCO_3$ (1%), and water

(1 : 1.1–1.3) and (2) saw dust (39%), wheat bran (49%), sugar cane (1%), $CaCO_3$ (1%), and water (1 : 1.1–1.3) (He et al., 2000). Similarly to other *Pleurotus* spp., *P. tuber-regium* sclerotia have also been found to grow (five to six weeks after spawning) on a variety of cellulosic waste materials, such as cotton waste (biological efficiency 30.11%), rice straw (29.51%), corn cobs (22.85%), and banana leaves (13.58%) (Fasidi and Ekuere, 1993; Garcha et al., 1984; Jandaik, 1974) and on the moist drill dust from the wood of *Daniella oliveri* and *Elaeis guineensis* trees after 65 and 71 days inoculation, respectively (Okhuoya and Okogbo, 1990). As suggested by Fasidi and Ekuere (1993), the high saprophytic ability of *P. tuber-regium* can most likely be attributed to its capability of secreting a wide range of hydrolyzing and oxidizing enzymes, as other members of the genus do (Isikhuemhen and Nerud, 1999; Kadiri and Fasidi, 1990; Toyama and Ogawa, 1974; Ulezlo et al., 1975;). Furthermore, recent studies (Jiang et al., 2000) have reported that, in addition to a sufficient oxygen supply, the optimum conditions for the cultivation of *P. tuber-regium* sclerotia are a temperature of 23–28°C, a pH of 7.5–8, a culture medium-to-water ratio of 1 : 2.2, and corn starch, wheat bran, and cow dung carbon and nitrogen sources.

4.5.2 Sclerotia of *Polyporus rhinocerus* Cooke

Polyporus rhinocerus (Chinese common name, hurulingzhi) is a type of white-rot fungus that is mainly distributed in China, Malaysia, Sri Lanka, the Philippines, Australia, and East Africa (Huang, 1999a). Very limited information on this mushroom has been reported in the literature.

The sclerotium of *P. rhinocerus* is subterranean with a spherical, oval, or even irregular shape (about 4–5 cm in diameter). The rough and wrinkly surface (rind) of the sclerotia (which is white to pale brown in color), on which oval-shaped fruiting bodies (that are tea brown in color, ciliated, and depressed in the center) are grown, is thin, and the internal structure is white and powdery (Huang, 1999b). The *P. rhinocerus* sclerotium is an expensive folk medicine used by Chinese physicians to treat liver cancer, chronic hepatitis, and gastric ulcers. Because of its successful cultivation, the scientific name and taxonomy of this mushroom have recently been further identified and confirmed by conventional morphology characterization as *P. rhinocerus* Cooke or *Lignosus rhinocerus* Cooke Ryv., belonging to Eumycota, Basidiomycotina, Hymenomycetes, Aphllophorales, and Polyporaceae (Huang, 1999a, b).

In contrast to *P. tuber-regium*, information concerning the cultivation of *P. rhinocerus* sclerotia is very limited. Nevertheless, the successful cultivation of *P. rhinocerus* sclerotia using substrate bags was reported by Huang (1999b). The ingredients of the compost used were saw dust (80%), wheat bran (18%), sugar cane (1%), $CaCO_3$ (1%), and water (1 : 1–1.4). Briefly, all of the compost ingredients were well mixed and filled into transparent polypropylene plastic bags (170 × 350 × 380 × 0.05mm) to two-thirds height. After putting them on plastic rings, the open ends of the plastic bags were plugged with cotton plugs, wrapped in paper, and tied up with rubber bands prior to sterilization with an autoclave.

It is interesting to note that loosely packed compost was found to cause wrinkles on the surface of the resulting sclerotia (Huang, 1999b). When the plastic bags had cooled down to about 30°C, each substrate bag was aseptically inoculated with spawn (young *P. rhinocerus* mycelia with sclerotium-forming ability, grown on a sawdust–wheat bran medium for 30–40 days at 20–25°C), followed by incubation at 20–26°C in incubation rooms. After about one and a half months, the mycelia had fully colonized the substrates in the plastic bags, and sclerotia had begun to form. At this stage, the sclerotia could either have been kept growing in the substrate bags or buried in soil (without the plastic bags) under broad-leaf trees and covered with loose soil and withered leaves. When the sclerotia maintained with the compost in the substrate bags had shrunk drastically, become softened, and undergone exudation (at about six months), they were mature enough for harvesting. All of the harvested sclerotia were washed clean and sun or oven dried prior to further processing for edible or medicinal use.

4.5.3 Sclerotia of *Wolfiporia cocos* (Schw.) Ryv. Et Gilbn [*Poria cocos* (Schw.) Wolf]

Wolfiporia cocos is also known as Fu Ling or Hoelen (the Chinese names for its sclerotium) and is mainly distributed in the southern provinces of China such as Yunnan and Fujian (Bi et al., 1993). The sclerotium of *W. cocos* is subterranean with a spherical, oval, or even irregular shape and ranges in diameter from 10 to 30 cm (Keys, 1976; Ooi, 2000). When it is fresh, the sclerotium of *W. cocos* is slightly soft, but it becomes very hard when it is dry (Bi et al., 1993; Liu and Bau, 1980). *Wolfiporia cocos* sclerotia can be collected all year round, especially in August and September (Liu and Bau, 1980).

The sclerotium of *W. cocos* is one of the earliest and most commonly used fungi in Chinese medicine. The rough and wrinkly surface (rind) of the sclerotia (which are brownish yellow to dark brown in color) is mainly used as a diuretic (that regulates the K and Na balance) (Keys, 1976; Xu and Wang, 2002) and as a decoction for coughs. Umbrella-shaped fruiting bodies (that are white or pale yellow when fresh, with nearly no stem) are resupinately grown on the sclerotia and form a thin layer (Bi et al., 1993; Ooi, 2000). Their white or pink interiors are powdery and used as a cardiotonic agent and to relieve the uneasiness that arises from pregnancy. The sliced or whole sclerotium is often applied to treat jaundice and to induce menstruation (Ooi, 2000). The polysaccharides extracted from the sclerotium of *W. cocos*, such as debranched pachyman, exhibit strong antitumor and immunomodulatory effects (Chihara et al., 1970; Ding et al., 1998), whereas its low-molecular-weight tetracyclic triterpenes have been found to have immunostimulating, antiviral, tumor inhibitory, and cytotoxic properties (Hobbs, 1995; Kaminaga et al., 1996; Ukiya et al., 2002; Wang et al., 1995). Other folk medicinal functions of *W. cocos* sclerotia include the treatment of diarrhea, spleen dampness, and insomnia. They are also used as a sedative to tranquilize the mind and refresh the spirit (Keys, 1976; Xu and Wang, 2002).

Wolfiporia cocos is a brown rot fungus that mainly grows in association with the roots of various conifers, especially the Chinese red pine and the Taiwanese pine, and oaks (Keys, 1976; Ooi, 2000). At present, *W. cocos* sclerotia for medicinal use are primarily obtained from cultivation. In China, *W. cocos* sclerotia are mainly cultivated on the surface of pine logs that are buried in caverns after the spawn (young mycelia with sclerotium-forming ability) are inoculated. As the procedures for and information on this cultivation technique (including the season for cultivation, preparation of the spawn, type and method of inoculation, management after inoculation, and harvesting method) have been comprehensively reviewed in the recent literature [Central Agricultural Broadcasting and Television School and National Farmer's Science and Technology Education Training Centre (CABTS/NFSTTC), 2006], they will not be described in detail here. Igari et al. (2000) reported that 10 strains of *W. cocos* sclerotia (about 5 cm in diameter; 62 kg/m^3 pine log) were successfully cultivated on the surface of pine logs buried in a field after spawn inoculation for 21 months and that their quality (in terms of ash content, TLC patterns, and amount of pachymic and debydropachymic acids in the diluted ethanol extracts) was highly comparable to that of commercial strains. Recently, an indoor cultivation technique (without soil) for *W. cocos* sclerotia was developed and reported by Kubo et al. (2006). Briefly, *W. cocos* sclerotia were cultivated in mushroom culture bottles that contained three pine logs (*Pinus densiflora* SIEB. *Et* Zucc.; 5 cm in diameter and 10 cm in length), and the caps of the bottles were equipped with commercially available cloth air filters. After spawn [*W. cocos* mycelia with proven sclerotium-forming ability, grown on a sawdust–rice bran medium (3 : 1 v/v) with a moisture content of 70% at 30°C for one month] inoculation on the pine logs, all of the culture bottles were incubated in the dark at 25°C for 24 weeks to facilitate sclerotia formation. Compared with the efficiency of field cultivation (21 kg/m^3 DW, 21 months), the indoor cultivation of *W. cocos* sclerotia exhibited a remarkably faster growth rate (14 weeks) and higher productivity (110 kg/m^3 DW, about 15 cm in length). In addition, both cultivated and commercial *W. cocos* sclerotia share similar TLC patterns and pachymic and debydropachymic acid content in their methanol extracts. Furthermore, the progressive decay (mainly hemicellulose and cellulose) of the pine logs by the *W. cocos* also markedly increased its alkaline solubility in 1% aqueous NaOH (based on the weight of the decayed wood and the percentage weight loss due to wood decay), which is consistent with the main characteristic of brown rot fungi. As this indoor technique successfully cultivated *W. cocos* sclerotia with a high yield in a short period of time, further study on its applicability to other mushroom sclerotia, such as those of *P. tuber-regium* and *P. rhinocerus*, would be worthwhile.

4.6 BIOCHEMICAL, NUTRITIONAL, AND TECHNOLOGICAL CHARACTERISTICS OF MUSHROOM SCLEROTIA

4.6.1 Biochemical Components of Mushroom Sclerotia

4.6.1.1 Cell Walls Similarly to the fungal cell wall of most Basidiomycetes, the cell wall of sclerotial hyphae usually contains chitin and β-glucans (with 1,3 and 1,6 linkages at different degrees) as its major structural and matrix components (Backhouse and Willetts, 1984; Bullock et al., 1980; Jones et al., 1972; Kohn and Grenville, 1989a, b; Willetts et al., 1990). Protein has also been detected in the hyphal walls of some sclerotia (Chet et al., 1967; Kohn and Grenville, 1989a, b). Melanins, or oxidized phenolic pigments, are found to be deposited in large amounts in the rind walls and sometimes in the walls of the outer cortical cells (Bullock et al., 1980; Kohn and Grenville, 1989a, b). In addition to reducing the permeability of the cell wall of the peripheral hyphae, melanins also contribute to its resistance to radiation and biological degradation (Willetts, 1971). When forming a complex with the chitin of the fungal wall, melanins also act as inhibitors of the polysaccharide-hydrolytic enzymes of the fungus itself and its antagonistic microorganisms (Bull, 1970). This finding could well explain why the rind remains intact even when the medullary hyphae have been completely lysed by the activities of the antagonistic microorganisms (Coley-Smith, 1980) and even after the sclerotia have been degraded during germination (Backhouse and Willetts, 1985; Bullock et al., 1983).

4.6.1.2 Extracellular Matrix In various mushroom sclerotia, the major chemical composition of their extracellular matrix, which consists of highly hydrated materials expanding and filling the interhyphal spaces within the sclerotia, is found to be similar to a structure that is composed mainly of β-1,3-glucan backbone with β-1,6-linked side branches (Backhouse and Willetts, 1984; Bullock and Willetts; 1996; Bullock et al., 1980; Dubourdieu et al., 1981). In addition to morphogenesis and the storage and supply of water to withstand unfavorable environmental conditions (e.g., drought), one of the most important functions of the extracellular matrix is to provide a large energy reserve of carbohydrates during sclerotial germination (Backhouse and Willetts, 1985; Bullock et al., 1983; Ueno et al., 1980).

4.6.1.3 Cytoplasmic Reserves The identification of sclerotial cytoplasmic reserves by histochemistry has been reported in numerous studies (Backhouse and Willetts, 1984; Bullock et al., 1980; Moore et al., 1991; Willetts et al., 1990), and the main cytoplasmic reserves detected in mushroom sclerotia include glycogen, protein polyphosphate, and lipids (Moore, 1995; Willetts and Bullock, 1992).

Glycogen. Glycogen is present throughout the cytoplasm of cortical and medullary hyphae at all stages of sclerotial differentiation, whereas the hyphal cells in the rind region contain a much lower quantity of glycogen than do those of other regions, even at the beginning of rind differentiation. As reported by Bullock et al. (1980), granular glycogen deposits, which fill the spaces between other storage bodies and organelles, were observed in the medulla hyphae of the sclerotia of *S. minor* by a transmission electron microscopy. Glycogen decreases in fully grown sclerotia and is probably utilized in the synthesis of other reserves, such as proteins.

Polyphosphates. Polyphosphate granules are present in vegetative hyphae and sclerotial initials, and their quantity increases during the growth of sclerotia, particularly in the outer medulla and cortex (the main storage region at maturity) (Kohn and Grenville, 1989a, b; Willetts et al., 1990). The presence of phosphates in the metachromatic granules and tissue of the sclerotia of *Paxillus involutus* was confirmed by Moore et al. (1991) using energy-dispersive X-ray microanalysis. Polyphosphates play several possible roles in sclerotia, including energy storage, the regulation of soluble phosphate levels, and phosphorus storage (Harold, 1966; Kulaev, 1975).

Protein. The size and number of protein bodies increase throughout sclerotial development, particularly at the stage at which the sclerotium has grown to almost its full size and is not yet fully pigmented. Under a transmission electron microscope, protein bodies are round or elongated membrane-bound structures and are completely filled with moderately electron-dense materials (Willetts and Bullock, 1992; Willetts et al., 1990). ER, in the form of long and parallel cisternae, becomes much more abundant in sclerotial hyphae when the protein bodies are being formed, which suggests its possible role in protein synthesis, as occurs in other animal and plant systems (Gunning and Steer, 1975; Jorgensen et al., 1977). In the sclerotia of *P. involutus* (Moore et al., 1991) and *T. incarrnata* (Willetts et al., 1990), polyphosphates embedded in the matrix of the protein bodies were observed. In mature sclerotia, protein bodies are the major cytoplasmic storage reserve.

Lipids. Rather than special storage bodies, Backhouse and Willetts (1984) suggested that the lipid bodies in the sclerotia of *Botrytis cinerea* and *Botrytis fabae* are normal constituents of the hyphae, as they resemble both the sclerotia and vegetative hyphae of the *Botrytis* spp. However, remarkable amounts of lipids were reported in the sclerotia of *Monilinia fructicola* (Kohn and Grenville, 1989a, b) and *V. dahliae* (Willetts and Bullock, 1992). In addition to being species dependent, the lipid content of mushroom sclerotia may also be partially attributed to the culture medium and the particular isolate used (Kohn and Grenville, 1987).

4.6.2 Nutritional Evaluation of Mushroom Sclerotia

4.6.2.1 Proximate Composition Our previous studies (Wong et al., 2003) have found that the sclerotia of *P. tuber-regium, P. rhinocerus*, and *W. cocos* exhibit similar patterns of proximate composition, with a substantial amount of carbohydrates [ranging from 90.5 to 98.1% dry matter (DM)] and an extremely low lipid content (ranging from 0.02 to 0.14% DM). This indicates that all three sclerotia may belong to the carbohydrate-rich sclerotia type, as was previously suggested by Coley-Smith and Cooke (1971). Similar results were reported previously by Cheung (1997), Fasidi and Ekuere (1993), Nwokolo (1987), and Ude et al. (2001). The crude protein (0.67–6.71% DM) and ash contents (ranging from 1.09 to 2.78% DM) of the three sclerotia were low and were significantly different from one another (*P. tuber-regium* was the highest and *W. cocos* the lowest). The crude protein content (6.71% DM) of the sclerotia of *P. tuber-regium* was similar to that of cultivated specimens grown on different cellulosic waste materials (ranging from 6.32 to 8.24% DM; Fasidi and Ekuere, 1993) but was much lower than that (10.8% DM) reported by Ude et al. (2001). Nevertheless, these data would be comparable if a protein conversion factor of 4.38, rather than 6.25, was used in the latter case (7.58% DM). In fact, Basidiomycetes such as *P. tuber-regium* possess an appreciable amount of nonprotein N (mainly from chitin/chitosan), which must be corrected for before the crude protein content can be estimated from the N content and the protein conversion factor (Nwokolo, 1987; Kurasawa et al., 1991). The ash content of *P. tuber-regium* (2.78% DM) lies within the range for cultivated *P. tuber-regium* sclerotia (ranging from 0.54 to 4.00 DM) in the previous literature (Fasidi and Ekuere, 1993; Nwokolo, 1987; Ude et al., 2001). All of the air-dried sclerotia possessed a notably high level of moisture, with that of *P. tuber-regium* being significantly the lowest. Although the moisture content of the *P. tuber-regium* sclerotium (12.9% DW) was consistent with that of specimens grown in Nigeria (14.9% DW) (Ude et al., 2001), a much higher moisture content level (23.7% DW) for this air-dried sclerotium was reported by Nwokolo (1987). Except for the moisture content, all of the proximate compositions of *W. cocos* were in agreement with those of a previous study (Cheung, 1997).

4.6.2.2 Sclerotial Dietary Fiber Extensive research over the past three decades has demonstrated that the intake of sufficient dietary fiber (DF) has benefits for health maintenance and disease prevention [American Diabetic Association (ADA), 2002]. In general, DF can be divided into soluble (SDF) and insoluble (IDF) fractions based on its solubility in an aqueous medium. The viscosity of SDF is responsible for slower digestion and the absorption of nutrients, which helps to attenuate blood cholesterol and glucose levels. In contrast, IDF is characterized by its ability to increase fecal bulk (nonfermentable or partially fermentable in colonic microflora) and decrease intestinal transit time, thus promoting laxation (Potty, 1996). The increased awareness of the potential health benefits of DF among consumers has undoubtedly encouraged food manufacturers to explore new DF sources and develop fiber-enriched or fiber-fortified food products such as snack foods, beverages, cookies, and canned meats (McKee and Latner, 2000; Sloan, 2001). Today, most fiber supplements are obtained from the by-products that result from the processing (e.g., milling) of cereals, fruits, vegetables, and legumes (McKee and Latner, 2000). Because of the highly competitive market for fiber-enriched food products, there is an urgent need to explore new sources of DF. Because chitin and β-linked glucose-based polysaccharides cannot be digested or absorbed in the human intestine, mushroom sclerotia obviously contain an abundant amount of cell wall and extracellular matrix materials that can be classified as DF and may thus serve as an alternative source of DF in the food industry (Cheung, 1997; Wasser and Weis, 1999).

Our previous studies (Wong et al., 2003; Wong and Cheung, 2005a) have shown that the sclerotia of *P. tuber-regium, P. rhinoceros*, and *W. cocos* possess remarkably high levels of IDF content (ranging from 77.4 to 94.6% DM) and exceptionally small levels of SDF content (ranging from 1.45 to 2.50% DM), whereas their TDF content is (ranging from 81.7 to 96.3% sample DM) comparable to that of some commercial DF-rich supplements [HUMAMIL (glucomannan) 82.9% DM; FYBOGEL (Ispaghula husk) 88.5% DM; FIBRAPLAN (soluble fiber from algae, seeds, flour, and nonspecified plants) 86.6% DM] (Goñi and Martin-Carrón, 1998). This finding suggests that the DF of all three sclerotia has great potential to act as an alternative source of high fiber in the food industry. In addition, the three types of sclerotial DF have notably high levels of NSP (86.6–94.3% sclerotial TDF DM), in which the predominant glucose residues (89.7–94.5% NSPs DM), together with the glucosamine content (1.83–6.28% NSP DM), collectively ensure that β-glucans and chitin are the main matrix and fibrillar components of the fungal cell wall polysaccharide in the three sclerotia (Wong et al., 2003; Willetts and Bullock, 1992). Other minor sugar residues found in the three types of sclerotial DF include mannose, galactose, rhamnose, and uronic acids, which may indicate the presence of small amounts of mannan, galactan, and polyuronides. The three sclerotia may possess glucuronic acids, as the presence of glucuronic acids in other edible fungi such as *Tremella aurantia* and *Tremella fuciformis* has also been reported (Gao et al., 1996; Kiho et al., 2000). Furthermore, scanning electron micrographs have shown fragments of interwoven hyphae and insoluble materials in the three sclerotial IDF fractions, but only the amorphous structure of soluble materials was observed in the SDF fractions (Wong et al., 2003).

All three types of sclerotial DF exhibited very low levels (<30.0 μg/g of sclerotial DF) of five nutritionally important divalent minerals—calcium (Ca), magnesium (Mg), copper (Cu), iron (Fe), and zinc (Zn)—when compared with those of common DF sources such as cereals (wheat bran, rice bran, and oats: Ca 701–1904 μg/g; Mg 771–8825 μg/g), fruits (apples and oranges: Mg 519–879 μg/g; Zn 9–16 μg/g), legumes (butter beans, broad beans, lentils: Ca 977–1730 μg/g; Mg 286–424 μg/g; Fe 180–390 μg/g; Cu 11.1–30 μg/g; Zn 29–62 μg/g), vegetables (tomato and sugar beet fiber: Mg 1530–3475 μg/g; Zn 13–41 μg/g) (Elhardallou and Walker, 1999; Idouraine et al., 1995; Thibault and Ralet, 2001), and some commercial DF supplements (e.g., Citrucel, Fiber One, All-Bran, Metamucil: Ca 333–6340 μg/g) (Luccia and Kunkel, 2002). This finding indicates that a considerable amount of these five minerals was most likely lost during the enzymatic preparation of sclerotial DF, as considerably higher amounts (Ca 6200–21000 μg/g; Mg 5800–15100 μg/g; Fe 100–500 μg/g; Cu 100–500 μg/g; Zn 100–500 μg/g) were previously reported in cultivated *P. tuber-regium* sclerotia, in addition to an abundant amount of potassium (24,500–95, 600 μg/g) and small amounts of phosphorus (8200–18, 300 μg/g) and manganese (100–1000 μg/g) (Fasidi and Ekuere, 1993). Furthermore, by using an energy-dispersive X-ray microanalyzer, trace amounts of Ca-oxalate, Si, and Al were detected in the crystalline substances of *W. cocos* sclerotia

(Tanaka, 1990). As suggested by Fasidi and Ekuere (1993), the mineral content in mushroom sclerotia is highly dependent on their cultivated or natural environments.

4.6.3 Physicochemical and Functional Properties of Mushroom Sclerotial DF

In addition to its physiological benefits, DF has desirable functional properties, such as providing texture, gelling, thickening, emulsification, and stabilization in DF-enriched foods (Nelson, 2001; Dreher, 1987). Therefore, DF research, particularly in the growing nutraceutical industry, has gained a lot of attention recently (Jalili et al., 2000; Thebaudin et al., 1997). DF of different origins possesses different structures, chemical compositions, and physicochemical properties that exhibit different nutritional, technological, and physiological benefits (Blackwood et al., 2000; Guillon and Champ, 2000; Nelson, 2001; Thebaudin et al., 1997). The preparation of sclerotial DF from *P. tuber-regium, P. rhinocerus*, and *W. cocos* using industrial food-grade glycolytic and proteolytic enzymes has recently been carried out in our laboratory (Wong and Cheung, 2004). To evaluate their potential for developing fiber-enriched products with a high level of consumer acceptance, some of their physicochemical and functional properties [such as color, pH, water-binding capacity (WBC), oil-holding capacity (OHC), emulsifying activities (EA), and emulsion stability (ES)] were investigated and compared with those of a commercial fiber-rich barley ingredient (Wong and Cheung, 2005a). We found that the pH of the suspension of all three types of sclerotial DF was slightly acidic (ranging from 5.59 to 6.11), which was consistent with their low levels of uronic acids (0.51–2.14% DM). In addition, their pH values were significantly lower ($p < 0.025$) than those of the barley DF and higher than those of the DF concentrates prepared from peaches (ranging from 3.63 to 3.86) (Grigelmo-Miguel and Martin-Belloso, 1997) and oranges (ranging from 3.85 to 3.93) (Grigelmo-Miguel and Martin-Belloso, 1999).

Compared with the color of the barley DF control, both the *P. tuber-regium* and *P. rhinocerus* DF possessed significantly higher ($p < 0.025$) values of lightness (L^*) but smaller increments of redness (a^*) and yellowness (b^*). Such a high degree of whiteness is a technological advantage for these two types of sclerotial DF when they are added to such bakery products as white bread and sugar-type cookies because their incorporation is not likely to produce an off color (darker than desired) (Good, 2002; Nelson, 2001). The color of the *W. cocos* DF was characterized by similar values of L^* and a^*, but a significantly lower ($p < 0.025$) value of b* when compared with the color of the barley control. Among the three types of sclerotial DF, the value of the total color difference (ΔE^*) between the *W. cocos* DF and the barley DF control was the lowest (9.84; $p < 0.025$), indicating their relatively high similarity in color compared to the other two types of sclerotial DF. The pinkish brown color of the *W. cocos* DF suggests that its incorporation into a food system may affect the color of the final product. Color is influenced by many factors, including species variety, the maturity of the sample, and the processing method (e.g., drying) (Grigelmo-Miguel and Martin-Belloso, 1999).

The WBC of a fiber measures the amount of water that it retains after it is subject to a stress such as centrifugation (Nelson, 2001). This hydration property of a DF ingredient is crucial to its successful application in food that will be subject to physical stress (e.g., the extrusion of cereals). The *W. cocos* DF had the highest ($p < 0.05$) value of WBC (6.26 g/g DW), which was highly comparable to that of some DF derived from cereal processing by-products [wheat bran 6.4–6.6 g/g DW (Adams et al., 1986; Ralet al., 1990); oat bran 5.5 g/g DW (Cadden, 1987)], fruits [apple DF 6.3–6.9 g/g DW and pear DF 6.8 g/g DW (Grigelmo-Miguel and Martin-Belloso, 1997)], and some commercial DF-rich supplements [AGIOLAX (Ispaghula seed and husk, cassia fruit) 6.6 g/g DW (Goñi and Martin-Carrón, 1998); FIBREX (sugar beet) 4.56 g/g DW (Abdul-Hamid and Luan, 2000)]. The WBC of *P. tuber-regium* (2.78 g/g DW) and *P. rhinocerus* DF (2.72 g/g DW) did not differ significantly ($p < 0.025$) from that of the barley DF (2.54 g/g DW), but their WBC levels were also consistent with those of various high-fiber ingredients from apple pulp (2.3 g/g DW), wheat bran (2.6 g/g DW), corn bran (2.5 g/g DW), and soy bran (2.4 g/g DW) (Dreher, 1987). The remarkably high WBC of the *W. cocos* DF suggests that this material could be used as a functional ingredient to avoid syneresis (weeping) and to improve the rubbery texture of formulated products such as cheese (Nelson, 2001), in addition to reducing calories by the total or partial substitution of high-energy ingredients. The methods of measurement and food system environments (such as pH, ionic strength, concentration, and presence of other water binding materials) are crucially important factors that affect the WBC of high-fiber ingredients (Auffret et al., 1994; Fleury and Lahaye, 1991; Nelson, 2001).

Only the OHC of *P. rhinocerus* DF (1.87 g/g DW) was comparable to that of barley DF (1.88 g/g DW) and wheat DF (2.3 g/g DW) (Thebaudin et al., 1997). Although the OHC values of the DF obtained from *P. tuber-regium* and *W. cocos* were significantly lower ($p < 0.025$) (1.36–1.37 g/g DW), they were comparable to those of orange DF concentrate (0.86–1.28 g/g) (Grigelmo-Miguel and Martin-Belloso, 1999) and the commercial DF-rich supplement FIBREX (1.29 g/g DW) (Abdul-Hamid and Luan, 2000).

The ability of a fiber to bind oil is more a function of the porosity of the fiber structure than the affinity of the fiber molecule for oil (Nelson, 2001). Other factors, such as the number of lipophilic sites, overall hydrophobicity, and capillary attraction (Kinsella, 1976), may also contribute to the variations of OHC in sclerotial DF. A high-fiber ingredient with a high OHC allows the stabilization of high fat content and emulsion in formulated food products such as comminuted or emulsified meat by retaining the fat. In low-fat meat applications, the OHC of the high-fiber ingredients can also retain the low amount of fat present, which aids in the flavor, texture, and juiciness of the final cooked product.

The emulsion formed by all of the DF samples was generally good, as their EA values (56.7–71.9%) were >50% (Wang and Kinsella, 1976; Yasumatsu et al., 1972). In addition, the EA of all of the sclerotial DF was notably higher ($p <$ 0.0.5) than that of the barley DF control (56.7%), rice bran DF (14.4%), and the commercial fiber-rich supplement FIBREX (3.46%) (Fleury and Lahaye, 1991), thus suggesting their great potential to act as an emulsifier in formulated food. Furthermore, the emulsions formed by all of the DF samples were very stable, as evidenced by their similarly high percentage of ES after incubation at 80°C for 30 minutes.

The three novel types of sclerotial DF appear to be versatile low-calorie food ingredients with several technological advantages (natural origin, high DF content, and good physicochemical functional properties) that are of interest to the food ingredient market, and they could thus be incorporated into a wide range of formulated foods such as bakery products, noodles, and snacks. Further investigation of their potential role as a functional food fiber or nutraceutical via the assessment of some of their physiological benefits would be interesting.

4.7 BIOPHARMACOLOGICAL VALUES OF MUSHROOM SCLEROTIA OF *P. tuber-regium, P. rhinocerus,* AND *W. cocos*

4.7.1 In Vitro Mineral Binding Capacity

The recommendation for an increase in DF intake has raised questions about the possible negative effects on mineral bioavailability, particularly in high-risk population groups such as the elderly, infants, and pregnant women (Idouraine et al., 1995, 1996). Because the electrostatistic binding and/or trapping of minerals within DF particles is one of the main factors that determines the undesirable effects of DF on mineral bioavailability, the in vitro mineral binding capacity of DF is believed to be a crucial parameter for predicting its effect on mineral bioavailability in humans (Laszlo, 1989). To predict the possible effects of sclerotial DF prepared from *P. tuber-regium, P. rhinocerus*, and *W. cocos* on mineral bioavailability in the human gastrointestinal tract, their in vitro mineral binding capacity on five nutritionally important divalent minerals—calcium (Ca), magnesium (Mg), copper (Cu), iron (Fe), and zinc (Zn)—under sequential simulated physiological conditions of the human stomach, small intestine, and colon was investigated and compared (Wong and Cheung, 2005b). In addition to releasing most of their endogenous Ca (ranging from 96.9 to 97.9% removal) and Mg (ranging from 95.9 to 96.7% removal), the simulated physiological conditions of the stomach also attenuated the possible adverse binding effects of the three types of sclerotial DF to the exogenous minerals by lowering their cation exchange capacity (ranging from 20.8 to 32.3%) and removing a substantial amount of their potential mineral chelators, including protein (ranging from 16.2 to 37.8%) and phytate (ranging from 58.5 to 64.2%). This finding suggests that when the three sclerotial DF types reach the stomach, most of their endogenous Ca and Mg will be readily released, and about half of their endogenous Cu, Fe, and Zn will remain bound. The in vitro mineral binding capacity of the three sclerotial types of DF under the simulated physiological conditions of the small intestine was found to be low, especially for Ca (ranging from 4.79 to 5.91% binding) and Mg (ranging from 3.16 to 4.18% binding) and was highly correlated ($r > 0.97$) with their residual protein content. The three types of partially demineralized sclerotial DF from the stomach could only rebind a limited amount of the five nutritionally important minerals in the small intestine and may not have a detrimental effect on mineral bioavailability compared with other fibers (Harland, 1989; Kelsay, 1986; Munoz and Harland, 1993; Reinhold et al., 1976). Under the simulated physiological conditions of the colon with a slightly acidic pH (5.80), only bound Ca was readily released (ranging from 34.2 to 72.3% releasing) from the three types of sclerotial DF. This finding indicates the potential physiological benefits of the three sclerotial DF types on Ca bioavailability. On reaching the human colon, part of the small intestinal condition-treated sclerotial DF would be fermented by the anaerobic microflora in the large intestine, thus releasing some of their bound minerals. If the fermentability of the three types of small intestinal condition-treated sclerotial DF was high enough to create an acidic colonic environment (pH $\leqslant$ 5.80), then this not only would release an appreciable amount (34.2–72.3%) of bound Ca from the three types of nonfermented sclerotial DF but also might promote their ionization together with the already released and unabsorbed minerals. As a result, passive mineral absorption, especially Ca, in the large intestine might be enhanced and the overall Ca bioavailability might then be improved. The additional absorption of Ca in the colon is especially important in elderly people and postmenopausal women who have insufficient Ca intake or insufficiently active Ca absorption from the small intestine. Nevertheless, the main criterion in determining the potential enhancing effect of the three types of sclerotial DF on passive Ca absorption in the human large intestine is their fermentability.

4.7.2 In Vitro Fermentability

DF escapes digestion and absorption in the human small intestine and constitutes the main substrate for colonic fermentation (Cummings, 1982). The fermentative breakdown of DF in the human colon by anaerobic saccharolytic microflora leads to the production of certain gases (CO_2, CH_4, and H_2), microbial biomass, and short-chain fatty acids (SCFAs), which considerably influences the physiological functions of humans (Topping and Clifton, 2001; Tungland and Meyer, 2002). The rate and extent of fiber fermentation depend on two main categories of factors: (1) host-specific factors, such as the activities and composition of the colonic microflora and gastrointestinal tract transit time, and (2) substrate-specific factors, including the physicochemical properties (e.g., particle size, solubility, and cell wall architecture) of the fiber source and the chemical composition (monosaccharide profile) and structural arrangement (degree of branching and linkages between monosaccharide) of the fiber constituents (Auffret et al., 1993; McBurney and Thompson, 1989; Titgemeyer et al., 1991). The SCFAs produced during fermentation are rapidly absorbed by the colonic mucosa, stimulating water and sodium absorption and peristalsis, which, in turn, aid the bowel function (Cummings et al., 1987; Ruppin et al., 1980). An appreciable amount of SCFAs resulting from highly fermentable DF would also lower the colonic pH, thus modulating the composition of the colonic microflora by inhibiting the growth of pathogenic bacteria (non–acid tolerant) but stimulating the growth of those, such as *Lactobacillus*, that are beneficial (Cummings et al., 2001; Gibson and Roberfroid, 1995). To predict the fate of the sclerotial DF prepared from *P. tuber-regium, P. rhinocerus*, and *W. cocos* in the human colon after consumption as a food ingredient, their fermentability was evaluated in vitro by using human fecal microflora and comparing it to a cellulose control (Wong et al., 2005).

Briefly, all of the DF samples (0.5 g each) were fermented in vitro with a human fecal homogenate (10 mL) in a batch system (total volume 50 mL) under strictly anaerobic conditions (using an oxygen-reducing enzyme under argon atmosphere) at 37°C for 24 h. All three types of novel sclerotial DF exhibited notably higher DM disappearance (*P. tuber-regium* 8.56%; *P. rhinocerus* 13.5%; *W. cocos* 53.4%) and organic matter disappearance (*P. tuber-regium* 9.82%; *P. rhinocerus* 14.6%; *W. cocos* 57.4%) when compared with the cellulose control. Nevertheless, only the *W. cocos* DF was remarkably degraded to produce a considerable amount of total SCFAs (5.23 mmol/g DF on an organic matter basis, with a relatively higher molar ratio of propionate) that lowered the pH of the nonfermented residue to a slightly acidic level (5.89). These findings suggest that during the fermentation of the *W. cocos* DF by the human colonic bacteria, the high fermentability may result in a sufficient amount of SCFAs to acidify the colonic pH, which may in turn promote the ionization of the unabsorbed minerals and enhance their passive absorption in the colon, as in other highly fermented DF reported previously (Coudray et al., 1997; Morohashi et al., 1998; Tahiri et al., 2001; Younes et al., 2001). From a physiological point of view, the relatively higher level of propionate produced by the readily fermented *W. cocos* DF implies that it may have hypoglycemic (probably by increasing hepatic glucose utilization or maximizing the insulin response) and/or hypocholesterolemic effect(s) on humans (probably by suppressing the synthesis of hepatic cholesterol or redistributing cholesterol from the plasma to the liver), as has been proposed in previous human (Todesco et al., 1991) and animal (Chen et al., 1984) studies. The *P. tuber-regium* and *P. rhinocerus* DF remains nonfermented in the human colon, which, in turn, contributes to the fecal bulking capacity and the bacterial biomass. In addition to diluting the carcinogenic and toxic substances by providing a bulkier stool, the nonfermented *P. tuber-regium* and *P. rhinocerus* DF may also decrease the transit time of the stool through the colon and lower the chance of exposure to carcinogens similar to other cereal polysaccharides (Karppinen et al., 2000). Interestingly, the fermentation of the β-1,4-glucan-rich samples, the *P. tuber-regium* DF, and the cellulose control exhibited similar SCFA profiles with a relatively higher molar ratio of butyrate (acetate–propionate–butyrate in *P. tuber-regium* DF 2.1 : 1 : 1.4; cellulose 1.86 : 1 : 1.14). This finding is also comparable to that of the β-1,4 linkage-rich glucan from cereals with a relative molar ratio of 2.09 : 1 : 1.84 (Botham et al., 1998). In the case of the DF of both *P. rhinocerus* and *W. cocos*, which was rich in β-1,3 linkages, it had a relatively higher molar ratio of propionate (*P. rhinocerus* DF 4.75 : 2.25 : 1; *W. cocos* DF 2.82 : 1.46 : 1) after 24 hours in vitro fermentation, and the SCFA profiles of both were also similar. The production of a relatively higher molar ratio of propionate has also been reported in the fermentation of curdlan (3.72 : 1.41 : 1) (Shimizu et al., 2001) and laminarian (1.7 : 1.25 : 1) (Michel et al., 1996). An obvious structure–function relationship between the three types of sclerotial DF and their in vitro fermentability was present, and the variations in their in vitro fermentability may mainly be attributed to the different amounts of interwoven hyphae present (different amounts of the enzyme-inaccessible cell wall component) and to the possibly different structural arrangements (linkage and degree of branching) of their β-glucan components. Thus, further investigation of the possible enhancing effect of the sclerotial DF of *W. cocos* on passive mineral absorption in the large intestine by using an animal model would be interesting.

4.7.3 In Vivo Ca and Mg Absorption

Previous studies have shown that DF, especially its insoluble fraction, can bind strongly to Ca and form unabsorbable complexes owing to its anionic nature (Platt and Clydesdale, 1987). As a result, it has been proposed that DF may impair Ca absorption. However, during the past decade, there has been substantial evidence to indicate that Ca absorption is not affected by the fiber component per se (Harrington et al., 2001; Kennefick and Cashman, 2000). Many recent studies have also shown that fermentable DF, including oligosaccharides (e.g., fructo-oligosaccharides and inulin) and polysaccharides (e.g., resistant starch), even improves overall Ca absorption in both humans (Coudray et al., 1997; Tahiri et al., 2001) and rats (Morohashi et al., 1998; Younes et al., 2001).

According to Campbell. (1997), the beneficial effect of DF on overall Ca absorption depends on its fermentability, the dosage used, and the duration of the animal experiments. Although the detailed mechanisms of the enhancing effect of fermentable DF on overall Ca absorption remain unclear, it is widely accepted that the microbial degradation of fermentable DF in the large intestine is the most important factor (Ohta et al., 1995; Shiga et al., 1998). The fermentation by-products, SCFAs, are also believed to be the major contributor (Kishi et al., 1999; Lutz, 1991) to increasing the concentration of ionized Ca and promoting its absorption in the large intestine (Mineo et al., 2001; Younes et al., 1996).

Compared with cellulose control, the effects of the sclerotial DF prepared from *P. tuber-regium*, *P. rhinocerus*, and *W. cocos* on apparent Ca and Mg absorption were evaluated in ovariectomized (OVX) rats fed sclerotial DF-based and low-Ca (0.3%) diets for 14 days (Wong et al., 2006). All three of the sclerotial DF-based diet groups possessed significantly higher ($p < 0.025$) molar concentrations of total SCFAs (ranging from 80.3 to 204 μmol/g of cecal content) in their cecums, with the *W. cocos* DF group being the highest ($p < 0.025$). However, only the cecal pH (about 5.88) and cecal content (1.15 g) of the *W. cocos* DF group were significantly lower ($p < 0.025$) than those of the cellulose control group. These findings suggest that the ingestion of *W. cocos* DF leads to greater cecal fermentation and produces significantly higher ($p < 0.025$) amounts of total SCFAs that lower cecal pH to a slightly acidic level. The slightly acidic environment established in the *W. cocos* DF group also led to the remarkable ($p < 0.025$) augmentation of the cecal-soluble Ca (2.56-fold) and Mg (1.22-fold) concentrations.

Compared with the cellulose control group, the apparent Ca and Mg absorption of the *W. cocos* DF group was also notably ($p < 0.025$) enhanced (Ca 16.5%; Mg 15.3%), together with a significant ($p < 0.025$) elevation of their serum Ca (3.61 mmol/L) and Mg (1.07 mmol/L) concentrations but suppression of their serum PTH levels (57.1 pg/mL). These findings illustrate that the detrimental effects induced by both ovariectomy and a low-Ca diet can be alleviated by the ingestion of *W. cocos* DF. Its enhancing effect on Ca and Mg absorption in the cecum of OVX rats is of particular interest for populations with inefficiently active Ca absorption, such as elderly people and postmenopausal women. Fermentability is the main factor that determines the enhancing effect of nondigestible carbohydrates on mineral absorption. Therefore, to explore the three types of novel sclerotial DF as functional food ingredients for the enhancement of overall Ca and Mg absorption, their fermentability should be improved (Cashman, 2003), probably by isolating their main DF component, β-glucan-rich polysaccharides, or even preparing some novel β-glucose-based oligosaccharides from the three types of sclerotial DF using partial acid or enzymatic hydrolysis.

4.7.4 Antitumor and Immunomodulatory Activities

β-Glucans, the major structural component of fungal cell walls (Bartnicki-Garcia, 1970), have been found to stimulate both the innate and adaptive immunity of the host, followed by a wide range of immunopharmacological activities, particularly antitumor activities, via their cytokine production and signaling cascade (Bohn and BeMiller, 1995; Moradali et al., 2007). Mushroom sclerotia (a dried compact biomass of fungal hyphae) have been found to possess substantial amounts of β-glucans (>80% on a DM basis; Wong et al., 2003), which have exhibited remarkable immunomodulatory and antitumor activities in numerous previous studies (Wong et al., 2007; Zhang et al., 2006a). Our research team has been actively studying the immunomodulatory and antitumor activities of both native and chemically modified sclerotial β-glucans for the past six years (Chau et al., 2007; Lai and Cheung, 2004; Lai, 2005; Lai et al., 2005; Tao et al., 2006; Wong et al., 2007; M. Zhang et al., 2001, 2004a, b, 2006a, b). We have found that sclerotial β-glucan fractions isolated from *P. tuber-regium* (hot alkaline soluble, hot water soluble, ultrasonic, sulfated, and carboxymethylated fractions) and *P. rhinocerus* (hot water soluble and ultrasonic fractions) possess remarkable immunomodulatory and antitumor activities when they are administered intraperitoneally on BALB/c mice bearing sarcoma 180 (allogeneic solid tumor cells) (M. Zhang et al., 2001, 2004a, b; Lai, 2005; Lai et al., 2005; Tao et al., 2006). These sclerotial β-glucan fractions also exhibited a direct cytotoxic effect on various mammalian cancer cell lines (such as HL-60, MCF-7, and HepG2) but were noncytotoxic to normal kidney cells from monkey (VERO) cells (M. Zhang et al., 2001, 2004a, b; Lai, 2005; Tao et al., 2006). Furthermore, flow cytometric analysis showed that the sclerotial β-glucan fractions isolated from *P. tuber-regium* (carboxymethylated fractions) and *P. rhinocerus* (hot water-soluble fractions) not only arrested the cell cycle progression of the MCF-7 (with down regulation of cyclin D1 and cyclin E expressions) and HL-60 cells, respectively, at the G_1 phase but also induced their apoptosis with decreased expression of Bcl-2 and increased expression of the Bax/Bcl-2 ratio (Lai, 2005; Zhang et al., 2006a). In the case of *W. cocos*, its sclerotial β-glucan fractions (1% sodium carbonate soluble) not only significantly induced nitric oxide (NO) production by the peritoneal macrophages of B6 C3F1 mice but also inducible NO synthase (iNOS) transcription of the murine macrophage-like cell line, RAW 264.7, via activation of the transcription factor, namely nuclear factor-κB/Rel (NF-κB/Rel) (Lee and Jeon, 2003). Its methanol extract and isolated triterpene acids were also found to inhibit 12-*O*-tetradecanoylphorbol-13-acetate-induced ear edema and tumor promotion in mouse skin (Kaminaga et al., 1996).

Despite the fact that sclerotial β-glucans exhibited the aforementioned remarkable immunomodulatory and antitumor activities, their underlying in vivo mechanisms are still not fully understood, even though this issue has been of interest to many scientists over the past five decades (Ooi and Liu, 2000). The most controversial issue is how these macromolecules, fungal β-glucans, act on or are recognized by the innate immunity prior to triggering the acquired immune response and exerting their antitumor effects.

Recently, a novel cell surface receptor, Dectin-1, that recognizes the yeast β-glucan (zymosan) and mediates its immunomodulatory and antitumor effects, has been found on the surface of various innate immune cells [such as macrophages, natural killer (NK) cells, and dendritic cells] in both humans and mice (Brown and Gordon, 2001; Brown, 2006; Heinsbroek et al., 2006; Taylor et al., 2002). Although the discovery of β-glucan receptors on the surface of innate immune cells would undoubtedly provide a stepping stone for an investigation of the underlying mechanisms of the immunomodulatory and antitumor effects of fungal β-glucans, nearly all related studies have been limited to yeast-derived β-glucans, zymosan (Goodridge et al., 2007; Olsson and Sundler, 2007), but not in the case of other important fungi such as mushroom sclerotia. Thus, it would be valuable to determine whether there is a unique cell surface receptor(s) for the β-glucans of other important fungi, especially mushroom sclerotia.

Our latest in vivo immunophenotyping studies on the peritoneal exudate cells and hepatic mononuclear cells of healthy BALB/c mice have discovered that sclerotial β-glucan fractions isolated from *P. tuber-regium* (hot water-soluble fraction) and *P. rhinocerus* (hot water-soluble and ultrasonic fractions) exhibit a remarkable stimulatory effect on both the NK cells ($CD56^+$) and macrophages ($Mac-1^+$) of the innate immunity (unpublished data). In addition to a significant increase in the weight of their spleens, the levels of various cytokines (including interleukins IL-12 and IL-13) and macrophage inflammatory proteins were also notably elevated in the serum of the mice pretreated with the hot water-soluble β-glucans that were isolated from *P. rhinocerus* (Lai et al., 2005).

Using this research as background, the underlying mechanisms of the innate antitumor immunity that is mediated by these previously proven immunopotentiating mushroom sclerotial β-glucan fractions could be further investigated using athymic nude mice with human tumor xenografts, as this animal model is T-cell deficient to support the growth of human tumors without immune rejection. Innate immune cells pretreated with these mushroom sclerotial β-glucan fractions could also be isolated from healthy athymic nude mice followed by an assessment of their functional activities. Furthermore, the β-glucan receptor(s) that are specific to these mushroom sclerotial β-glucan fractions could be identified on the surface of innate human and murine primary cells (isolated from human peripheral blood and athymic nude mice, respectively) by a molecular technique termed "phage display." These findings would be a major breakthrough in antitumor studies of mushroom sclerotia, as they would provide a new perspective to explain their immunomodulatory and antitumor effects in terms of the kind of innate immune cells involved, the type of cytokines induced, and the possible cell surface β-glucan receptor(s) identified. By increasing our knowledge of the interaction between mushroom sclerotial β-glucans and innate immunity, a more effective utilization of these macromolecules as antitumor agents can be made possible.

4.8 CONCLUSION

The detailed mechanistic actions of the bioactive components in mushroom sclerotia are still not completely understood. Mushroom sclerotia thus remain underutilized at the moment. It is anticipated that, with advances in molecular biology and biotechnology, the content and the structure of the bioactive components of sclerotia, especially β-glucans, can be manipulated to produce "tailor-made" functional food products in the future.

REFERENCES

Abdul-Hamid, A. and Luan, Y. S. (2000). Functional properties of dietary fiber from defatted rice bran. *Food Chemistry*, *68*, 15–19.

Adams, R. G., Evans, A. J., Oakenfull, D. G., and Sidhu, G. S. (1986). Fruit processing wastes as dietary fiber supplements. *Proceedings of the Nutrition Society of Australia*, *11*, 115.

Akobundu, E. N. T. and Eluchie, G. U. (1992). Quality characteristics of pork sausage containing mushroom (*Pleurotus tuber-regium*) and local spices. *Journal of Food Technology*, *29*, 159–161.

American Diabetic Association (ADA) (2002). Position of the American Dietetic Association: Health implication of dietary fiber. *Journal of the American Dietetic Association*, *102*, 993–1000.

Auffret, A., Barry, J. L., and Thibault, J. F. (1993). Effect of chemical treatments of sugar beet fibres on their physico-chemical properties and on their *in vitro* fermentation. *Journal of the Science of Food and Agriculture*, *61*, 95–203.

Auffret, A., Ralet, M. C., Guillon, F., Barry, J. L., and Thibault, J. F. (1994). Effect of grinding and experimental conditions on the measurement of hydration properties of dietary fibers. *Lebensmittel-Wissenschaft and Technologie*, *27*, 166–172.

Backhouse, D. and Willetts, H. J. (1984). A histochemical study of sclerotia of *Botrytis cinerea* and *Botrytis fabae*. *Canadian Journal of Microbiology*, *30*, 171–178.

Backhouse, D. and Willetts, H. J. (1985). Histochemical changes during conidiogenic germination of sclerotia of *Botrytis cinerea. Canadian Journal of Microbiology, 31*, 282–286.

Bartnicki-Garcia, S. (1970). Cell wall composition and other biochemical markers in fungal phylogeny. In *Phytochemical Phylogeny*. Harborne, J. B., editor. London: Academic, pp. 81–103.

Bi, Z., Zheng, G. Y., and Li, T. H. (1993). *The Macrofungus Flora of China's Guangdong Province*. Hong Kong: The Chinese University Press. pp. 185–186.

Blackwood, A. D., Salter, J., Dettmar, P. W., and Chaplin, M. F. (2000). Dietary fiber, physico-chemical properties and their relationship to health. *Journal of the Royal Society for the Promotion of Health, 120*, 242–247.

Bohn, J. A. and BeMiller, J. N. (1995). $(1 \rightarrow 3)$-β-D-Glucans as biological response modifiers: A review of structure-functional activity relationships. *Carbohydrate Polymers, 28*, 3–14.

Botham, R. L., Ryden, P., Robertson, J. A., and Ring, S. G. (1998). Structural features of polysaccharides and their influence on fermentation behaviour. In *Functional Properties of Nondigestible Carbohydrates*. Guillon, F., editor. Nantes: National Institute for Agricultural Research, pp. 46–49.

Brefeld, O. (1877). *Botanische Untersuchungen über Schimmelpilze*, Vol. III. Leipzig: Felxi.

Brown, G. D. (2006). Dectin-1: A signaling non-TRL pattern-recognition receptor. *Nature Reviews. Immunology, 6*, 33–43.

Brown, G. D. and Gordon, S. (2001). Immune recognition: A new receptor for beta-glucans. *Nature, 413*, 36–37.

Bull, A. T. (1970). Inhibition of polysaccharases by melanin: Enzyme inhibition in relation to mycolysis. *Archives of Biochemistry and Biophysics, 137*, 345–356.

Bullock, S. and Willetts, H. J. (1996). Ultrastructural and histochemical studies on mycelial germination of sclerotia of *Sclerotinia minor*. *Mycological Research, 100*, 561–570.

Bullock, S., Willetts, H. J., and Ashford, A. E. (1980). The structure and histochemistry of sclerotia of *Sclerotinia minor* Jagger. I. Light and electron microscope studies on sclerotial development. *Protoplasma, 104*, 315–331.

Bullock, S., Willetts, H. J., and Ashford, A. E. (1983). The structure and histochemistry of sclerotia of *Sclerotinia minor* Jagger. III. Changes in ultrastructure and loss of reserve materials during carpogenic germination. *Protoplasma, 117*, 214–225.

Cadden, A. M. (1987). Comparative effects of particle size reduction on physical structure and water binding properties of several plant fibers. *Journal of Food Science, 52*, 1595–1599.

Campbell, J. M., Fahey, G. C., and Wolf, B. W. (1997). Selected indigestible oligosaccharides affect large bowel mass, cecal and fecal short chain fatty acids, pH, and microflora in rats. *Journal of Nutrition, 127*, 130–136.

Carlisle, M. J., Watkinson, S. C., and Gooday, G. W. (2001). *The Fungi*, 2nd ed. London: Academic, pp. 57–83, 171–172.

Cashman, K. (2003). Prebiotics and calcium bioavailability. *Current Issues in Intestinal Microbiology, 4*, 21–32.

Central Agricultural Broadcasting and Television School and National Farmer's Science and Technology Education Training Centre (CABTS/NFSTTC) (2006). Yao Yong Zhen

Jun Gao Xiao Sheng Chan Xin Ji Shu. Beijing: Zhong Guo Nong Ye Chu Ban She, pp. 102–111.

Chang, S. T. (1999). World production of cultivated edible and medicinal mushrooms in 1997 with emphasis on *Lentinus edodes* (Berk.) Sing. in China. *International Journal of Medicinal Mushrooms, 1*, 291–300.

Chau, A. H. Y., Cheung, P. C. K., Lai, C. K. M., and Chau, C. F. (2007). The effect of particle size reduction on the extraction, chemical characteristics and antitumor activity of mushroom polysaccharides. Poster 137-02. *Abstract of IFT Annual Meeting*. Chicago: IFT, p. 175.

Chen, W. J. L., Anderson, J. W., and Jennings, D. J. (1984). Propionate may mediate the hypocholesterolemic effects of certain soluble plant fibers in cholesterol-fed rats. *Proceedings of the Society for Experimental Biology and Medicine, 175*, 215–218.

Chet, I. and Henis, Y. (1968). X-ray analysis of hyphal and sclerotial walls of *Sclerotium rolfsii* Sacc. *Canadian Journal of Microbiology, 14*, 815–817.

Chet, I. and Henis, Y. (1975). Sclerotial morphogenesis in fungi. *Annual Review of Phytopathology, 13*, 169–192.

Chet, I., Henis, Y., and Mitchell, R. (1967). Chemical composition of hyphal and sclerotial walls of *Sclerotium rolfsii* Sacc. *Canadian Journal of Microbiology, 13*, 137–141.

Cheung, P. C. K. (1997). Dietary fiber content and composition of some edible fungi determined by two methods of analysis. *Journal of the Science of Food and Agriculture, 73*, 255–260.

Cheung, P. C. K. and Lee, M. Y. (1998). Comparative chemical analysis of fiber material prepared by enzymatic and chemical methods from two mushrooms (*Pleurotus sajor-caju* and *Pleurotus tuber-regium*). *Journal of Agricultural and Food Chemistry, 46*, 4854–4857.

Cheung, P. C. K. and Lee, M. Y. (1999). Isolation and characterization of mushroom dietary fiber (non-starch polysaccharides) as potential nutraceuticals. Poster 15-6. *Abstract of IFT Annual Meeting*. Chicago: IFT, pp. 27–28.

Cheung, P. C. K. and Lee, M. Y. (2000). Fractionation and characterization of mushroom dietary fiber (nonstarch polysaccharides) as potential nutraceuticals from sclerotia of *Pleurotus tuber-regium* (Fries) Singer. *Journal of Agricultural and Food Chemistry, 48*, 3148–3151.

Cheung, P. C. K. and Wong K. H. (2004). Non-digestible carbohydrates prepared from mushroom sclerotia: Relationship of fermentability and chemical structure. Poster P-20. *Abstract of Vahouny-ILSI Japan International Symposium on Non-Digestible Carbohydrate*. Tokyo: ILSI.

Chihara, G., Hamuro, J., Maeda, Y., Arai, Y., and Fukuoka, F. (1970). Antitumor polysaccharide derived chemically from natural glucan (pachyman). *Nature, 225*, 943–944.

Coley-Smith, J. R. (1980). Sclerotia and other structures in survival. In *The Biology of Botrytis*. Coley-Smith, J. R., Verhoeff, V., and Jarvis, W. R., editors. London: Academic, pp. 85–114.

Coley-Smith, J. R. and Cooke, R. C. (1971). Survival and germination of fungal sclerotia. *Annual Review of Phytopathology, 9*, 65–92.

Colotelo, N. (1978). Fungal exudates. *Canadian Journal of Microbiology, 24*, 1173–1181.

Cooke, R. C. (1969). Changes in soluble carbohydrates during sclerotium formation by *Sclerotinia sclerotiorum* and *S. trifoliorum. Transactions of the British Mycological Society, 53*, 77–86.

Cooke, R. C. (1970). Physiological aspects of sclerotium growth in *Sclerotinia sclerotiorum. Transactions of the British Mycological Society, 54*, 364–365.

Cooke, R. C. (1971). Physiology of sclerotium of *Sclerotinia sclerotiorum* during growth and maturation. *Transactions of the British Mycological Society, 56*, 51–59.

Cooke, R. C. (1983). Morphogenesis of sclerotia. In *Fungal Differentiation, a Contemporary Synthesis*. Smith, J. E., editor. New York: Marcel Dekker, pp. 397–418.

Coudray, C., Bellanger, J., Castiglia-Delavaud, C., Rémésy, C., Vermorel, M., and Ravssignuir, Y. (1997). Effect of soluble or partly soluble dietary fibers supplementation on absorption and balance of calcium, magnesium, iron and zinc in healthy young men. *European Journal of Clinical Nutrition, 51*, 375–380.

Cummings, J. H. (1982). Consequences of the metabolism of fiber in the human large intestine. In *Dietary Fiber in Health and Disease*. Vahouny, G. V. and Kritchevsky, D., editors. New York: Plenum, pp. 9–22.

Cummings, J. H., Macfarlane, G. T., and Englyst, H. N. (2001). Prebiotic digestion and fermentation. *American Journal of Clinical Nutrition, 73* (Suppl.), 415–420.

Cummings, J. H., Pomare, E. W., Branch, W. J., Naylor, C. P. E., and MacFarlane, G. T. (1987). Short chain fatty acids in human large intestine, portal, hepatic and venous blood. *Gut, 28*, 1221–1227.

De Bary, A. (1887). *Comparative Morphology and Biology of the Fungi, Mycetozoa and Bacteria*. London: Clarendon.

Deng, C. H., Yang, X. L., Gu, X. M., Wang, Y., and Zhou, J. Y. (2000). A β-glucan from the sclerotia of *Pleurotus tuber-regium* (Fr.) Sing. *Carbohydrate Research, 328*, 629–633.

Ding, Q., Zhang, L. N., and Zeng, F. B. (1998). Influence of molecular weight and periodate-modification of beta D-glucan from *Poria cocos* sclerotium on antitumor activities. *Chinese Journal of Polymer Science, 16*, 62–66.

Dreher, M. L. (1987). *Handbook of Dietary Fiber: An Applied Approach*. New York: Marcel Dekker.

Dubourdieu, D., Ribéreau-Gayon, P., and Fournet, B. (1981). Structure of the extracellular β-D-glucan from *Botrytis cinerea. Carbohydrate Research, 93*, 294–299.

Elhardallou, S. B. and Walker, A. F. (1999). The effect of multi-mineral mix (Fe, Zn, Ca and Cu) on magnesium binding to starchy legumes under simulated gastrointestinal conditions. *Food Chemistry, 67*, 113–121.

Elliott C. G. (1994). *Reproduction in Fungi: Genetical and Physiological Aspects*. London: Chapman & Hall.

Fasidi, I. O. and Ekuere, U. U. (1993). Studies on *Pleurotus tuber-regium* (Fries) Singer: Cultivation, proximate composition and mineral content of sclerotia. *Food Chemistry, 48*, 255–258.

Fasidi, I. O. and Olorunmaiye, K. S. (1994). Studies on the requirements for vegetative growth of *Pleurotus tuber-regium* (Fr.) Singer, a Nigerian mushroom. *Food Chemistry, 50*, 397–401.

Fleury, N. and Lahaye, M. (1991). Chemical and physico-chemical characterization of fibers from *Laminaria digitata* (Kombu Breton): A physiological approach. *Journal of the Science of Food and Agriculture, 55*, 389–400.

Gao, Q., Seljelid, R., Chen, H., and Jiang, R. (1996). Characterisation of acidic heteroglycans from *Tremella fuciformis* Berk with cytokine stimulating activity. *Carbohydrate Chemistry, 288*, 135–142.

Garcha, H. S., Dhanda, S., and Khanna, P. (1984). Evaluation of various organic residues for the cultivation of *Pleurotus* (Dhingri) spp. *Mushroom Newsletter for the Tropics*, *5*, 13–16.

Garrett. S. D. (1970). *Pathogenic Root-Infecting Fungi*. Cambridge: Cambridge University Press.

Gibson, G. R. and Roberfroid, M. B. (1995). Dietary modulation of the human colonic microbiota: Introducing the concept of prebiotics. *Journal of Nutrition*, *125*, 1401–1412.

Goñi, I. and Martin-Carr6n, N. (1998). *In vitro* fermentation and hydration properties of commercial dietary fiber-rich supplements. *Nutrition Research*, *18*, 1077–1089.

Good, H. (2002). Measurement of color in cereal products. *Cereal Foods World*, *47*, 5–6.

Goodridge, H. S., Simmons, R. M., and Underhill D. M. (2007). Dectin-1 stimulation by *Candida albicans* yeast or zymosan triggers NFAT activation in macrophages and dendritic cells. *Journal of Immunology*, *178*, 3107–3115.

Grigelmo-Miguel, N. and Martín-Belloso, O. (1997). A dietary fiber supplement from fruit and vegetable processing waste. Poster 73G. *1997 Conference of Food Engineering (CoFE'97)*. Los Angeles: American Institute of Chemical Engineers.

Grigelmo-Miguel, N. and Martín-Belloso, O. (1999). Characterization of dietary fiber from orange juice extraction. *Food Research International*, *31*, 355–361.

Guillon, G. and Champ, M. (2000). Structural and physical properties of dietary fibers, and consequences of processing on human physiology. *Food Research International*, *33*, 233–245.

Gunning, B. E. S. and Steer, M. G. (1975). *Ultrastructure and the Biology of Plant Cells*. London: Edward Arnold.

Hall, F. C., Beute, M. K., and Smith, T. (1982). Effect of soil pH and volatile stimulants from remoistened peanut leaves on germination of sclerotia of *Sclerotinia minor*. *Plant Disease*, *66*, 223–224.

Harland, B. F. (1989). Dietary fiber and mineral bioavailability. *Nutrition Research Reviews*, *2*, 133–147.

Harold, F. M. (1966). Inorganic polyphosphorus in biology: Structure, metabolism and function. *Bacteriological Reviews*, *30*, 772–794.

Harrington, M. E., Flynn, A., and Cashman, K. D. (2001). Effects of dietary fiber extracts on calcium absorption in the rat. *Food Chemistry*, *73*, 263–269.

He, P. X., Meng, L., Wang, Z. H., Xu, G. F., Yuan, G. R., and Yang, H.Z. (2000). Ming Te Xin Shi Yong Jun San Shi Zhong. Beijing: Zhong Guo Nong Ye Chu Ban She, pp. 144–147.

Heinsbroek, S. E. M., Taylor, P. R., Rosas, M., Willment, J. A., Williams, D. L., Gordon, S., and Brown, G. D. (2006). Expression of functionally different dectin-1 isoforms by murine macrophages. *Journal of Immunology*, *176*, 5513–5518.

Hobbs, C. (1995). *Medicinal Mushrooms: An Exploration of Tradition, Healing and Culture*. Santa Cruz: Botanica.

Huang, H. C. (1991). Induction of myceliogenic germination of sclerotia of *Sclerotinia sclerotiorum* by exposure to sub-freezing temperatures. *Plant Pathology*, *40*, 621–625.

Huang, N. L. (1999a). Identification of the scientific name of Hurulingzhi. *Acta Edulis Fungi*, *6*, 30–32.

Huang, N. L. (1999b). A new species of medicinal mushroom—Cultivation of Hurulingzhi. *Edible Fungi China*, *18*, 8–9.

Huang, N. L. (2000). Current and future perspectives on production of edible mushroom in China. *Edible Fungi China, 4*, 3–5.

Idouraine, A., Hassani, B. Z., Claye, S. S., and Weber, C. W. (1995). *In vitro* mineral binding capacity of various fiber sources for magnesium, zinc and copper. *Journal of the Agricultural and Food Chemistry, 43*, 1580–1584.

Idouraine, A., Khan, M. J., Kohlhepp, E. A., and Weber, C. W. (1996). *In vitro* mineral binding capacity of three fiber sources for Ca, Mg, Cu, and Zn by two different methods. *International Journal of Food Sciences and Nutrition, 47*, 285–293.

Igari, N., Minoura, S., Sano, T., Kubo, T., Higuchi M., Sasaki, H., and Okada, M. (2000). Cultivation of sclerotia of *Wolfiporia cocos. Natural Medicines, 54*, 237–240.

Imolehin, E. D., Grogan, R. G., and Duniway, J. M. (1980). Effects of temperature and moisture tension on growth, sclerotial production, germination and infection by *Sclerotinia minor. Phytopathology, 70*, 1153–1157.

Isikhuemhen, O. S. and Nerud, F. (1999). Preliminary studies on the ligninolytic enzymes produced by the tropical fungus *Pleurotus tuber-regium* (Fr.) Sing. *Antonie van Leeuwenhoek, 75*, 257–260.

Iwuagwu M. A. and Onyekweli (2002). Preliminary investigation into the use of *Pleurotus tuber-regium* powder as a tablet disintegrant. *Tropical Journal of Pharmaceutical Research, 1*, 29–37.

Jalili, T., Wildman, R. E. C., and Medeiros, D. M. (2000). Nutraceutical roles of dietary fiber. *Journal of Nutraceuticals, Functional and Medical Foods, 2*, 19–34.

Jandaik, C. L. (1974). Artificial cultivation of *Pleurotus sajorcaju. Mushroom Journal, 22*, 4405.

Jennings, D. H. (1987). Translocation of solutes in fungi. *Biological Reviews, 62*, 215–243.

Jiang, Z., Li, S. S., Zheng, Y. B., and Tang, P. Y. (2000). Studies on the biological characteristics of Pleurotus tuber-regium. *Acta Edulis Fungi, 7*, 11–17.

Jones, D. (1970). Ultrastructure and composition of the cell walls of *Sclerotinia sclerotiorum. Transactions of the British Mycological Society, 54*, 351–360.

Jones, D., Farmer, V. C., Bacon, J. S. D., and Wilson, M. J. (1972). Comparison of ultrastructure and chemical components of cell walls of certain plant pathogenic fungi. *Transactions of the British Mycological Society, 54*, 351–360.

Jorgensen, L. B., Beinke, H. D., and Mabry, T. J. (1977). Protein-accumulating cells and dilated cisternae of the endoplasmic reticulum in three glucosinolate-containing genera: *Armaracia, Capparis, Drypetes*, *Planta, 137*, 215–224.

Kadiri, M. and Fasidi, I. O. (1990). Studies on enzyme activities of *Pleurotus tuber-regium* Heim at various fruit-body stages. *Nahrung, 34*, 695–699.

Kaminaga, T., Yasukawa, K., Takido, M., Tai, T., and Nunoura, Y. (1996). Inhibitory effect of *Poria cocos* on 12-O-tetradecanoylphorbol-13-acetate-induced ear oedema and tumor promotion in mouse skin. *Phytotherapy Research, 10*, 581–584.

Karppinen, S., Liukkonen, K., Aura, A. M., Forssell, P., and Poutanen, K. (2000). *In vitro* fermentation of polysaccharides of rye, wheat, and oat brans and inulin by human faecal bacteria. *Journal of the Science of Food and Agriculture, 80*, 1469–1476.

Kelsay, J. L. (1986). Update on fiber and mineral availability. In *Dietary Fibers, Basic and Clinical Aspects*. Vahouny, G. and Kritchevsky, D., editors. New York: Plenum, pp. 361–372.

Kennefick, S. and Cashman, K. D. (2000). Inhibitory effect of wheat fibre extract on calcium absorption in Caco-2 cells: Evidence for a role of associated phytate rather than fiber *per sec*. *European Journal of Nutrition*, *39*, 12–17.

Keys, J. D. (1976). *Chinese Herbs—Their Botany, Chemistry, and Pharmacodynamics*. Hong Kong: Swindon Book Company, pp. 21–22.

Khan, S. R. and Aldrich H. C. (1973). Conidiogenesis in *Termitaria snyderi* (Fungi Imperfecti). *Canadian Journal of Botany*, *51*, 2307–2314.

Kiho, T., Kobayashi, T., Morimoto, H., Usui, S., Ukai, S., Hirano, K., Aizawa, K., and Inakuma, T. (2000). Structural features of an anti-diabetic polysaccharides (TAP) from *Tremella aurantia*. *Chemical and Pharmaceutical Bulletin*, *48*, 1793–1795.

Kinsella, J. E. (1976). Functional properties of protein in foods: A survey. *J. Food Sci. Nutr.*, *7*, 219–280.

Kishi, M., Fukaya, M., Tsukammoto, Y., Nagasawa, T., Takehana, K., and Nishizawa, N. (1999). Enhancing effect of dietary vinegar on the intestinal absorption of calcium in ovariectomized rats. *Bioscience, Biotechnology, and Biochemistry*, *63*, 905–910.

Kohn, L. M. and Grenville, D. J. (1987). Some factors influencing lipid storage in sclerotia of *Sclerotinia minor*. *Mycologia*, *79*, 907–909.

Kohn, L. M. and Grenville, D. J. (1989a). Anatomy and histochemistry of stromatal anamorphs in the Sclerotiniaceae. *Canadian Journal of Botany*, *67*, 371–393.

Kohn, L.M. and Grenville, D. J. (1989b). Ultrastructure of stromatal anamorphs in the Sclerotiniaceae. *Canadian Journal of Botany*, *67*, 394–406.

Kubo, T., Terabayashi, S., Takeda, S., Sasaki, H., Aburada, M., and Miyamoto, K. (2006). Indoor cultivation and cultural characteristics of *Wolfiporia cocos* sclerotia using mushroom culture bottles. *Biological and Pharmaceutical Bulletin*, *29*, 1191–1196.

Kulaev, I. S. (1975). Biochemistry of inorganic polyphosphates. *Reviews of Physiology, Biochemistry and Pharmacology*, *73*, 131–158.

Kurasawa, S., Sugahara, T., and Hayashi, J. (1991). Determination of dietary fiber in chitin-containing mushrooms by the enzymatic-gravimetric method. *Journal of Japanese Society of Nutrition and Food Science*, *44*, 293–303.

Lai, C. K. M. (2005). Antitumor effects of polysaccharides extracted from mushroom sclerotia: An *in vitro* and *in vivo* study. M.Phil. thesis. Hong Kong: The Chinese University of Hong Kong.

Lai, C. K. M. and Cheung, P. C. K. (2004). Preparation and characterization of antitumor polysaccharides from mushroom sclerotia. Poster 29-8. *Abstracts of IFT Annual Meeting*. Las Vegas: IFT.

Lai, C. K. M., Wong, K. H., Cheung, P. C. K., and Chiu, L. C. M. (2005). Immunomodulatory and antitumor effect of mushroom sclerotial polysaccharides. Poster 04.14. *Programme Abstract of ILSI's First International Conference on Nutrigenomics—Opportunities in Asia*. Singapore: ILSI.

Laszlo, J. A. (1989). Effect of gastrointestinal conditions on the mineral-binding properties of dietary fibers. In *Mineral Absorption in the Monogastric GI Tract*. Dintzis, F. R. and Laszlo, J. A., editors. New York: Plenum, pp.133–145.

Lee, K. Y. and Jeon, Y. J. (2003). Polysaccharide isolated from *Poria cocos* sclerotium induces NF-κB/Rel activation and iNOS expression in murine macrophages. *International Journal of Immunopharmacology*, *3*, 1353–1362.

Liu, B. and Bau, Y. S. (1980). *Fungi Pharmacopoeia (Sinica)*. Oakland, CA: Kinoko.

Luccia, B. H. D. and Kunkel, M. E. (2002). *In vitro* availability of calcium from sources of cellulose, methylcellulose, and psyllium. *Food Chemistry*, *77*, 139–146.

Luttrell, E. S. (1980). Host-parasite relationships and development of the ergot sclerotium in *Clavicrps purpurea*. *Canadian Journal of Botany*, *58*, 942–958.

Lutz, T. (1991). The effect of SCFA on Ca absorption by rat colon. *Experimental Physiology*, *76*, 615–618.

Marchant, R., Peat, A., and Banbury G. H. (1967). The ultrastructural basis of hyphal growth. *New Phytologist*, *66*, 623–629.

McBurney, M. I. and Thompson, L. U. (1989). Effect of human faecal donor on *in vitro* fermentation variables. *Scandinavian Journal of Gastroenterology*, *24*, 359–367.

McKee, L. H. and Latner, T. A. (2000). Under-utilized sources of dietary fiber: A review. *Plant Food for Human Nutrition*, *55*, 285–304.

Michel, C., Lahaye, M., Bonnet, C., Mabeau, S., and Barry, J. L. (1996). *In vitro* fermentation by human faecal bacteria of total and purified dietary fibers from brown seaweeds. *British Journal of Nutrition*, *75*, 263–280.

Mineo, H., Hara, H., and Tomita, F. (2001). Short chain fatty acids enhance diffusional Ca transport in the epithelium of the rat cecum and colon. *Life Science*, *69*, 517–526.

Moore, A. E. P., Ashfold, A. E., and Petersen R. L. (1991). Reserve substances in *Paxillus involutus* sclerotia: Determination by histochemistry and X-ray microanalysis. *Protoplasma*, *163*, 67–81.

Moore, D. (1984). Positional control of development in fungi. In *Positional Control in Plant Development*. Barlow, P. W. and Carr, D. J., editors. Cambridge: Cambridge University Press, pp. 107–135.

Moore, D. (1995). Tissue formation. In *The Growing Fungus*. Gow, N. A. R. and Gadd, G. M., editors. London: Chapman and Hall, pp. 424–435.

Moore, D. (2003). *Fungal Morphogenesis*. Cambridge, Cambridge University Press.

Moradali, M. F., Mostafavi, H., Ghods, S., and Hedjaroude, G. A. (2007). Immunomodulating and anticancer agents in the realm of macromycetes fungi (macrofungi). *International Journal of Immunopharmacology*, *7*, 701–724.

Morohashi, T., Sano T., Ohta, A., and Yamada S. (1998). True calcium absorption in the intestine is enhanced by fructo-oligosaccharide feeding in rats. *Journal of Nutrition*, *128*, 1815–1818.

Munoz, J. M. and Harland, B. F. (1993). Overview of the effect of dietary fiber on the utilization of minerals and trace elements. In *CRC Handbook of Dietary Fiber in Human Nutrition*. Spiller, G. A., editor. Boca Raton, FL: CRC Press, pp. 245–252.

Nelson, A. L. (2001). *High-Fiber Ingredients*. St. Paul, MN: Eagan.

Nwokolo, E. (1987). Composition of nutrients in the sclerotium of the mushroom *Pleurotus tuber-regium*. *Plant Food for Human Nutrition*, *37*, 133–137.

Ohta, A., Ohtuki, M., Baba, S., Adachi, T., Sakata, T., and Sakaguchi, E. I. (1995). Calcium and magnesium absorption from the colon and rectum are increased in rats fed fructo-oligosaccharides. *Journal of Nutrition*, *125*, 2417–2424.

Okhuoya, J. A. and Etugo, J. E. (1993). Studies of the cultivation of *Pleurotus tuber-regium* (Fr.) Sing. and edible mushroom. *Bioresource Technology*, *44*, 1–3.

Okhuoya, J. A. and Okogbo, F. O. (1990). Induction of edible sclerotia of *Pleurotus tuber-regium* (Fr) Sing. in the laboratory. *Annals of Applied Biology*, *117*, 295–298.

Olsson, S. and Sundler, R. (2007). The macrophage glucan receptor mediates arachidonate release induced by zymosan: Essential role for Src family kinase. *Molecular Immunology*, *44*, 1509–1515.

Ooi, V. E. C. (2000). Medicinally important fungi. In *Science and Cultivation of Edible Fungi*, Vol. *1*. Van Griensven, L. J. L. D., editor. Rotterdam: Balkema, pp. 41–49.

Ooi, V. E. C. and Liu, F. (2000). Immunomodulation and anticancer activity of polysaccharide-protein complexes. *Current Medicinal Chemistry*, *7*, 715–729.

Oso, B. A. (1975). Mushroom and the Yoruba people of Nigeria. *Mycologia*, *67*, 311–319.

Oso, B. A. (1977). *Pleurotus tuber-regium* from Nigeria. *Mycologia*, *69*, 271–279.

Platt, S. R. and Clydesdale, F. M. (1987). Mineral binding characteristics of lignin, guar gum, cellulose, pectin and NDF under simulated duodenal pH conditions. *Journal of Food Science*, *52*, 1414–1419.

Potty, V. H. (1996). Physico-chemical aspects, physiological functions, nutritional importance and technological significance of dietary fiber—A critical appraisal. *Journal of Food Science and Technology*, *33*, 1–18.

Ralet, M. C., Thibault, J. F., and Della Valle, G. (1990). Influence of extrusion cooking on the physico-chemical properties of wheat bran. *Journal of Cereal Science*, *11*, 249–259.

Reijnders, A. F. M. and Moore, D. (1985). Developmental biology of agarics—An overview. In *Developmental Biology of Higher Fungi*. Moore, D., Casselton, L. A., Wood, D. A., and Frankland, J. C., editors. Cambridge: Cambridge University Press, pp. 581–595.

Reinhold, J. G., Faradji, B., Abdi, P., and Ismail-Beigo, F. (1976). Decreased absorption of calcium, magnesium, zinc and phosphorus by human due to increased fiber and phosphorus consumption as wheat bread. *Journal of Nutrition*, *106*, 493–503.

Ruppin, H., Barmeir, S., Soergel, K. H., Wood, C. M., and Schmidt, M. G. (1980). Absorption of short chain fatty acids by the colon. *Gastroenterology*, *78*, 1500–1507.

Shiga, K., Hara, H., and Kasai, T. (1998). The large intestine compensates for insufficient calcium absorption in the small intestine in rats. *Journal of Nutritional Science and Vitaminology*, *44*, 737–744.

Shimizu, J., Tsuchihashi, N., Kudoh, K., Wada, M., Takita, T., and Innami, S. (2001). Dietary curdlan increases proliferation of bifidobacteria in the cecum of rats. *Bioscience, Biotechnology, and Biochemistry*, *65*, 466–469.

Singer, R. (1961). *Mushrooms and Truffles*. London: Leonard Hill (Books) and New York: Interscience Publishers, p. 272.

Sloan, A. E. (2001). Dietary fiber moves back into mainstream. *Food Technology*, *55*, 18.

Tahiri, H. M., Tressol, J. C., Arnaud, J., Bornet, F., Bouteloup-Demange, C., Feillet-Coudray, C., Ducros, V., Repin, D., Brouns, F., Roussel, A. N., Rayssiguier, Y., and Coudray, C. (2001). Five-week intake of short-chain fructo-oligosaccharides increases intestinal absorption and status of magnesium in postmenopausal women. *Journal of Bone and Mineral Research*, *16*, 2152–2160.

Tanaka, Y. (1990). Microscopic studies on powdered crude drugs (1) crystals of fungi drugs. *Shoyakugaku Zasshi*, *44*, 5–10.

Tao, Y. Z., Zhang, L., and Cheung, P. C. K. (2006). Physicochemical properties and antitumor activities of water-soluble native and sulfated hyperbranched mushroom polysaccharides. *Carbohydrate Research, 341*, 2261–2269.

Taylor, P. R., Brown, G. D., Reid, D. M., Willment, J. A., Martinez-Pomares, L. Gordon, S., and Wong, S. Y. (2002). The beta-glucan receptor, Dectin-1, is predominantly expressed on the surface of cells of the monocyte/macrophage and neutrophil lineages. *Journal of Immunology, 169*, 3876–3882.

Thebaudin, J. Y., Lefebvre, A. C., Harrington, M., and Bourgeois, C. M. (1997). Dietary fibers: Nutritional and technological interest. *Trends in Food Science and Technology, 8*, 41–48.

Thibault, J. F. and Ralet, M. C. (2001). Pectins, their origin, structure and functions. In *Advanced Dietary Fiber Technology*. McCleary, B. V. and Prosky, L., editors. Oxford: Blackwell Science, pp. 369–379.

Titgemeyer, E. C., Bourquin, L. D., Fahey, G. C., and Garleb, K. A. (1991). Fermentability of various fiber sources by human faecal bacteria *in vitro*. *American Journal of Clinical Nutrition, 53*, 1418–1424.

Todesco, T., Rao, A. V., Bosello, O., and Jenkins, D. J. A. (1991). Propionate lowers blood glucose and alters lipid metabolism in health subjects. *American Journal of Clinical Nutrition, 54*, 860–865.

Topping, D. L. and Clifton, P. M. (2001). Short chain fatty acids and human colonic function: Roles of resistant starch and non-starch polysaccharides. *Physiological Reviews, 81*, 1031–1064.

Townsend, B. B. and Willetts, H. J. (1954). The development of sclerotia of certain fungi. *Transactions of the British Mycological Society, 37*, 213–221.

Toyama, N. and Ogawa, K. (1974). Comparative studies on cellulotic and oxidizing enzyme activities of edible and inedible wood rotters. *Mushroom Science, 9*, 745–760.

Tungland, B. C. and Meyer, D. (2002). Nondigestible oligo-saccharides (dietary fiber): Their physiology and role in human health and food. *Comprehensive Reviews in Food Science and Food Safety, 1*, 73–92.

Ude, C. M., Ezenwugo, A. E. N., and Agu, R. C. (2001). Composition and food value of sclerotium (Osu) and edible mushroom (*Pleurotus tuber-regium*). *Journal of Food Science and Technology, 38*, 612–614.

Ueno, Y., Hachisuka, Y., Esaki, H., Yamaguchi, R., and Kato, K. (1980). Structural analysis of extracellular polysaccharides of *Sclerotinia libertiana*. *Agricultural and Biological Chemistry, 44*, 353–359.

Ukiya, M., Akihisa, T., Tokuda, H., Suzuki, H., Mukainaka, T., Ichiishi, E., Ken Yasukawa, K., Kasahara, Y., and Nishino, H. (2002). Inhibition of tumor-promoting effect by poricoic acids G and H and other lanostane-type triterpenes and cytotoxic activity of poricoic acids A and G from *Poria cocos*. *Journal of Natural Products, 65*, 462–465.

Ulezlo, I. V., Upcrova, T. N., and Feniksova, R. V. (1975). Oxidative enzymes of the lignin-degrading fungus, *Pleurotus ostreatus*. *Prikladnaia Biokhimiia i Mikrobiologiia, 11*, 535–538.

Wang, J. C. and Kinsella, J. E. (1976). Functional properties of novel protein: Alfalfa leaf protein. *Journal of Food Science, 41*, 286–290.

Wang, S. X., Wen, Y. Y., and Hu, C. X. (1995). Immunoactivities of the polysaccharides from *Morus alba, Chlamydomonas mexicana* and *Poria cocos*. *Phytotherapy Research, 9*, 448–451.

Wasser, S. P., Nevo, E., Sokolov, D., Reshetnikov, S., and Timor-Tismenetsky, M. (2000). Dietary supplements from medicinal mushrooms: Diversity of types and variety of regulations. *International Journal of Medicinal Mushrooms*, *2*, 1–19.

Wasser, S. P. and Weis, A. L. (1999). Therapeutic effects of substances occurring in higher Basidiomycetes mushrooms: A modern perspective. *Critical Reviews in Immunology*, *19*, 65–96.

Wilcoxson, R. D. and Sudia, T. W. (1968). Translocation in fungi. *The Botanical Review*, *34*, 32–52.

Willetts, H. J. (1971). The survival of fungal sclerotia under adverse environmental conditions. *Biological Reviews*, *46*, 387–407.

Willetts, H. J. (1972). The morphogenesis and possible evolutionary origins of fungal sclerotia. *Biological Reviews*, *47*, 515–536.

Willetts, H. J. (1978). Sclerotium formation. In *The Filamentous Fungi—Developmental Mycology*. Smith, J. E. and Berry, D. R., editors. London: Edward Arnold, pp. 197–213.

Willetts, H. J. (1997). Morphology, development and evolution of stromata/sclerotia and macroconidia of the Sclerotiniaceae. *Mycological Research, 101*, 939–952.

Willetts, H. J. and Bullock, S. (1982). Studies on the ontogeny and ultrastructure of the sclerotium of *Botrytis cinerea* Pers. ex Nocca and Balbis. *Canadian Journal of Microbiology*, *68*, 2083–2091.

Willetts, H. J. and Bullock, S. (1992). Developmental biology of sclerotia. *Mycological Research*, *96*, 801–816.

Willetts, H. J., Bullock, S., Begg, E., and Matsumoto, N. (1990). The structure and histochemistry of Typhula incarnate. *Canadian Journal of Botany*, *68*, 2083–2091.

Willetts, H. J. and Wong, A. L. (1971). Ontogenetic diversity of sclerotia of *Sclerotinia sclerotiorum* and related species. *Transactions of the British Mycological Society*, *57*, 515–524.

Wong, A. L. and Willetts, H. J. (1974). Polyacrylamide-gel electrophoresis of enzymes during morphogenesis of sclerotia of *Sclerotinia sclerotiorum*. *Journal of General Microbiology*, *81*, 101–109.

Wong, K. H. and Cheung, P. C. K. (2004). Preparation of non-digestible carbohydrates from mushroom sclerotia using industrial enzymes. Poster P-21. *Abstract of Vahouny-ILSI Japan International Symposium on Non-digestible Carbohydrate*. Tokyo: ILSI.

Wong, K. H. and Cheung, P. C. K. (2005a). Dietary fiber from mushroom sclerotia: 1. Preparation, physico-chemical and functional properties. *Journal of Agricultural and Food Chemistry*, *53*, 9395–9400.

Wong, K. H. and Cheung, P. C. K. (2005b). Dietary fiber from mushroom sclerotia: 2. *In vitro* mineral binding capacity under sequential simulated physiological conditions of the human gastrointestinal tract. *Journal of Agricultural and Food Chemistry*, *53*, 9401–9406.

Wong, K. H. Cheung, P. C. K., and Wu, J. Z. (2003). Biochemical and microstructural characteristics of insoluble and soluble dietary fiber prepared from mushroom sclerotia of *Pleurotus tuber-regium, Polyporus rhinocerus* and *Wolfiporia cocos*. *Journal of Agricultural and Food Chemistry*, *51*, 7197–7202.

Wong, K. H., Katsumata, S., Masuyama, R., Uehara, M., Suzuki, K., and Cheung, P. C. K. (2006). Dietary fibers from mushroom sclerotia: 4. In vivo mineral absorption using ovariectomized rat model. *Journal of Agricultural and Food Chemistry*, *54*, 1921–1927.

Wong, K. H. Wong, K. Y., Kwan, H. S., and Cheung, P. C. K. (2005). Dietary fiber from mushroom sclerotia: 3. *In vitro* fermentability using human faecal microflora. *Journal of Agricultural and Food Chemistry*, *53*, 9407–9412.

Wong, S. M., Wong, K. K., Chiu, L. C. M., and Cheung, P. C. K. (2007). Non-starch polysaccharides from different developmental stages of *Pleurotus tuber-regium* inhibited the growth of human acute promyelotic leukemia HL-60 cells by cell-cycle arrest and/or apoptotic induction. *Carbohydrate Polymer*, *68*, 206–217.

Xu, L. and Wang, W. (2002). *Chinese Materia Medica—Combination and Applications*. London: Donica Publishing, pp. 217–219.

Yasumatsu, K., Sawada, K., Moritaka, S., Misaki, M., Toda, J., Wada, T., and Ishi, K. (1972). Whipping and emulsifying properties of soybean products. *Agricultural and Biological Chemistry*, *36*, 719–737.

Younes, H., Coudrays, C., Bellanger, J., Demigné, C., Rayssiguier, Y., and Rémésy, C. (2001). Effects of two fermentable carbohydrates (inulin and resistant starch) and their combination on calcium and magnesium balance in rats. *Journal of Nutrition*, *86*, 479–485.

Younes, H., Demigné, C., and Rémésy, C. (1996). Acidic fermentation in the caecum increases absorption of calcium and magnesium in the intestine of the rats. *British Journal of Nutrition*, *75*, 301–314.

Young, N. and Ashfold, A. E. (1992). Changes during development in the permeability of sclerotia of *Sclerotinia minor* to an apoplastic tracer. *Protoplasma*, *167*, 205–214.

Zadrazil, F. (1996). *Pleurotus* (*Lentinus*) *tuber-regium. Mushroom Information*, *109*, 24–29.

Zhang, L. N., Zhang, M., Dong, J., Guo, J., Song, Y. Y., and Cheung, P. C. K. (2001). Chemical structure and chain conformation of the water-insoluble glucan isolated from *Pleurotus tuber-regium. Biopolymers*, *59*, 457–464.

Zhang, M., Cheung, P. C. K., and Zhang, L. (2001). Evaluation of mushroom dietary fiber (nonstarch polysaccharides) from sclerotia of *Pleurotus tuber-regium* (Fries) Singer as a potential antitumor agent. *Journal of Agricultural and Food Chemistry*, *49*, 5059–5062.

Zhang, M., Zhang, L., Cheung, P. C. K., and Dong, J. (2003a). Fractionation and characterization of a polysaccharide from the sclerotia of *Pleurotus tuber-regium* by preparative size-exclusion chromatography. *Journal of Biochemical and Biophysical Methods*, *56*, 281–289.

Zhang, M., Zhang, L., Wang, Y., and Cheung, P. C. K. (2003b). Chain conformation of sulfated derivatives of β-glucans from sclerotia of *Pleurotus tuber-regium. Carbohydrate Research*, *338*, 2863–2870.

Zhang, M., Cheung, P. C. K., Zhang, L., Chiu, C. M., and Ooi, V. E. C. (2004a). Carboxymethylated β-glucans from mushroom sclerotium of *Pleurotus tuber-regium* as novel water-soluble anti-tumor agent. *Carbohydrate Polymer*, *57*, 319–325.

Zhang, M., Zhang, L., Cheung, P. C. K., and Ooi, V. E. C. (2004b). Molecular weight and anti-tumor activity of the water-soluble polysaccharides isolated by hot water and ultrasonic treatment from the sclerotia and mycelia of *Pleurotus tuber-regium. Carbohydrate Polymer*, *56*, 123–128.

Zhang, M., Cheung, P. C. K., Chiu, L. C. M., Wong, E. Y. L., and Ooi V. E. C. (2006a). Cell-cycle arrest and apoptosis induction in human breast carcinoma MCF-7 cells by carboxymethylated β-glucan from mushroom sclerotia of *Pleurotus tuber-regium. Carbohydrate Polymer*, *66*, 455–462.

Zhang, M., Cui, S. W., Cheung, P. C. K., and Wang, Q. (2006b). Polysaccharides from mushrooms: A review on their isolation process, structural characteristics and antitumor activity. *Trends in Food Science and Technology*, *18*, 4–19.

Zoberi, M. H. (1972). *Tropical Macrofungi—Some Common Species*. London: Macmillan Press, pp. 57–59, 158.

Zoberi, M. H. (1973). Some edible mushrooms from Nigeria. *Nigerian Field*, *38*, 81–90.

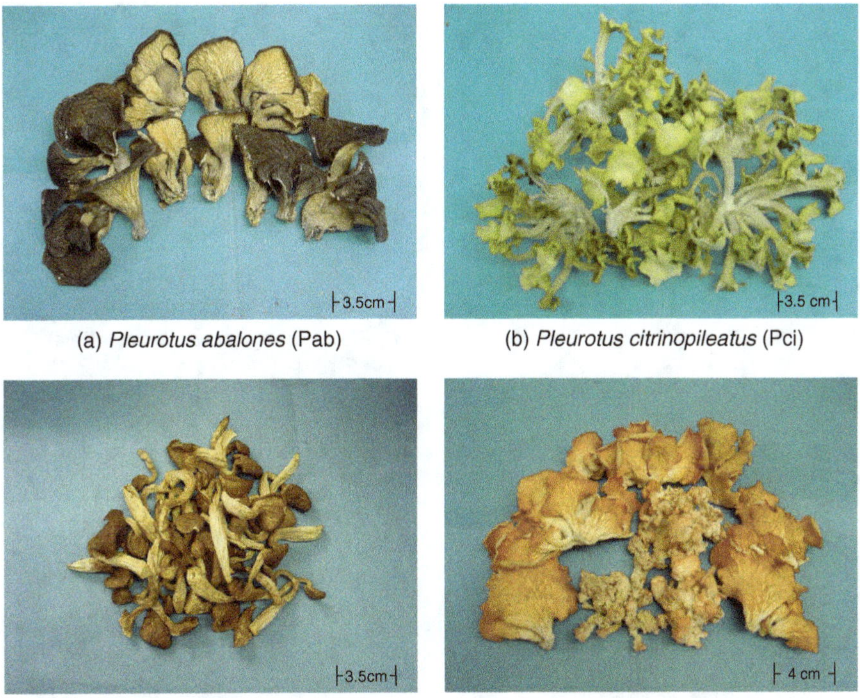

(a) *Pleurotus abalones* (Pab)
(b) *Pleurotus citrinopileatus* (Pci)
(c) *Pleurotus cornucopiae* (Pco)
(d) *Pleurotus djamor* (Pd)

Figure 3.2 Photos of dried samples of newly developed cultivated mushrooms.

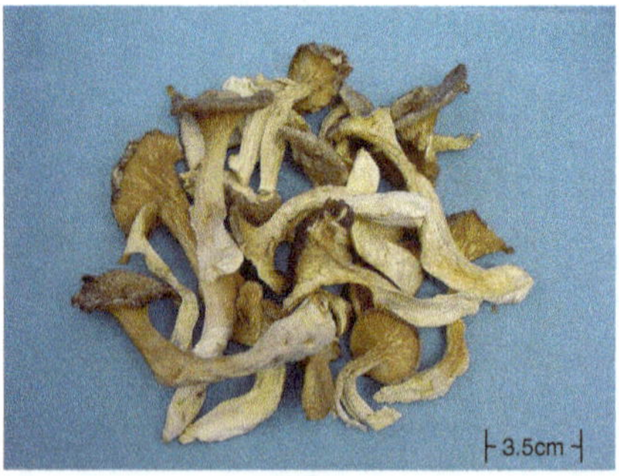

(e) *Pleurotus eryngii* (Pe)

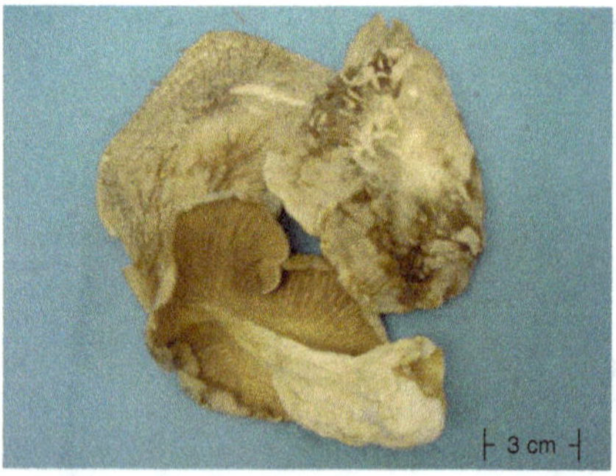

(f) *Pleurotus eryngii var. ferulae* (Pevf)

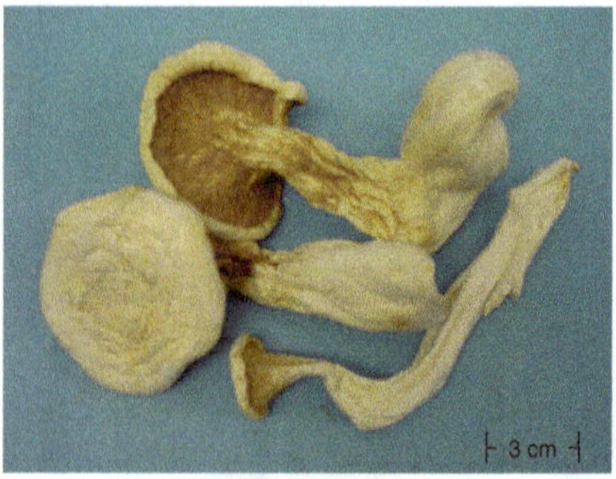

(g) *Pleurotus eryngii var. nebrodensis* (Pevn)

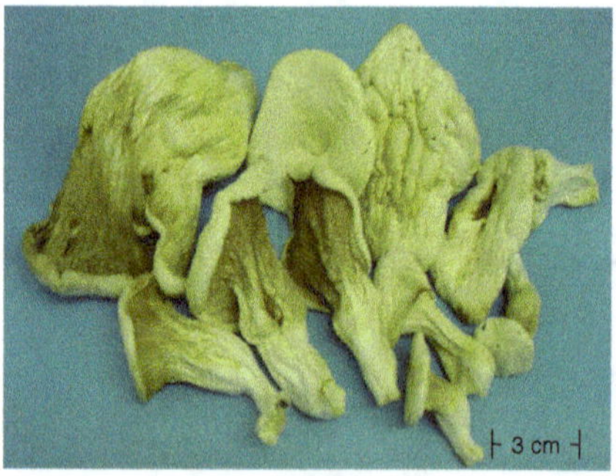

(h) *Pleurotus nebrodensis* (Pne)

Figure 3.2 *(Continued)*

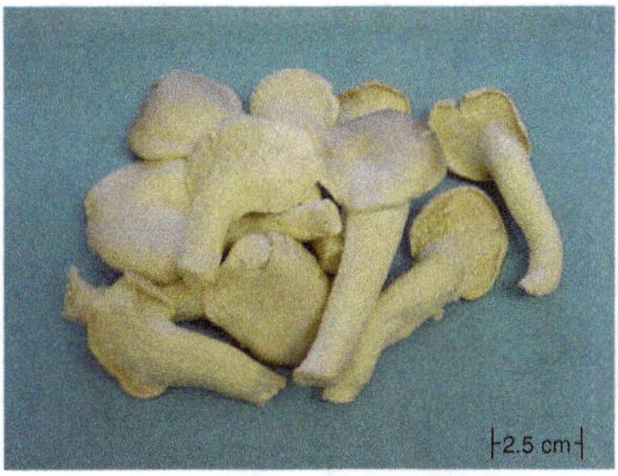

(i) *Pleurotus ostreatus* (Po)

(j) *Pleurotus plumonarius* (Pp)

(k) *Pleurotus sapidus* (Ps)

Figure 3.2 *(Continued)*

(l) *Agrocybe aegerita* (Aa)

(m) *Agaricus blazei* (Ab)

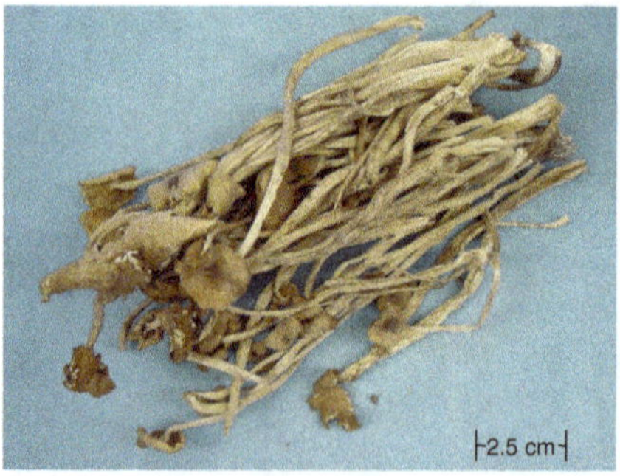

(n) *Agrocybe chaxinggu* (Ac)

Figure 3.2 (*Continued*)

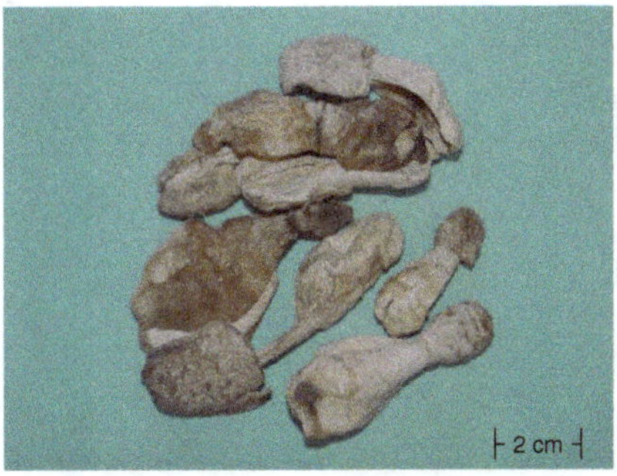

(o) *Coprinus comatus* (Cc)

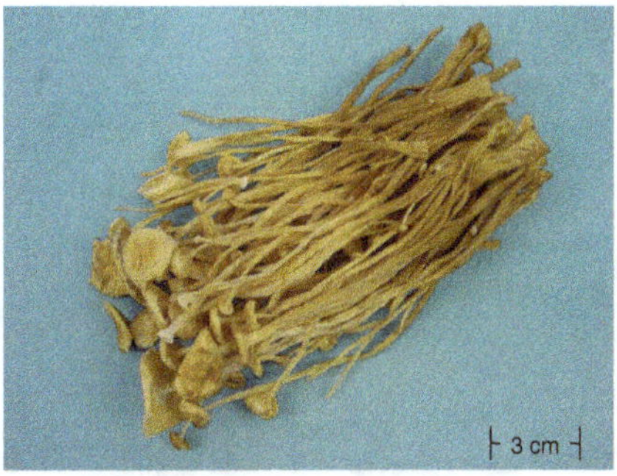

(p) *Flammulina velutipes* (Fv)

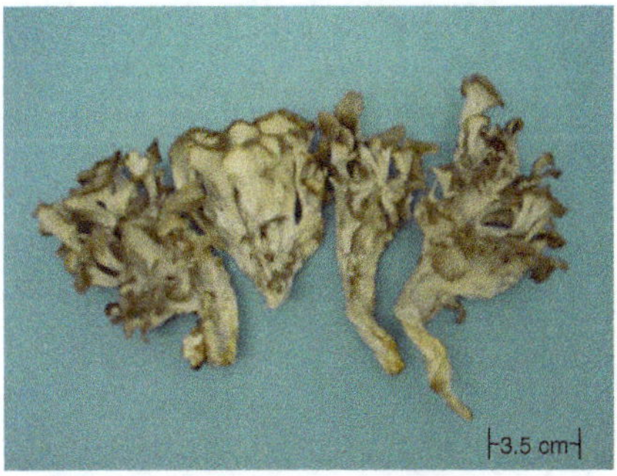

(q) *Grifola frondosa* (Gf)

Figure 3.2 *(Continued)*

(r) GK16

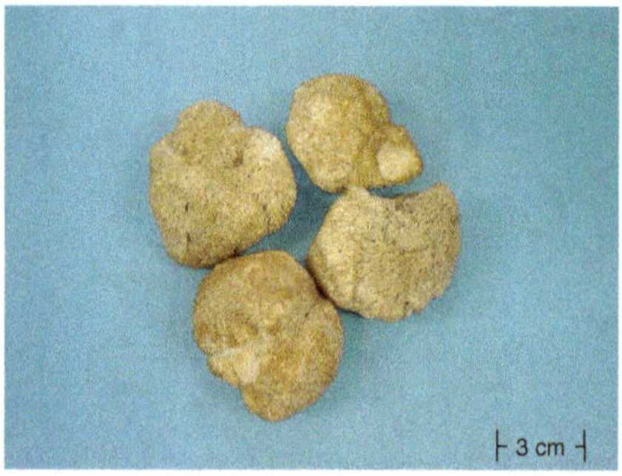

(s) *Hericium erinaceus* (He)

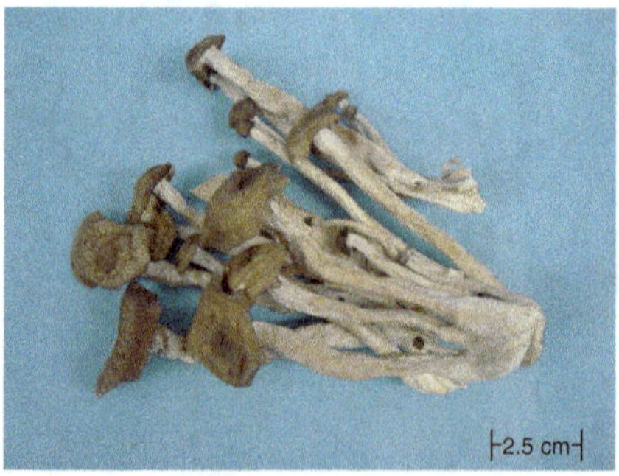

(t) *Hypsizigus marmoreus* (Hm)

Figure 3.2 *(Continued)*

(u) *Hericium ramosum* (Hr)

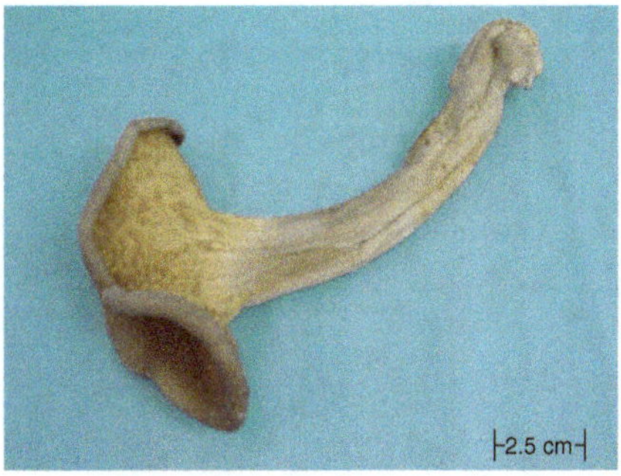

(v) *Lentinula giganteus* (Lg)

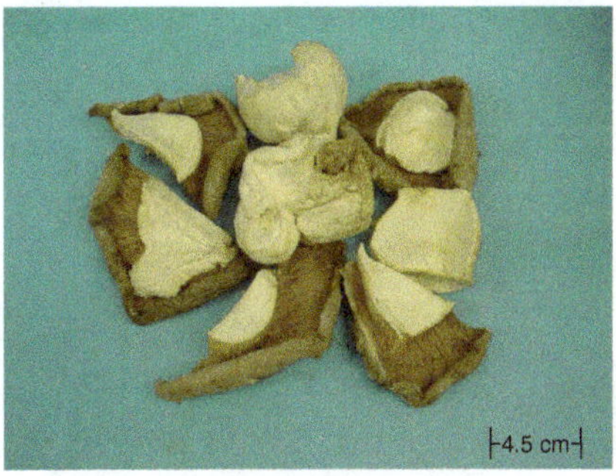

(w) LPK 15

Figure 3.2 (*Continued*)

(x) *Pholiota adipose* (Pa)

(y) *Pholiota nameko* (Pn)

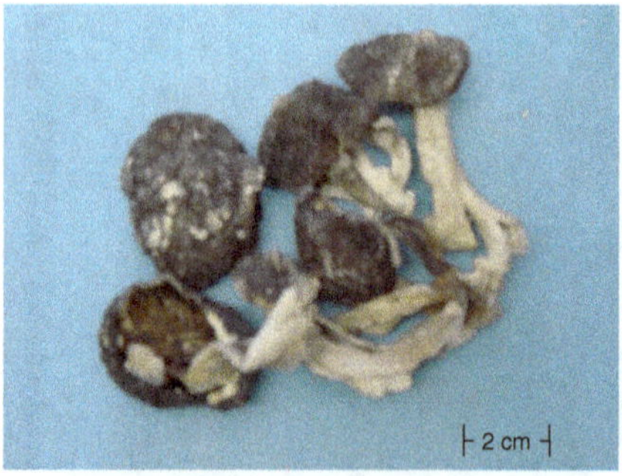

(z) *Stropharia rugoso-annulata* (Sra)

Figure 3.2 (*Continued*)

CHAPTER 5

Antitumor and Immunomodulatory Activities of Mushroom Polysaccharides

Vincent E. C. Ooi

Department of Biology and Institute of Chinese Medicine, The Chinese University of Hong Kong, Hong Kong, China

CONTENTS

5.1 INTRODUCTION

Mushrooms have a long history of medicinal application in addition to their nutritional value. They have been used as food and medicinal materials in many Oriental countries such as China, Japan, and Korea. In the past three to four decades, there has been an upsurge of interest in research on the medicinal value of mushrooms and their products. Chief among the most promising biopharmacological activities of mushrooms are their immunomodulation and antitumor effects. Polysaccharides are the best known mushroom-derived substances with potent antitumor and immunomodulatory properties (Ooi and Liu, 2000; Wasser, 2002; Moradali et al., 2007; Zhang et al., 2007). The beneficial uses of many mushroom polysaccharides as therapeutic adjuvants or dietary supplements have been extensively studied and their new wider usages and trials are further explored (Borchers et al., 2004;

Mushrooms as Functional Foods, Edited by Peter C. K. Cheung

Zekovic et al., 2005; Sullivan et al., 2006). Polysaccharides represent a structurally diverse class of biological macromolecules of relatively widespread occurrence in nature. Unlike proteins and nucleic acids, they contain repetitive structural features which are polymers of monosaccharide residues joined to each other by glycosidic linkages and can interconnect at several points to form a wide variety of branched or linear structures. Among these macromolecules, polysaccharides offer the highest capacity for carrying biological information because they have the greatest potential for structural variability. This enormous variability in polysaccharide structures gives the necessary flexibility for the precise regulatory mechanisms of various cell–cell interactions in higher organisms (Sharon and Lis, 1993). Polysaccharides or polysaccharide–protein complexes derived from mushrooms have attracted much attention of research because they are generally believed to be able to suppress the tumor growth of the host by restoring or enhancing the immune defense system, which is vitally important for the maintenance of homeostasis. They are often considered as host defense potentiators or biological response modifiers (BRMs) (Bohn and BeMiller, 1995; Ooi and Liu, 2000; Leung et al., 2006; Wasser, 2002; Moradali et al., 2007). In addition, these biomacromolecules have also been acclaimed to prevent carcinogenesis and tumor metastasis (Kim et al., 1999; Baek et al., 2002; Guterres et al., 2005; Lee et al., 2005). Although the mechanisms of antitumor action of polysaccharides are not completely clear, they can potentiate cell-mediated immune responses through the activation of specific immune cells to enhance a variety of cellular functions such as cytotoxic and phagocytic responses against tumor cells. They are considered as multicytokine inducers that are able to induce gene expression of various immunomodulatory cytokines by immunocompetent cells in the innate immunity (Ooi and Liu, 2000; Wasser, 2002; Moradali et al., 2007). Nonetheless, many of these macromolecules have also been documented to have direct cytotoxicity against cancer cell lines in vitro (Chen and Chang, 2004; Zaidman et al., 2005; M. Zhang et al., 2006a, b, 2007; Hui et al., 2005). The mechanistic action of antitumor polysaccharides in the regulation of cell cycle and in the activation of the cell death program (apoptosis) has recently received more attention of research (Fullerton et al., 2000; M. Zhang et al., 2006a, b; Fang et al., 2006; Lavi et al., 2006; Wong et al., 2007).

In the 1970s and 1980s, several antitumor polysaccharides, such as lentinan, schizophyllan, and polysaccharide–protein complexes (PSK, PSP), were isolated from *Lentinus edodes, Schizophyllum commune*, and *Coriolus (Trametes) versicolor*, respectively, and have since become very popular in Japan, China, and other Oriental regions (Mizuno et al., 1995b; Ooi and Liu, 1999, 2000). β-D-Glucans from *Grifola frondosa, Sparassis crispa, Agaricus blazei, Phellinus linteus*, and many others have also been widely used as dietary supplements or therapeutic adjuvants for cancer treatment. Several excellent review articles about the isolation, biological activity, chemical structure and modification, and drug development of antitumor polysaccharides have been published in the last few years (Ooi and Liu, 2000; Wasser, 2002; Zhang et al., 2007; Moradali et al., 2007). This present review focuses on the recent findings of the antitumor polysaccharides of mushrooms with an emphasis on the relationship between their structure and antitumor activity, elucidation of their antitumor and immunomodulatory mechanisms of action at the molecular and cellular level, and improvement of their various biological activities by chemical modifications.

5.2 ANTITUMOR POLYSACCHARIDES FROM MUSHROOMS (HIGHER FUNGI)

Mushrooms are considered as macrofungi with a distinctive fruiting body which is large enough to be seen with the naked eyes. Most macrofungi belong to the class Basidiomycetes, but there are also others from the class Ascomycetes (Chang and Miles, 1989). The life cycles of mushroom-like fungi are complex and may involve a number of different morphological forms, including mycelium, fruiting body, and sclerotium. The number of large filamentous fungi in the sense of this definition is at least 14,000 species and perhaps as many as 22,000 species (Hawksworth, 2001). It is estimated that more than 2000 are safe as edible mushrooms, and about 700 species are known to possess significant pharmacological properties (Wasser, 2002). Thus, mushrooms comprise a vast and yet largely untapped source of powerful new pharmaceutical products. For modern medicine, in particular, they represent an unlimited source of polysaccharides with antitumor and immunomodulatory properties. Many, if not all, mushrooms contain biologically active polysaccharides in fruiting bodies, cultured mycelia, sclerotia, and culture filtrates. Wasser (2002) reported that at least 651 species and 7 infraspecific taxa representing 182 genera of Basidiomycetes (mushroom-like fungi) contain antitumor and immunostimulatory polysaccharides. These polysaccharides may vary in their chemical composition, structure, and antitumor activity (Ooi and Liu, 2000; Wasser, 2002; Zhang et al., 2007). Many polysaccharides purified from mushroom fungi are mainly present as glucans with different types of glycosidic linkages, but some are true heteroglycans, while others mostly bind to protein residues as polysaccharide–protein complexes (Ooi and Liu, 2000). The main source of mushroom antitumor polysaccharides appears to be related to the cell walls, which consist of polysaccharides such as chitin, cellulose, $(1 \rightarrow 3, 1 \rightarrow 6)$-$\beta$-glucans, and $(1 \rightarrow 3)$-α-glucans or polysaccharide–protein complexes such as galactomannan–protein, glucuromannan–protein (Zhang et al., 2007). However, chitin and chitosan (fungal chitin) have not been reported to have any antitumor activity (Mizuno et al., 1995b).

Modern research with mushroom polysaccharides can be traced to the 1960s in Japan to the work of the Ikegawa group on the host-mediated antitumor activity of hot-water extracts of several edible mushrooms using sarcoma 180–bearing mice (Ikekawa et al., 1969) and that of Chihara et al. (1970b) on the antitumor polysaccharides isolated from *L. edodes*. Since then, numerous polysaccharides and polysaccharide–protein complexes have been isolated and characterized from mushrooms and used as a source of therapeutic agents for cancers.

In the last three to four decades, numerous polysaccharides and polysaccharide–protein complexes have been isolated from mushrooms and used as a source of nutraceuticals and therapeutic agents. Three antitumor mushroom polysaccharides, that is, lentinan, schizophyllan, and polysaccharide–protein complexes (PSK, PSP), have become very popular nutraceuticals in the Oriental countries. Lentinan and schizophyllan are pure β-glucans (Komatsu et al., 1969; Chihara et al., 1970b, 1989), whereas PSK (Krestin) is a β-glucan–protein complex containing 25–38% protein residues (Tsukagoshi et al., 1984). It is a $(1 \rightarrow 4)$-β-glucan with $(1 \rightarrow 6)$-β-glucopyranosidic side chains for every fourth glucose unit and has a structure with branches at the 3 and 6 positions in a proportion of one per every several residual groups of $1 \rightarrow 4$ bonds and associated with peptide moiety (Tsukagoshi et al., 1984). PSP (polysaccharopeptide) isolated from a strain of *C. (T.) versicolor* in China (Yang et al., 1992) is found to be quite similar in glucan structure to PSK in Japan. Lentinan from the fruiting body of *L. edodes* is a representative mushroom $(1 \rightarrow 3)$-β-glucan with effective antitumor and immunopotentiating activity. Its primary structure is a $(1 \rightarrow 3)$-β-glucan consisting of five $(1 \rightarrow 3)$-β-glucose residues in a linear linkage and two $(1 \rightarrow 6)$-β-glucopyranoside branches in side chains which result in a right-handed triple-helical structure (Chihara et al., 1989)]. Another highly potent antitumor polysaccharide, schizophyllan from *S. commune*, is also a $(1 \rightarrow 3)$-β-glucan having a β-glucopyranosyl group linked $1 \rightarrow 6$ to every third or fourth residue of the main chain. It is similar to lentinan in its triple-helix structure and biological activity but physicochemically unlike lentinan (Komatsu et al., 1969; Ohno et al., 1995).

More recent additions to the list of antitumor polysaccharides that are widely used include β-D-glucans isolated from *G. frondosa, A. blazei, S. crispa, P. linteus*, and many others. A bioactive β-glucan extracted from the maitake mushroom has a cytotoxic effect on prostatic cancer cells in vitro, leading to apoptosis (Fullerton et al., 2000). A D-fraction, a $(1 \rightarrow 3)$-branched $(1 \rightarrow 6)$-β-glucan extracted from the fruiting bodies of *G. frondosa*, has strong antiproliferative activity against human prostatic cancer PC-3 cells in vitro and shows a synergistic potentiation when coadministered with vitamin C or an anticancer agent carmustine (BCNU) (Konno et al., 2002). It can activate macrophages, dendritic cells, and T cells and inhibit the growth of tumor cells. The D-fraction enhances the cytotoxicity of natural killer (NK) cells through the production of interleukin (IL) 12 by macrophages activated by D-fraction (Kodama et al., 2005a). A 21-kDa heteropolysaccharide, coded as GFPS1b, obtained from the cultured mycelia of *G. frondosa* exhibits more potent antiproliferative activity on MCF-7 cells than other polysaccharide fractions. GFPS1b was an acidic polysaccharide with a backbone consisting of $(1 \rightarrow 4)$-α-linked D-galacopyranosyl and $(1 \rightarrow 3)$-α-linked D-glucopyranosyl residues (Cui et al., 2007).

Agaricus blazei is an edible mushroom that is widely considered to be medically important. Polysaccharide fractions prepared from cultured *A. blazei* mainly show antitumor activity against the solid form of sarcoma 180 in ICR mice. The highly branched $(1 \rightarrow 3)$-β-glucan segment forms the active center of the antitumor activity (Ohno et al., 2001; Chung et al., 2005). A $(1 \rightarrow 6)$-β-D-glucan extracted from

A. blazei has cytotoxic effect against human ovarian cancer HRA cells in vitro, promoting p38 MAPK activity for suppressing HRA cell proliferation and amplifying the apoptosis cascade. In mice, oral supplementation with β-glucan reduces pulmonary metastasis of 3LL cells and peritoneal disseminated metastasis of HRA cells and inhibits the growth of these metastatic tumors in lung or peritoneal cavity (Kobayashi et al., 2005). The mycelium polysaccharide and exopolysaccharide (EPS) of *Agaricus brasiliensis* also demonstrate a strong antitumor action against the solid form of sarcoma 180 in ICR mice (Fan et al., 2007).

Sparassis crispa Fr., an edible mushroom recently cultivable in Japan, contains a remarkably high content of 6-branched $(1 \rightarrow 3)$-β-D-glucan (SCG) showing antitumor activity (Ohno et al., 2000). The addition of recombinant murine granulocyte-macrophage colony-stimulating foctor (rMuGM-CSF) to spleen cell cultures from various strains of mice synergistically enhanced interferon gamma (IFN-γ), tumor necrosis factor alpha (TNF-α), and IL-12p70 in the presence of SCG (Harada et al., 2002, 2004). Acidic polysaccharide (PL) isolated from *P. linteus* exhibits antitumor activity against B16 melanoma cells in a dose-dependent manner through the up-regulation of nitric oxide (NO) and TNF-α production, suggesting that PL acts as an effective immunomodulator that enhances the antitumor activity (Kim et al., 2004a). Protein-bound polysaccharide from *P. linteus* induces G2/M phase arrest and apoptosis in SW480 human colon cancer cells (Li et al., 2004).

Ganoderma lucidum and other related *Ganoderma* species, including *G. lucidum, G. tsugae, G. capense*, and *G. applanatum*, are the most well known medicinal fungi in the Orient. The extract of *G. lucidum* (LZE) has antitumor activity with proapoptotic and anti-inflammatory functions, as well as inhibitory effects on cytokine expression during early inflammation in colonic carcinoma cells (Hong et al., 2004). It exerts its effect on cancer cells by multiple mechanisms and suppresses the growth of breast cancer cells through the inhibition of Akt/nuclear factor kappa B (NF-κB) signaling pathway (Jiang et al., 2004a). Various antitumor polysaccharide components such as β-glucan, glucuronoglycan, mannoglucan, and other active heteroglycans as well as polysaccharide–protein complexes have been isolated and purified from *Ganoderma* species for medicinal use (Mizuno et al., 1995b; Ooi and Liu, 2000; Gao et al., 2004). GLP, a polysaccharide fraction isolated from fruiting bodies of *G. lucidum*, significantly suppresses in vivo the growth of sarcoma 180 solid tumor. GLP induces a marked increase in the expression levels of IL-1α (two fold), IL-1β (3-fold), TNF-α (2-fold), IL-12 p35 (up to 6-fold), and IL-12 p40. In the macrophages, GLP promotes a remarkable increase in the expression levels of IL-1β (2.5–to 3-fold), TNF-α (up to 6-fold), and macrophage colony-stimulating factor (M-CSF) (up to 2-fold) (Ooi et al., 2002). Ganopoly, the refined polysaccharide extracted from *G. lucidum*, exhibits antitumor potential with a broad spectrum of immunomodulating activities, including significantly increased cytotoxic T-lymphocyte cytotoxicity and NK cell activity in mice (Gao et al., 2005). One β-glucan, named Lzps-1 (MW 8000), obtained from *G. lucidum* spore has antitumor activity against sarcoma 180 and Lewis lung cancer in mice and enhances the NK cell activity

(Jiang et al., 2005). Ganoderan (MW 20,000), an immunomodulatory β-glucan of *G. lucidum*, induces potent antitumor immunity in tumour-bearing mice (Han et al., 1995). Several β-glucans, heteroglycans, and glycan–protein complexes having high antitumor activity have been isolated from the fruiting body and mycelium of *G. applanatum*, with the molecular weight of the primary polysaccharides ranging from 30×10^3 to 1000×10^3, and the basic chemical structure is $(1 \rightarrow 3)$-β-glucopyranan having 1–15 $(1 \rightarrow 6)$-monoglucosyl side chains. It seems that the greater the molecular weight and the higher the water solubility of these polysaccharids, the higher the antitumor activity (Mizuno, 1995). Among the seven strong antitumor polysaccharide–protein complexes from *Ganoderma tsugae*, two are identified as protein-containing glucogalactans associated with mannose and fucose, and five are protein-containing $(1 \rightarrow 3)$-β-glucans. It is noteworthy that antitumor polysaccharides having high activity from fruiting bodies are mostly heteropolysaccharides with molecular weight of about 10,000 and consisting of galactose, glucose, mannose, and fucose, whereas the highly active polysaccharides from mycelia are mainly protein-containing glucans with molecular weight of 10,000 (Zhang et al., 1994a; Mizuno et al., 1995c; Gao et al., 2004; Peng et al., 2005). The active polysaccharides isolated from the fruiting bodies and mycelia of these three *Ganoderma* species are markedly different in their component monosaccharides, their protein moiety content, and their average molecular weight.

A highly branched $(1 \rightarrow 3)$-β-glucan with a pentasaccharide segment consisting of one nonreducing terminus on 3,6-O-substituted and three 3-mono-O-substituted β-glucopyranosyl side chains isolated from *Pleurotus ostreatus* was reported to have strong antitumor activity (Yoshioka et al., 1985). A newly identified low-molecular-weight α-glucan from the same mushroom exhibits antiproliferative and proapoptotic activities against HT-29 colon cancer cells in vitro via the induction of programmed cell death (Lavi et al., 2006). Several potent antitumor polysaccharide–protein complexes have been purified from fruiting bodies of a Chinese edible mushroom, *Pleurotus sajor-caju*, including (a) protein-containing xyloglucan (MW 280,000) with polysaccharide–protein ratio 76 : 24 (w/w); (b) protein-containing mannogalactan (MW 120,000) with polysaccharide–protein ratio 76 : 16; (c) protein-containing xylan (MW 200,00) with polysacchairde–protein ratio 62 : 21; (d) protein-containing glucoxylan (MW 90,000) with polysaccharide–protein ratio 71 : 15; and (e) protein-containing xyloglucan (MW 70,000) with polysaccharide–protein ratio 69 : 3 (Zhuang et al., 1993). Another interesting edible mushroom, *Pleurotus tuber-regium* (PTR), exists as a stage of mycelium, fruiting body, or sclerotium. The water-soluble nonstarch polysaccharides (NSPs) extracted from the fruiting body, mycelium, and culture medium (coded as HWE, EDP, and CEP, respectively) of *P. tuber-regium* have been shown to exert antiproliferative activity through the induction of apoptosis in HL-60 cells with an increase in the ratio of Bax/Bcl-2 (Wong et al., 2007). The mechanism for the antitumor activity of a water-soluble carboxymethylated β-glucan (CMPTR), partially synthesized from an insoluble native glucan isolated from the sclerotia of *P. tuber-regium*, is related to cell cycle arrest and apoptosis induction in human breast carcinoma MCF-7 breast cancer cells in vitro (Zhang et al., 2006a). An immunomodulating polysaccharide isolated from the aqueous extract of *Pleurotus florida* fruiting bodies exhibits significant macrophage activity through the release of nitric oxide (Rout et al., 2004). Three polysaccharides, namely, $(1 \rightarrow 3)$-β-D-glucan, $(1 \rightarrow 6)$-β-D-glucan, and $(1 \rightarrow 3, 1 \rightarrow 6)$-$\beta$-D-glucan, purified from fruiting bodies of *Lyophyllum decastes* Sing., a newly cultivated mushroom in Japan, show marked antitumor activity against sarcoma 180 (Ukawa et al., 2000). EA6 (a protein-bound polysaccharide) isolated from the culinary-medicinal mushroom *Flammulina velutipes* augments the antitumor immunity in combination with surgical excision (SE) in Meth-A fibrosarcoma-bearing mice, and the effect is mediated by CD4-positive T cells (Maruyama and Ikekawa, 2005).

The complete list of the antitumor polysaccharides and polysaccharide–protein complexes from mushroom fungi are presented separately in Tables 5.1 and 5.2.

5.3 MECHANISMS OF ANTITUMOR ACTION OF MUSHROOM POLYSACCHARIDES

Numerous polysaccharides or polysaccharide–protein complexes from mushrooms have been identified and shown to have antitumor activities (Ooi and Liu, 2000b; Wasser, 2002; Zhang et al., 2007). Although a complete answer to the mechanisms of antitumor action of polysaccharides or polysaccharide–protein complexes is not yet available, they are generally known as biological response modifiers which are able to restore or enhance various immune responses in vivo and in vitro. Their actions are predominantly considered to be host mediated. However, many of these macromolecules have been documented to also possess direct cytotoxic effects on cancer cells. It is possible that, in some instances, these two types of inhibitory action may be interwoven. Therefore, the possible modes of anticancer action may include both (1) direct cytotoxicity to cancer cells as shown in many in vitro studies and (2) indirect antitumor inhibition through immunomodulation of the body defense system.

5.3.1 Antiproliferation of Cancer Cells and Induction of Apoptosis

Various glucans such as lentinan (β-glucan of *L. edodes*) show no direct growth-inhibitory effects on tumor cell lines in vitro (Ooi and Liu, 2000b; Wasser, 2002). However, the indirect cytotoxicity of lentinan is observed to enhance the antitumor cytotoxic activity of peritoneal macrophages against human melanoma target cells in vitro (Ladanyi et al., 1993). The activity of lymphokine-activated killer cells (LAK) stimulated by IL-2 and lentinan against autologous tumour cells and K562 human erythroleukemia cells is greater than that stimulated by IL-2 alone in vitro showing the augmentation of cytotoxicity of LAK by lentinan (Tani et al., 1993). When the water-insoluble α-$(1 \rightarrow 3)$-D-glucan (L-FV-II) isolated from fruiting bodies of *L. edodes* is chemically modified to the water-soluble

TABLE 5.1 Antitumor Polysaccharides from Mushrooms

Mushroom Species	Polysaccharide Component	References
	Homoglucans	
Agaricus blazei	(1 → 4)-α- and (1 → 6)- β-Glucan	Fujimiya et al., 1998; Kobayashi et al., 2005
	(1 → 6)- α- and (1 → 4)-α-Glucan	Mizuno et al., 1998
	(1 → 6)- β- and (1 → 3)-β-Glucan	Mizuno et al., 1990a; Chung et al., 2005
Agaricus brasiliensis	(1 → 6)- β- and (1 → 3)-β-Glucan	Camelini et al., 2005; Angeli et al., 2006
Agrocybe cylindracea	(1 → 3)-β-Glucan (fruiting body)	Kiho et al., 1989; Yoshida et al., 1996
Amanita muscaria	(1 → 3)-α-Glucan	Kiho et al., 1994
	(1 → 3)-β-Glucan (fruiting body)	Kiho et al., 1992a
Armillariella tabescens	(1 → 3)-α-Glucan	Kiho et al., 1992b
Auricularia auricula	(1 → 3)-β-Glucan (fruiting body)	Misaki and Kakuta, 1995
Collybia dryophila	(1 → 3)-β-Glucan (fruiting body)	Pacheco-Sanchez et al., 2006
Cordyceps sinensis	Cordyglucan (1 → 3)-β-glucan	Yalin et al., 2005
Cryptoporus volvatus	(1 → 3)-β-Glucan (fruiting body)	Kitamura et al., 1994
Dictyophora indusiata	(1 → 3)- β-Glucan (fruiting body)	Hara et al., 1991
Flammulina velutipes	(1 → 3)-β-Glucan (fruiting body)	Smiderle et al., 2006; Leung and Fung, 1997
Ganoderma lucidum	Ganoderan [(1 → 3)-β-glucan]	Han et al., 1995
	(1 → 3)-β-Glucan (spore)	Bao et al., 2001
Ganoderma tsugae	(1 → 4)-α- and (1 → 3)-β-Glucan (mycelium)	Peng et al., 2005
Grifora frondosa	Grifola (1 → 3)-β-D-Glucan	Zhuang et al., 1994b
	(1 → 3)-β-Glucan (fruiting body, mycelium, medium product)	Kodama et al., 2002
Grifora umbellata	(1 → 3)-β-Glucan (sclerotium)	Ogawa and Kaburagi, 1982
Hericium erinaceus	(1 → 3)-β-Glucan	Dong et al., 2006
Hypsizigus marmoreus	(1 → 3)-β-Glucan (fruiting body)	Ikekawa et al., 1992
Lyophyllum decastes	(1 → 6)- β-Glucan	Ukawa et al., 2000
Lentinus edodes	Lentinan, (1 → 3)-β-glucan (fruiting body, mycelium)	Chihara et al., 1970b; Surenjav et al., 2005
Omphalia lapidescens	(1 → 3)-β-Glucan (fruiting body)	Ohno et al., 1993; Saito, et al., 1992
Phellinus linteus	(1 → 3)-β-Glucan (fruiting body)	Kim et al., 1996

TABLE 5.1 *(Continued)*

Mushroom Species	Polysaccharide Component	References
Pleurotus eryngii	(1 → 3)-β-Glucan	Carbonero et al., 2006
Pleurotus florida	(1 → 3)-β-Glucan	Rout et al., 2005
Pleurotus ostreatoroseus	(1 → 3)-β-Glucan	Carbonero et al., 2006
Pleurotus pulmonarius	(1 → 3)-β-Glucan	Gutierrez et al., 1996
Pleurotus tuber-regium	(1 → 3)-β-D-Glucan (Sclerotium)	Zhang et al., 2003; Tao and Zhang, 2006
Polyporus confluens	(1 → 3)-β-D-Glucan (fruiting body)	Mizuno et al., 1992
Poria cocos	Pachyman, (1 → 3)-β-D-glucan	Kanayama et al., 1986
Porodisculus pendulus	β-Glucan (medium products)	Ogawa and Kaburagi, 1982
Sclerotinia libertiana	β-Glucan (medium products)	Ogawa and Kaburagi, 1982
Sclerotium sclerotia	Scleroglucan, (1 → 3)-β-glucan	Palleschi et al., 2005
Schizophyllum commune	Schizophyllan, (1 → 3)-β-glucan	Komatsu et al., 1969; Ogawa and Kaburagi, 1982
Sparassis crispa	(1 → 3)-β-Glucan	Ohno et al., 2000; Harada et al., 2006
Termitomyces eurhizus	(1 → 3)-β-Glucan	Chakraborty et al., 2006
Trametes gibbosa	β-Glucan (fruiting body)	Czarnecki and Grzybek, 1995
Tricholoma giganteum	(1 → 3)-β-Glucan (fruiting body)	Mizuno et al., 1995a
Tylopilus felleus	β-Glucan (fruiting body)	Grzybek et al.,1990; Kohlmunzer et al., 1990
Volvariella volvacea	β-Glucan (fruiting body)	Kishida et al., 1989
	Heteroglucans	
Agaricus blazei	Mannogalactoglucan	Cho et al., 1999
	Riboglucan	Cho et al., 1999
Flammulina velutipes	Galactomannoglucan	Ikekawa et al., 1982
Fomitella fraxinea	Mannogalactoglucan	Cho et al., 1998
Ganoderma lucidum	(1 → 3)-β-Glucuronoglucan	Saito et al., 1989
Ganoderma tsugae	Arabinoglucan	Zhang et al., 1994a
Grifola frondosa	Mannofucoxyloglucan	Zhuang et al., 1994b; Mizuno and Zhuang, 1995
	Xyloglucan	Mizuno et al., 1986
Hericium erinaceus	Galactoxyloglucan	Mizuno et al., 1992b; Wang et al., 2004

TABLE 5.1 *(Continued)*

Mushroom Species	Polysaccharide Component	References
Hohenbuehelia serotina	Galactomannoglucan	Ma et al., 1991
Inonotus obliquus	Xylogalactoglucan	T. Mizuno et al., 1999
Leucopaxillus giganteus	Galactomannoglucan	Mizuno et al., 1995
Pleurotus cornucopiae	Mannogalactoglucan	Gutierrez et al., 1996
Pleurotus pulmonarius	Mannogalactoglucan	Gutierrez et al., 1996
	Xyloglucan	Gutierrez et al., 1996
Polyporus confluens	Xyloglucan	Mizuno et al., 1992
Tremella fuciformis	$(1 \rightarrow 3)$-β-Mannoglucan	Gao et al., 1996
	Heteroglycans	
Agariscu bisporus	Heterogalactan	Shida and Sakai, 2004
Agaricus blazei	$(1 \rightarrow 2)$-β- and $(1 \rightarrow 3)$-β-Glucomannan	M. Mizuno et al., 1999
	Glucoxylan (fruiting body)	Mizuno et al., 1990b
	Heterogalactan	Shida and Sakai, 2004
Collybia maculata	Galactomann	Lim et al., 2005
Dictyophora indusiata	Fucomannogalactan	Hara et al., 1991
Flammulina velutipes	Xylomannan	Smiderle et al., 2006
	Heterogalactan	Shida and Sakai, 2004
Fomitella fraxinea	$(1 \rightarrow 6)$-α-Mannofucogalactan	Cho et al., 1998
Ganoderma tsugae	Glucogalactan	Wang et al., 1993
Grifola frondosa	Mannogalactofucan	Zhuang et al., 1994a
	Heterogalactan	Shida and Sakai, 2004
Hericium erinaceus	Xylan	Mizuno et al., 1992b
	Rhamnoglucogalactan	Jia et al., 2004
	Glucogalactan	Wang et al., 2004
	Mannoglucoxylan	Mizuno et al., 1992b
	Fucogalactan	A. Zhang et al., 2006; Shida and Sakai, 2004
Hypsizigus marmoreus	Heterogalactan	Shida and Sakai, 2004
Inonotus obliquus	α-Linked fucoglucomannan	Kim et al., 2006
Lampteromycer japonicus	Mannan (fruiting body)	Fukuda et al., 1975
Lentinus edodes	Galactoglucomannan	Fujii et al., 1978
Pleurotus citrinopileatus	Arabinogalactan	Zhang et al., 1994b
Pleurotus eryngii	Heterogalactan	Shida and Sakai, 2004

TABLE 5.1 (*Continued*)

Mushroom Species	Polysaccharide Component	References
Pleurotus ostreatus	Heterogalactan	Shida and Sakai, 2004
Pleurotus sajor-caju	Galactoglucomannan	Pramanik et al., 2005
Polyporus confluens	β-Glucopyranan (mycelium)	Mizuno et al., 1992a
Sarcodon aspratus	Fucogalactan	Mizuno et al., 2000
Tremella fuciformis	β-Glucuronoxylomannan (fruiting body)	Misaki and Kakuta, 1995; Yui et al., 1995; Gao et al., 1996
Tremella mesenterica	β-glucuronoxylomannan	Wasser et al., 2002; Vinogradov et al., 2004
Tricholoma giganteum	Xyloglucomannan (fruiting body)	Mizuno et al., 1995a

sulfated α-(1 $\rightarrow$ 3)-D-glucan (SL-FV-II), it has potent antiproliferation action (52%) on human MCF-7 breast carcinoma cells (Zhang and Cheung, 2002).

Moreover, a (1 $\rightarrow$ 3)-β-glucan extracted from the maitake mushroom (*G. frondosa*), known as D-fraction, demonstrates a direct cytotoxicity (at a dosage of 480 mg/mL) to prostatic cancer PC-3 cells, inducing nearly complete cell death (>95%) in 24 hours, and has a synergistic potentiation when it is coadministered with vitamic C or an anticancer agent carmustine (BCNU), resulting in a drastic reduction in cell viability (Fullerton et al., 2000; Konno et al., 2002). GFPS1b, a 21-kDa heteropolysaccharide obtained from the cultured mycelia of *G. frondosa*, exhibits more potent antiproliferative activity on MCF-7 cells than other polysaccharide fractions (Cui et al., 2007). *Phellinus linteus* polysaccharides (PLs) render murine or human lung cancer cells susceptible to apoptosis, and in the process of PL-induced apoptosis, caspase 2 is induced in LNCaP cells, which express the androgen receptor (AR), but not in PC-3 cells, which lack AR, demonstrating the AR-dependent and independent apoptotic pathways (Zhu et al., 2007). PL has a synergistic effect with doxorubicin (Dox) to activate caspases in prostate cancer LNCaP cells, suggesting that PL has therapeutic potential to augment the magnitude of apoptosis induced by antiprostate cancer drugs (Collins et al., 2006).

A newly identified low-molecular-weight α-glucan of *P. ostreatus* has promising antitumorigenic properties and demonstrates its direct effect on HT-29 colon cancer cell proliferation via the induction of programmed cell death (Lavi et al., 2006). A comparative study shows that the water-soluble NSPs extracted from the fruiting body, mycelium, and culture medium (coded as HWE, EDP, and CEP, respectively) of a novel edible mushroom PTR can induce apoptosis in HL-60 cells with an increase in the ratio of Bax/Bcl-2 (Wong et al., 2007). Among all PTR NSPs, HWE (a heteropolysaccharide–protein complex from the fruiting body, MW1.86 $\times 10^6$) has the strongest antiproliferative activity in

TABLE 5.2 Antitumor Polysaccharide–Protein Complexes from Mushrooms

Mushroom Species	Active Component	References
Agaricus blazei	(1 → 6)-β-D-Glucan–protein complex	Mizuno et al., 1990b; Kawagishi et al., 1990; Gonzaga et al., 2005; Hong and Choi, 2007
	Polysaccharide–protein complex (ATOM)	Ito et al., 1997
	Xyloglucan–protein	Mizuno et al., 1990b
Armillariella tabescens	Protein-containing heteroglycan (fruiting body)	Kiho et al., 1992b
Collybia confluens	Protein-bound polysaccharide (mycelium)	Kim et al., 1993
Cordyceps ophioglossoides	Protein-bound polysaccharide (medium product)	Ohmori et al., 1988a,
Coriolus (Trametes) versicolor	PSK, protein-bound polysaccharide (mycelium)	Tsukagoshi et al., 1984; Kanazawa et al., 2005
	PSP, polysaccharide–peptide complex (mycelium)	Yang et al., 1992; Wang et al., 1996a; Ho et al., 2004
Flammulina velutipes	Protein-bound glucan (fruiting body, mycelium)	Ikekawa et al., 1982; Ohkuma et al., 1982
Fomes fomentarius	Protein-containing polysaccharide (medium product)	Ito et al., 1976
Fomitella fraxinea	Protein-containing galactomannoglucan (fruiting body)	Cho et al., 1995
Ganoderma lucidum	Proteoglycan	J. Zhang et al., 2002
	Proteoglycan	Baek et al., 2002
Ganoderma tsugae	(1 → 3)-β-Glucan–protein complex (mycelium)	Wang et al., 1993
	Glucogalactan–protein complex (mycelium)	Zhang et al., 1994a; Peng et al., 2005
Grifola frondosa	Heteroglycan–protein complex	Zhuang et al., 1994a
Hebeloma crustuliniforme	Polysaccharide–protein complex	Cho and Chung, 1999
Lentinus edodes	Polysaccharide–peptide complex	Liu et al., 1998
Laetiporus sulphureus	Protein–polysaccharide (fruiting body)	Gasiorowski et al., 1993
Phellinus linteus	Polysaccharide–protein complex	Song et al., 1995; Kim et al., 2006
Pleurotus citrinopileatus	Protein-containing heteroglycan (fruiting body)	Zhang et al., 1994b
Pleurotus ostreatus	Proteoglycan	Sarangi et al., 2006
Pleurotus sajor-caju	Heteroglycan–protein complex (fruiting body)	Zhuang et al., 1993

TABLE 5.2 (*Continued*)

Mushroom Species	Active Component	References
Polyporus confluens	Xyloglucan–protein (fruiting body)	Mizuno et al., 1992a
Tremella fuciformis	Heteroglycan–protein (mycelium)	Cho et al., 2006
Tricholoma giganteum	Polysaccharide–protein complex (fruiting body)	Mizuno et al., 1995a
Tricholoma lobayense	Polysaccharide–protein complex (PSPC) (medium product)	Liu and Ooi, 1995; Liu et al., 1996a; Liu et al., 1996b
Tricholoma matsutake	α-Glucona–protein complex	Hoshi et al., 2005
Tricholoma mongolicum	Polysaccharide–peptide complex (mycelium)	Wang et al., 1996b

vitro against human acute promyelocytic leukemia cells (HL-60) whereas CEP (a mannose-rich polysaccharide, MW 44,000) has the least cytotoxicity. EDP (a glucose-rich polysaccharide from the mycelium of PTR, MW 509,000) can cause G_2/M arrest in HL-60 cells by lowering the Cdk1 expression. HWE exerts S-phase arrest in the HL-60 cells by a depletion of Cdk2 and an increase in cyclin E expression (Wong et al., 2007). CMPTR, partially synthesized from an insoluble native glucan isolated from the sclerotia of PTR, can inhibit the cell proliferation of MCF-7 by arresting the G phase of its cell cycle, which is associated with the down-regulation of cyclin D-1 and cyclin E expressions in breast cancer cells. In addition, CMPTR-treated MCF-7 cancer cells exhibit a decreased expression of antiapoptotic Bcl-2 protein and an increased expression of Bax/Bcl-2 ratio (Zhang et al., 2006a).

A neutral polysaccharide fraction isolated from *Poria cocos* (PC) shows a potent activity in suppressing the proliferation of human leukemic cells, U937 and HL-60 cells, and induces more than 50% of U937 cells and HL-60 cells to differentiate into mature monocytes/macrophages, which also markedly express surface antigens of CD11b, CD14, and CD68 (Chen and Chang, 2004). A recent study provides the preliminary insights into the mode of direct antitumor activity in vitro of a β-glucan from the mycelium of PC on human breast carcinoma MCF-7 cells via cell cycle arrest and apoptosis induction (Zhang et al., 2006).

Unlike most glucans, polysaccharide–protein complexes (PSK or PSP) from *C. (T.) versicolor* and other mushrooms usually have both direct and indirect cytotoxic effects on tumor cell lines. *Coriolus versicolor* (CV) extract is able to selectively and dose dependently inhibit the proliferation of lymphoma and leukemic cells possibly via an apoptosis-dependent pathway (Lau et al., 2004). The CV extract exerts antiproliferative activity through the induction of apoptosis, differentially dependent of p53 and Bcl-2 expression in human breast cancer cells (Ho et al., 2005). The protein-bound polysaccharide of *C. versicolor* (PSK) could suppress cell proliferation and induce subsequent cellular apoptosis in the Burkitt lymphoma cell line (Namalwa), out of 33 hematological malignant cell lines tested, indicating the initial evidence of the direct cytotoxic activity of PSK in a cancer cell line (Hattori et al., 2004). PSK could also inhibit cell growth and DNA synthesis in various cell lines such as L1210 leukemia, P388 leukemia, Ehrlich carcinoma, Yoshida sarcoma, AH-13, human hepatoma CHC-20, human choriocarcinoma GCH-1 and GCH-2, and human breast cancer cell MCF-7 (Tsukagoshi et al., 1984). Similarly, the polysaccharide peptide of *C. versicolor* (PSP) extracted from a strain of *C. versicolor* has a wide range of antitumor activities in vitro, inhibiting Ehrlich ascites tumor, leukemia P388, sarcoma 180, and four types of human cancer cell lines, namely human gastric cancer cells, human lung cancer cells, mononuclear leukemia cells, and human skin histiocytic lymphoma cells (Yang et al., 1992; Cui and Chisti, 2003). PSP is also effective in inhibiting cell proliferation through apoptosis. Cells treated with PSP show a significant reduction in cell proliferation with the induction of apoptosis via the up-regulation of p21 and down-regulation of cyclin D-1 (Chow et al., 2003). PSP, used in combination therapy, has the ability to lower the cytotoxicity of certain antileukemic drugs through their interaction with cell cycle–dependent and apoptotic pathways. Induction of S-phase cell arrest and caspase activation by PSP of *C. versicolor* enhances the cell cycle–dependent activity and apoptotic cell death of doxorubicin and etoposide but not cytarabine in HL-60 cells (Hui et al., 2005). PSK, PSP, and most β-D-glucans such as lentinan show different modes of antitumor action in vitro (Sakagami et al., 1991; Ooi and Liu, 2000).

Protein-bound polysaccharide from *P. linteus* induces G_2/M phase arrest and apoptosis in SW480 human colon cancer cells (Li et al., 2004). The antitumor polysaccharide–protein complex (PSPC) purified and characterized from the culture filtrates of *Tricholoma lobayense* exhibits strong antitumor activity in both ICR and BALB/c mice. PSPC has the ability to restore the phagocytic function of the peritoneal exudate cells (macrophages) and the mitogenic activity of T cells of tumor-bearing mice (Liu and Ooi, 1995; Liu et al., 1996b). PSPC also exerts indirect cytotoxic activity against P815 mastocytoma cells and L929 mouse fibroblast cells by activating macrophages to release reactive nitrogen intermediates (RNIs) and TNF-α, which are shown to increase significantly after PSPC treatment in the tumor-bearing mice (Liu and Ooi, 1995). PSPC both induces the various immune responses in vivo and exhibits cytotoxicity against tumor cell lines in the presence of PSPC in vitro (Liu et al., 1996b). Similarly, the antitumor activity of ATOM, a polysaccharide–protein complex prepared from the cultured mycelia of *A. blazei*, is highly effective against four kinds of established tumors in mice, namely subcutaneously implanted sarcoma 180, Ehrlich ascites carcinoma, Shionogi carcinoma 42, and Meth A fibrosarcoma (Itoh et al., 1994, 1997). Heteroglycan–protein complexes from *G. frondosa* are shown to depress tumor growth by activating the immune system as a biological response modifier (Cun et al., 1994). The heteropolysaccharide–protein complexes (mainly $(1 \rightarrow 3)$-α-D-glucan-bound protein with the presence of mannose and galactose) from *P. cocos* mycelia exhibit significant cytotoxic effect on the proliferation of HL-60 cells in vitro and the antitumor activity against sarcoma 180 in vivo (Jin et al., 2003c). Among all PTR NSPs, HWE (a heteropolysaccharide–protein complex, MW 1.86 $\times 10^6$) exerts the strongest anti-proliferative activity in vitro against human acute promyelocytic leukemia cells (HL-60) and causes S-phase arrest in the HL-60 cells by a depletion of Cdk2 and an increase in cyclin E expression (Wong et al., 2007). It is thus suggested that polysaccharide–protein/polysaccharide–peptide complexes of many mushrooms may have some unique structural features, possibly generated from the involvement of protein portions and/or unique structural configurations, including sugar to sugar linkages, that contribute to their immunomodulatory and cell cycle–dependent antitumor actions as well as direct cytotoxicity (Ooi and Liu, 2000b).

5.3.2 Immunomodulation

Immunomodulators are substances from a variety of sources which have the ability to augment the immune system in multiple ways. They are often considered pharmacologically as BRMs and have been reported to have antitumor activity among many biopharmacological properties. The most prominent BRMs are polysaccharide BRMs, especially $(1 \rightarrow 3)$-β-D-glucans from mushrooms (higher fungi), which occur widely in nature (Bohn and BeMiller, 1995; Ooi and Liu, 2000b; Wasser, 2002; Lull et al., 2005).

Many mushroom polysaccharides or polysaccharide–protein complexes have distinct antitumor activities in murine allogeneic, syngeneic, and autochthonous hosts (Tsukagoshi et al., 1984; Bohn and BeMiller, 1995). The preliminary determination of antitumor activity has often relied on a bioassay system normally based on an allogeneic tumor in mice. There are a number of factors such as strain of mice, type of tumor, suitable dosage, and strictly planned timing of drug administration that are essential to achieve the antitumor effect of polysaccharides (Chihara, 1992; Kerekgyarto et al., 1996). Hundreds of polysaccharides or polysaccharide–protein complexes have been screened for their antitumor activity, and three of them, namely schizophyllan, lentinan, and protein-bound polysacharides (PSK and PSP), have been used widely for more than 30 years (Chihara, 1992; Yang et al., 1992).

Several polysaccharides such as lentinan (β-glucan) and PSK (β-glucan–protein) have been shown to have effective antitumor action against a variety of transplantable experimental animal tumors and have been successfully used in clinical treatments (Chihara, 1992; Kobayashi et al., 1993). The enhancement or potentiation of host defense mechanisms has been recognized as a possible means of inhibiting tumor growth without harming the host. The body's defense against spontaneously arising malignant tumors involves a concerted interplay of innate and acquired immune responses. Innate immunity comprises cellular elements such as macrophages, cytotoxic lymphocytes, NK cells, and dendritic cells (DCs). Both cell-mediated immune response against the target cells initiated by macrophage–lymphocyte interactions and cytotoxicity induced by antibodies to target cells are believed to contribute to the elimination of target tumor cells (Chihara et al., 1989). This immune system is regulated by chemical mediators or cytokines and by activation of inflammatory responses. Mushroom polysaccharides can activate macrophages or NK cells to produce various cytokines such as interleukins, interferons, and other mediators so that they are targeted toward destroying tumor cells (Chihara, 1992). NK cells have two relevant functions related to the natural immune response against pathogens or tumor cells: (1) cytotoxicity that is mediated by the recognition and lysis of target cells, such as virus- and bacteria-infected or tumor cells, and (2) production of cytokines that can modulate natural and specific immune responses. Furthermore, DCs and macrophages activated by D-fraction produce cytokines such as IL-12 that stimulate NK cells to rapidly produce other cytokines (including IFN-γ, TNF-α, and GM-CSF) and enhance the cytotoxicity of NK cells (Kodama et al., 2002, 2005a).

The detailed mechanisms of action of polysaccharide immunomodulators or BRMs are not yet fully known, but it is generally accepted that they act on different immunocompetent cells which may initiate a cascade of signal transduction pathways that are responsible for the immune responses. The first step of the polysaccharide BRM in the modulation of cellular activity is the recognition of BRM and the binding of it to the specific immune cell receptors. Some evidence shows that there are pattern recognition receptors (PRRs) available for the molecular reception of polysaccharide BRMs (Lowe et al., 2001). The binding of ligands to PRRs may initiate Rel/NF-κB-mediated signaling events, which leads to the induction of gene expression and specific cellular functions of the innate immunity (Leung et al., 2006). It has been reported that some groups of PPRs can recognize the polysaccharide BRMs, for example, complement receptor 3 (CR3 or CD11b/CD18) (Ross et al., 1987), dectin-1 (Zimmerman et al., 1998; Taylor et al., 2002), and toll-like receptors (TLR-2 and TLR-4) (Shao et al., 2004). CR3 plays an important role as both membrane β-glucan receptor and adhesion molecule and occurs on cell membranes of monocytes, macrophages, neutrophils, NK cells, and DCs. CR3 may mediate a variety of cellular functions as it has the ability to bind various ligands such as intercellular adhesion molecule-1 (ICAM-1), iC3b, β-glucan, and others (Thornton et al., 1996; Lowe et al., 2001). β-Glucans can thus mediate cytotoxic and phagocytic responses through CR3 by binding them to these sites of the immune cells (Moradali et al., 2007). Mueller et al. (2000) suggests that the triple-helical conformation, molecular weight, and charge of the glucan polymer may be important determinants in CR3–ligand interaction. Dectin-1 is the β-glucan receptor and is mainly expressed on the surface of the monocytes, macrophages, neutrophils, and DCs (Taylor et al., 2002). The biological response transduced by dectin-1 depends on the TLR pathway, for example, dectin-1 is required to cooperate with TLR-2 for the activation of NF-κB and induction of TNF-α (Moradali et al., 2007). A proteoglycan isolated from *P. linteus* (PL) is shown to induce the phenotypic and functional maturation of DCs through the TLR-2- and/or TLR-4- mediated NF-κB signal pathways (Kim et al., 2004b). TLR-4 is also involved in *G. lucidum* polysaccharide (GLPS)–mediated macrophage activation (Shao et al., 2004). It is noteworthy that a thorough understanding of the recognition of mushroom polysaccharides (particularly β-glucan) by certain receptors on the immune cells and activation of signal transduction pathways appears to be a major challenge of future immunomodulation research.

5.3.2.1 Effects of Mushroom Polysaccharides on Macrophages and Spleen Cells Mushroom polysaccharides can enhance the natural immune system through the activation of monocytes/macrophages, splenocytes (including lymphocytes), NK cells, and DCs. The effects of mushroom polysaccharides on macrophages have been extensively studied in vitro and in vivo. Macrophages and spleen cells can be induced by mushroom polysaccharides to release several cytokines, such as IL-1, IL-1β, IL-6, IL-8, IL-10, IL-12, IL-18, TNF-α, and GM-CSF. Some of these cytokines are able to directly promote cytotoxicity of macrophages. The production of cytokines from the immune cells can be considered as a key event in the initiation and regulation of an immune response (Lull et al., 2005).

Ganoderma lucidum polysaccharide (GLP) significantly suppresses in vivo the growth of sarcoma 180 solid tumor, exhibiting antitumor activity, although GLP shows no direct antiproliferative effect as evaluated in vitro using several cancer cell lines, such as breast cancer MCF-7, lung cancer SPC-A, and hepatoma SMMC-7721 cells. A study on the immunomodulatory action of GLP as elucidated through analyzing the induced expression profile of cytokines in the treated mice using primers of specific cytokines, total RNA, and reverse-transcription polymerase chain reaction (RT-PCR) in the male inbred BALB/c mice showed that 7 out of 17 cytokine mRNAs were detected in the splenocytes and peritoneal exudate cells (macrophages) from the control and treated mice (Ooi et al., 2002). Among the seven detectable cytokine genes in the splenocytes, GLP induced a marked increase in the expression levels of $\mathrm{IL}-1\alpha$ (2-fold), $\mathrm{IL}-1\beta$ (3-fold), TNF-α (2-fold), IL-12 p35 (up to 6-fold), and IL-12 p40. In the macrophages, GLP promoted a remarkable increase in the expression levels of $\mathrm{IL}-1\beta$ (2.5- to 3-fold), TNF-α (up to 6-fold), and M-CSF (up to 2-fold). These results indicate that antitumor GLP can induce a cascade of immunomodulatory cytokines, but the potentiation of their gene expression and interaction seems quite complicated. The potency of TNF-α induction in macrophage is up-regulated after the challenge of GLPO (MW$<$12,000) more than that of GLPI (MW$>$12,000), suggesting that molecular size might be one of the factors in determining the structure–function relationship of this polysaccharide (Ooi et al., 2002). Lee et al. (2003) reported that polysaccharide purified from the mycelium of *G. lucidum* GLP(AI) is the major component to show the in vivo antitumor effect on fibrosarcoma growth in C3H mice. It could stimulate blood mononuclear cells to secrete cytokines, such as TNF-α and IFN-γ, which might induce apoptosis and differentiation in the treated leukemic U937 and HL-60 cells in vitro. The polysaccharide from *G. lucidum* (PS-G) is also reported to enhance phagocytic activity of human primary neutrophils and neutrophilic phenotype cells differentiated from all-trans retinoic acidy–treated HL-60 cells. Moreover, chemotactic action of PS-G requires the activities of phosphatidylinositol 3-kinase (PI3K), p38 MAPK, Src tyrosine kinases, and protein kinase C (PKC), demonstrating the abilities of PS-G to enhance neutrophil function in phagocytosis and chemotaxis (Hsu et al., 2003). It was reported that after the treatment of macrophages with a polysaccharide from fruiting bodies of *G. lucidum*, the levels of IL-1β, TNF-α, and IL-6 were, respectively, 5.1-, 9.8-, and 29-fold higher than in cultures of untreated cells, and the release of INF-γ from T lymphocytes was also greatly enhanced in the presence of this polysaccharide (Wang et al., 1997). The data suggest that the immunomodulating effects of *G. lucidum* polysaccharides include the activation of macrophages, splenocytes, NK cells, and DCs as well as the production of cytokines promoting antitumor activity (Lin, 2005).

SCG is a major 6-branched $(1 \rightarrow 3)$-β-D-glucan isolated from *S. crispa* Fr. with strong antitumor activities. The splenocytes from the naive DBA/1 and DBA/2 mice strongly react with SCG to produce IFN-γ. The addition of GM-CSF to spleen cell cultures in the presence of SCG synergistically enhances the production of IFN-γ, TNF-α, and IL-12p70, which would be significantly inhibited by neutralizing GM-CSF with anti-GM-CSF monoclonal antibody (mAb). It is concluded that GM-CSF plays a key role in regulating cytokine induction by $(1 \rightarrow 3)$-β-D-glucan SCG in DBA/2 mice in vitro (Harada et al., 2004). A bioactive $(1 \rightarrow 3, 1 \rightarrow 4)$-$\beta$-D-glucan from *Collybia dryophila* polysaccharide (CDP) strongly inhibits NO production in activated macrophages in a dose-dependent manner without affecting cell viability (Pacheco-Sanchez et al., 2006). The inhibition of NO by CDP is consistent with the decreases in both inducible nitric oxide synthase (iNOS) protein and mRNA expression, suggesting that CDP exerts its effect by inhibiting iNOS gene expression. CDP also significantly increases prostaglandin E [PGE(2)] production in LPS- and IFN-γ-induced macrophages when compared to the control (Pacheco-Sanchez et al., 2007). Three polysaccharides, namely, $(1 \rightarrow 3)$-β-D-glucan, $(1 \rightarrow 6)$-β-D-glucan, and $(1 \rightarrow 3, 1 \rightarrow 6)$-$\beta$-D-glucan or a mixture of both polysaccharides, purified from the newly cultivated mushroom *Lyophyllum decastes* Sing. show marked antitumor activities against sarcoma 180 and increased number of peritoneal macrophages with the complement (C3)–positive fluorescent cells in mice treated with $(1 \rightarrow 3)$-β-D-glucan (Ukawa et al., 2000).

Similarly, a purified polysaccharide fraction extracted from the mycelial culture and fruiting bodies of *A. blazei* Murill (ABM) could induce bigger increases in NO secretion than the others, mainly due to an increase in cytokine mRNAs or NO synthase mRNA (Sorimachi et al., 2001). ABM could also promote TNF-α and IL-8 secretion by macrophages derived from rat bone marrow. Thus ABM contains certain components which activate macrophages contributing to the immune response in vitro. Different doses of the polysaccharides of ABM synergically enhance the antitumor effect of cyclophosphamide (CP) in S-180-treated mice, and its mechanism is associated with the induced apoptosis of tumor cells and the increased expressions of IL-2 by macrophages and suppressive gene P27 (Liu et al., 2006).

Lentinan can restore and augment responsiveness of host cells but has no direct cytotoxicity against tumor cells. Interestingly, the antitumor activity of lentinan is also inhibited by pretreatment with antimacrophage agents. Thus, the various effects of lentinan are thought to be due to potentiation of the response of precursor Tcells and macrophages to cytokines produced by certain classes of lymphocytes after specific recognition of tumor cells. In addition, the induction of a marked increase in the amounts of cerebrospinal fluid (CSF), IL-1, and IL-3 by lentinan results in maturation, differentiation, and proliferation of the immunocompetent cells for host defense mechanisms (Chihara et al., 1989). Lentinan is able to restore the suppressed activity of helper T cells in the tumor-bearing host to their normal state, leading to the complete restoration of humoral immune responses (Maeda et al., 1988). The oral administration of lentinan to AKR mice exerts strong antitumor activity, resulting in the raised level of lymphocyte secretion of cytokines such as IFN-γ, TNF-α, IL-2, and IL-1α (Yap and Ng, 2003). Lentinan has been clinically applied as an antitumor and antimetastatic drug and has been reported to prevent both chemical and viral carcinogenesis. It is known that lentinan affects the tumorous vascular system resulting in the induction of hemorrhagic necrosis which is dependent on T cells in the tumor, (Mitamura et al., 2000). The polysaccharide L-II consisting of D-glucopyranose purified from the fruiting body of *L. edodes* demonstrates the antitumor activity in sarcoma 180–bearing mice mediated by the induction of T cells and macrophage-dependent immune system responses. It causes also a significant increase in TNF-α and IFN-γ but not in IL-2. It can also raise NO production and catalase activity in macrophages (Ruan et al., 2005).

Granulocytes/macrophages seem to be the major target cell type responsive to PG101 (a water-soluble extract from *Lentinus lepideus*) in irradiated mice. PG101 interacts with macrophages or related cells resulting in the activation of the transcription factor NF-κB, which sets off a series of reactions producing a variety of proinflammatory and anti-inflammatory cytokines (TNF-α, IL-1β, IL-10, IL-12, GM-CSF, IL-18) in a sequential manner. Despite its significant biological effect on various cytokines, PG101 remains nontoxic in both rats and human peripheral blood mononuclear cells (hPBMCs) even at a biological concentration approximately 20 times greater (Jin et al., 2003b).

Schizophyllan is similar to lentinan in the composition and biological activity, and its mechanism of antitumor action appears to be quite similar (Jong et al., 1991). The antitumor effect of schizophyllan is diminished in mice neonatally thymectomized and treated with antithymus globulin. Schizophyllan restores and enhances cellular immunity in the tumor-bearing host by functioning as a T-cell adjuvant and macrophage activator (Okazaki et al., 1995). Mechanistic study of the antitumor activity of schizophyllan suggests that macrophages may incorporate β-glucans through certain $(1 \rightarrow 3)$-β-D-glucan-specific mechanisms and/or other endocytosis pathways and that the glucan-mediated immunopharmacological activities are dependent on the helical conformation (Okazaki et al., 1995).

Grifolan, a $(1 \rightarrow 3)$-β-glucan isolated from *G. frondosa*, induces the release of IL-1, IL-6, and TNF-α from macrophages but appears to require a high-molecular-mass soluble form of grifolan for TNF-α production (Adachi et al., 1994; Okazaki et al., 1995; Ishibashi et al., 2001). D-fraction, another $(1 \rightarrow 3)$-β-glucan purified from *G. frondosa*, could induce a Th-2 dominant response via the activation of macrophages, resulting in the enhancement of humoral immunity rather than cell-mediated immunity. In addition, it could promote an increase in the percentage ratio of CD69 and CD89 expression on major histocompatibility complex II+ cells, which revealed activation of APCs 4 hours after D-fraction administration (Kodama et al., 2004). D-fraction is suggested as a novel inducer for iNOS which contributes at least in part to the antitumor activity of D-fraction through iNOS-mediated NO production in RAW264.7 macrophages (Sanzen et al., 2001). These results suggest that D-fraction can decrease the effective dosage in tumor-bearing mice by increasing the proliferation, differentiation, and activation of immunocompetent cells and thus provide a potential clinical benefit for patients with cancer (Kodama et al., 2005a).

A galactomannan isolated from a polar extract of *Morchella esculenta*, a highly prized mushroom in the world, exhibits immunostimulatory activity by enhancing NF-κB-directed luciferase expression in THP-1 human monocytic cells (macrophages) (Duncan et al., 2002). Similarly, a fucogalactan isolated from *Sarcodon aspratus* elicits the immunomodulating activity by the release of TNF-α and NO in macrophages of mice in vitro (Mizuno et al., 2000). The endopolysaccharide extracted from the mycelium of *Inonotus obliquus* is also an immunostimulating agent which can specifically activate B cells and macrophages. It enhances nitrite production and expression of IL-1β, IL-6, TNF-α, and iNOS in macrophages (Kim et al., 2005, 2006). An immunomodulating polysaccharide isolated from *P. florida* fruiting bodies exhibits significant macrophage activity through the release of No (Rout et al., 2004). The tumoricidal activity of peritoneal macrophages (PMs) cultured with acidic polysaccharide (PL) isolated from *Phellinus linteus* against B16 melanoma cells was enhanced in a dose-dependent manner (Kim et al., 2003). PL exerts cytotoxicity against Yac-1 cells through the up-regulation of NO and TNF-α production and enhances the expression of costimulatory molecules, CD80 and CD86, and major histocompatibility complex (MHC) molecules II in PM. Such properties of PL may be related to its ability to induce the production of the tumoricidal effector molecule NO mediated via the activity of protein tyrosine kinase (PTK) and protein kinase C (PKC) (Kim et al., 2003, 2004a) and through the enhancement of IFN-γ secretion by T lymphocytes (Oh et al., 2006). A novel polysaccharide–protein complex (PPC) extracted from *P. linteus* is able to increase the production of cytokines and NO from macrophages and enhance the lytic death of NO-sensitive tumor cells, B16 melanoma. PPC is a potent immunomodulator which can stimulate the tumoricidal activities of macrophages and NK cells and induce the proliferation of B cells in vitro (Kim et al., 2006).

PSK (a protein bound-polysaccharide K) prepared from *C. (T.) versicolor* in Japan has the ability to restore immune potential to the normal level after the host has been depressed by tumor burden or anticancer chemotherapeutic agents (Tsukagoshi et al., 1984; Kobayashi et al., 1993). The oral administration of PSK can improve the impaired antitumor CD4$^+$ T-cell response in gut-associated lymphoid tissue of specific-pathogen-free mice (Harada et al., 1997). PSK enhances the cytotoxic activity of peripheral blood lymphocytes (PBLs) in vivo and in vitro.

It may accelerate interaction of PBL with tumor cells such as T24 human urinary bladder tumors when both effector cells and target cells are exposed to PSK simultaneously. It is reported that PSK induces gene expression of some cytokines such as TNF-α, IL-1, IL-8, and IL-6 in vivo or in vitro (Kato et al., 1995; Liu et al., 1996a). These cytokines produced by monocytes, macrophage, and various other immune cell types mediate multiple biological effects by direct stimulation of cytotoxic T cells against tumors, enhancement of antibody production by B lymphocytes, and induction of IL-2 receptor expression on T lymphocytes. The induction of TNF-α by PSK would contribute, in part, to potent tumoricidal effects of this agent since the administration of neutralizing antibody against TNF-α significantly attenuates the antitumor activity of PSK in the murine model (Kato et al., 1995). Similarly, PSP prepared from *C. versicolor* in China may activate peritoneal macrophages to produce TNF-α and other cytokines. It is also able to suppress the growth of various human cancer cell lines and reverse tumor-induced immunodeficiencies in mice by increasing immunoglobulin (Ig)G and C3 complement levels (Yang et al., 1992; Liu et al., 1993; Cui and Chisti, 2003).

The antitumor PSPC purified and characterized from the culture filtrates of *T. lobayense* exhibits strong antitumor activity in both ICR and BALB/c mice (Liu et al., 1996b). PSPC has the ability to restore the phagocytic function of the peritoneal exudate cells (PEC) and the mitogenic activity of T cells of tumor-bearing mice. PSPC also exhibits indirect cytotoxic activity against P815 mastocytoma cells and L929 mouse fibroblast cells by activating PEC to release reactive nitrogen intermediates (RNIs) and TNF-α, which are shown to increase significantly after PSPC treatment in the tumor-bearing mice (Liu et al., 1996b). Similarly, the antitumor activity of ATOM, a polysaccharide–protein complex prepared from the cultured mycelia of *A. blazei*, is highly effective against four kinds of established tumors, that is, subcutaneously implanted sarcoma 180 in mice, Ehrlich ascites carcinoma, Shionogi carcinoma 42, and Meth A fibrosarcoma (Itoh et al., 1994; Ito et al., 1997). A heteroglycan–protein complex from *G. frondosa* is shown to depress tumor growth by activating the immune system as a biological response modifier (Cun et al., 1994).

5.3.2.2 Effects of Mushroom Polysaccharides on NK Cells Natural killer cells are a form of cytotoxic lymphocytes which constitutes a major component of the innate immune system. NK cells can be activated in response to interferons or macrophage-derived cytokines. Their activity is tightly regulated. NK cells can recognize the surface changes that occur on a variety of tumor cells and virally infected cells (Miller, 2002). They express a cascade of activating cell surface receptors that can trigger cytolytic programs and produce cytokines such as IFN-γ, TNF-α, and GM-CSF, which modulate natural and specific immune responses (Yokoyama et al., 2004; Vivier et al., 2004).

Maitake D-fraction, a $(1 \rightarrow 3)$-β-glucan extrated from *G. frondosa*, exerts its antitumor effect in tumor-bearing mice by enhancing the immune system through the activation of macrophages, T cells, and NK cells. Kodama et al. (Kodama et al. (2002) monitored the level of NK cell cytotoxic activity in cancer patients receiving

D-fraction and found that the elevated levels of cytotoxic activity were maintained for one year. D-fraction is capable of enhancing and maintaining peripheral blood NK cell activity in patients with lung and breast cancer. It stimulates the natural immunity related to the activation of NK cells indirectly through IL-12 produced by macrophages and DCs in normal mice (Kodama et al., 2003). Thus, NK cells are responsible not only for the early effects of D-fraction on tumor growth but also for the long-term tumor-suppressive effects of D-fraction through the increased release of IL-12 by macrophages. It appears to repress cancer progression primarily through the stimulation of NK activity (Kodama et al., 2005a). Thus immunomodulation effected by the binding of a $(1 \rightarrow 3)$-β-glucan molecule or particle probably includes activation of cytotoxic macrophages, helper T cells, and NK cells and promotion of T cell differentiation.

AC-PS, a unique polysaccharide component purified from the mycelium of *Antrodia camphorata*, has pronounced antitumor effects on both in vitro and in vivo model. AC-PS alone does not show any direct cytotoxic effect to human leukemic U937 cells, even at high concentrations (200 mg/mL). However, AC-PS could inhibit the proliferation of U937 cells via the activation of mononuclear cells and elicits its antitumor effect by promoting a Th1-dominant state and NK cell activity (Liu et al., 2004). Ganopoly, the polysaccharide fraction of *G. lucidum*, is reported to enhance the host immune functions (e.g., enhanced NK cell activity) in patients with advanced lung cancer. Administration of Ganopoly for 12 weeks results in a significant increase in the mitogenic reactivity of lymphocytes to concanavalin A, CD3 percentage, and NK cell activity; a marginal increase in the CD4 percentage and CD4/CD8 ratio; but a marginal reduction of CD8 (Gao et al., 2003). A well-characterized glycoprotein fraction containing fucose residues in the extract of *G. lucidum* polysaccharide (EORP) is reported to enhance CD14 endocytosis of LPS and promote TLR4 signal transduction of cytokine expression. EORP increases the surface expression of CD14 and TLR4 within murine macrophages J774A.1 cells in vitro (Hua et al., 2007).

FWE, the water extract containing mainly polysaccharides from five medicinal mushrooms, *C. versicolor, Cordyceps sinensis, L. edodes, A. blazei*, and *G. lucidum* in equal amounts, has the activity to enhance phagocytosis of peritoneal macrophages and NK activity in mice and suppress the growth of B-16 melanoma. FWE is able to activate NK cells to directly kill tumor cells and to secrete the cytotoxic agents to elicit the apoptotic pathway of tumor cells or other signaling pathways (W. Y. Zhang et al., 2004).

5.3.2.3 Effects of Mushroom Polysaccharides on DCs DCs are antigen-presenting cells (APCs) with a unique ability to induce primary immune responses. DCs not only induce the activation and polarization of T cells but also directly activate naive and memory B cells (Banchereau et al., 2000). Mushroom polysaccharides as BRMs may promote the cytotoxic activity of various effector cells by the induced production of multiple cytokines and suppression of immunosuppressive factors. DCs at different stages of differentiation can regulate effector cells of the innate immunity, such as NK cells and cytotoxic T cells, and initiate the induction of tumor immunity (Ogihara et al., 2004; Lull et al., 2005).

PS-G, the polysaccharide component with a branched (16)-β-D-glucan moiety of *G. lucidum*, exerts antitumor activity and can effectively promote the activation and maturation of immature human monocyte-derived DCs. The treatment of DCs with PS-G results in the enhanced expression of membrane molecules such as CD80, CD86, CD83, CD40, CD54, and IL-12p70, p40, and IL-10 and also IL-12p35, p40, and IL-10 mRNA. PS-G promotes the inhibitor of kappa B (I-κB) kinase and NF-κB activity and also IκBa and p38 mitogen-activated protein kinase (MAPK) phosphorylation. The data demonstrate that PS-G can effectively promote the activation and maturation of immature DCs, suggesting it has a potential in regulating immune responses (Lin et al., 2005). In examining the effects of PS-G on human monocyte-derived DCs with microarray analysis by Human Genome U133 Plus 2.0 GeneChip, the results also reveal that PS-G induces changes of gene expression of various immunomodulatory and proinflammatory cytokines in human DCs and promotes the immune response of Th-1 in BALB/c mice (Lin et al., 2006). The data from microarray analysis could be correlated with the in vivo effect of the immune-enhancing compound. In another study, Cao and Lin (2002) reported that Gl-PS, a fraction of *G. lucidum* polysaccharides, could promote not only the maturation of cultured murine bone marrow–derived DCs but also the initiation of immune response induced by DCs. GL-M, the extract of *G. lucidum* (GL) mycelia, could activate the proliferation of PBMCs and monocytes and stimulate Th-1 and Th-2 cytokine mRNA expression. GL-M enhances the maturation of DCs in terms of up-regulation of CD40, CD80, and CD86, and reduces endocytosis (Chan et al., 2005).

PG, a proteoglycan isolated from *Phellinus linteus*, strongly inhibits the MCA-102 tumor growth and proliferation in vivo through a mechanism leading to a Th-1-dominant immune state and the activation of CD11c^{+} CD8^{+} DCs. It induces the phenotypic and functional maturation of bone marrow–derived DCs in vitro via TLR-2- and TLR-4-mediated NF-κB, ERK, and p38 MAPK signal pathways. PG enhances the production of IL-12 and IFN-γ and surface molecules, including CD80 and CD86 in MCA-102 tumor-bearing mice, as well as the proliferation of CD4(+) and CD8(+) T cells (Kim et al., 2004c). Furthermore, the acidic PL isolated from *P. linteus* can also promote the maturation of DCs. PL significantly increases the membrane surface molecules, including MHC classes I and II, CD80, and CD86, and IL-12 in DCs. PL markedly reduces the endocytic activity of DCs and augments their capacity to promote the proliferation of naïve allogeneic T cells (Park et al., 2003).

PSK (a protein-bound polysaccharide) from the cultured mycelium of *C.* (*T.*) *versicolor* is reported to improve the immunosuppressed state and might be associated with DC maturation directly. PSK promotes both the phenotypic and functional maturation of DCs derived from human CD14^{+} mononuclear cells (Kanazawa et al., 2004). It can overcome the defective maturation of DCs exposed to tumor-derived factors in vitro (Okuzawa et al., 2002). In a clinical trial involving 6 normal adults and 14 patients with gastric or colorectal cancers, the Th1/Th2 and DC1/DC2 balance becomes Th2 and DC2 dominant in the cancer-bearing state. PSK therapy results in a shift of the Th1/Th2 and DC1/DC2 balance toward Th1 and DC1 dominance (Kanazawa et al., 2005). It might be possible to combine DC vaccination therapy with oral PSK to promote the induction of T cell and DC differentiation in cancer patients (Kanazawa et al., 2005).

5.3.2.4 Effects of Mushroom Polysaccharides on Hematopoietic Stem Cells MBG (a β-glucan isolated from the fruiting body of *G. frondosa*) can enhance the blood-forming process (hematopoiesis) of mouse bone marrow cells (BMCs) in vitro and protects BMCs from doxorubicin (DOX) toxicity on fresh human umbilical cord blood (CB) cells. MBG treatment significantly raises the response of the colony-formation unit of granulocytes-macrophages (CFU-GM) over the whole dose range of 12.5–100 mg/mL ($P < 0.05$). The addition of MBG to DOX-treated CB cells significantly protectes granulocyte-macrophage colony formation from the toxicity of DOX, which otherwise produces strong hematopoietic repression. The data show that MBG induces granulocyte colony-stimulating factor (G-CSF) production in CB CD33$^+$ monocytes, but adult peripheral blood monocytes did not produce a significant G-CSF response to MBG. They have the potential to reduce hematopoietic suppression induced by chemotherapy (Lin et al., 2007). Recently, Lin et al. reported that maitake MD-fraction (obtained by further purification of D-fraction from the fruiting body of *G. frondosa*) causes direct enhancement of the response of CFU-GM of BMC progenitors and enhances the recovery of the CFU-GM response after DOX-induced hematopoietic suppression (H. Lin et al., 2004). These studies demonstrate that MBG or MD-fraction may induce the proliferation of hematopoietic stem cells and differentiation of CFU-GM in umbilical CB cells and act directly to induce the production of G-CSF.

PG101, a water-soluble extract containing protein-bound polysaccharides isolated from cultured mycelia of *L. lepideus*, is a potential biological response modifier that activates selective cytokines in vitro, mainly by controlling cellular transcription factor NF-κB (Jin et al., 2003b). It was reported that in irradiated mice given PG101 by gavage daily for 24 days the number of colony-forming cells, including CFU-GM) and erythroid burst-forming units (BFU-E), increased to almost the levels seen in the nonirradiated control as early as 8 days after irradiation. PG101 increases the number of granulocytes (ER-MP12$^-$20$^+$) and myeloid progenitors (ER-MP12$^+$20$^+$) in the bone marrow cell population and raises the serum levels of GM-CSF, IL-6, and IL-1β. The level of TNF-α, elevated by irradiation in the control mice, maintains at the background level in the PG101-treated mice. The results suggest that PG101 might induce differentiation of progenitor cells to granulocytes and/or proliferation of the committed cells and might effectively suppress TNF-α-related pathological conditions (Jin et al., 2003a).

A branched β-glucan isolated from fruiting bodies of *S. crispa* (SCG) is able to promote the hematopoietic response in cyclophosphamide (CY)–induced leukopenic mice by prior or postadministration. Monocytes and granulocytes in the peritoneal cavity, liver, spleen, and bone marrow recover faster than in the control group. The ratio of NK cells and T cells in the liver, spleen, and peritoneal cavity is also increased. The cotreatment of CY+SCG in the culture of peritoneal exudate cells (macrophages), spleen cells, and bone marrow cells induces the production of high amounts of IL-6 than that of the CY-treated group. IL-6 might be a key cytokine for the enhanced hematopoietic response by SCG (Harada et al., 2002).

5.3.3 Antimetastasis

Many polysaccharides or polysaccharide–protein complexes from mushrooms also exhibit antimetastatic effects. Lentinan, a purified $(1 \rightarrow 3)$-β-D-glucan from *L. edodes*, not only markedly prevents chemical and viral carcinogenesis (Suga et al., 1984), but also suppresses cancer metastasis and recurrence in animal models (Suga et al., 1989). The administration of lentinan prominently inhibits colony formation of metastasis in the lung after surgical resection of the tumors implanted subcutaneously into the mouse footpad, with the inhibition ratios of 95% in DBA/2.MC.CS-T sarcoma and 83% in Lewis lung carcinoma. The results suggest that lentinan may be effective in the prevention of tumor recurrence and/or metastasis (Suga et al., 1989).

PSK (a protein-bound polysaccharide from *C. versicolor*) has also been shown, that once the progression of carcinogenesis is initiated, to exhibit significant preventive action on cancer metastasis such as the suppression of pulmonary metastasis of methylcholanthrene-induced sarcomas, human prostate cancer DU145M, and lymphatic metastasis of mouse leukemia P388 in the spontaneous metastatic models. PSK also inhibits the metastasis of rat hepatoma AH60C, mouse colon cancer 26, and mouse leukemia RL male 1 in artificial metastatic models (Kobayashi et al., 1995). PSK could prevent distant metastases and improve the survival rates by 10–20% in colorectal cancer (Yoshikawa et al., 2004). Kobayashi et al (1995) suggested that PSK is able to influence the following steps of cancer metastasis: (a) through the inhibition of tumor invasion, adhesion and production of cell matrix–degrading enzymes; (b) by suppression of tumor cell attachment to endothelial cells; (c) by suppression of tumor cell motility and thus cell migration after extravasation; (d) through the inhibition of tumor angiogenesis and growth; (e) through the modulation of cytokine production and the augmentation of effector cell functions; and (f) by suppression of malignant progression of tumor cells through superoxide trapping. PSK may thus suppress cancer metastasis at any one step, and its primary mechanism of action can be ascribed to direct action on the tumor cell as well as to immunomodulation (Kobayashi et al., 1995).

In a study on the effects of natural polysaccharides isolated from *Phellinus gilvus* (PG) on human gastric cancer, Bae et al. (2006) showed that PG reduces cell proliferation and increased cell apoptosis in a dose-dependent manner in vitro and markedly suppresses tumor growth (peritoneal carcinomatosis) in a nude mouse model. The data show that PG significantly inhibits tumor growth and metastasis in an orthotopic model of human gastric adenocarcinoma without detectable adverse effects (Bae et al., 2006).

An exopolysaccharide fraction (EPSF) prepared from cultivated *C sinensis* (Chinese caterpillar fungus) has immunomodulatory and antitumor activity. EPSF enhances phagocytosis capacity of peritoneal macrophages and proliferation of spleen lymphocytes in the treated B16-bearing mice. EPSF significantly inhibits the metastasis of B16 melanoma cells to the lung and liver (Zhang et al., 2005).

M. Zhang et al. (2004) reported that FWE (the water extract composed mainly of polysaccharides from five medicinal mushrooms, i.e., *C. versicolor, C. sinensis, L. edodes, A. blazei*, and *G. lucidum*) significantly promoted the NK cell activity of treated mice and depressed the levels of bcl-2 and P53 protein in the liver and lung cells. The minimum inhibition rates (MIRs) of lung metastasis of B16 melanoma cells by low dose and high dose of FWE were 15.5% and 72.7%, respectively. These results indicate that FWE not only promotes the mouse host-mediated immunity but also inhibits tumor metastasis (M. Zhang et al., 2004).

5.3.4 Antiangiogenesis

Angiogenesis is the process of new blood vessel formation from the preexisting microvascular networks (Folkman, 1995). It is crucial to the progressive growth and metastasis of solid tumors, and thus antiangiogenesis is considered an important strategy for therapeutic intervention of tumor proliferation (Chen et al., 1995). The formation of new blood vessels involves the angiogenic switch in the balance of angiogenic and antiangiogenic factors and the interactions of tumor cells, endothelial cells, and extracellular matrix, leading to endothelial proliferation, migration, and tube formation (Hanahan and Folkman, 1996). Cancer cells can produce several angiogenic factors such as vascular endothelial growth factor (VEGF) and few inflammatory cytokines.

PSK, a protein-bound polysaccharide immunomodulating agent from *C. versicolor*, was reported to have an antiangiogenic effect (Wada et al., 2002). PSK inhibits the proliferation of human umbilical vein endothelial cells (HUVECs) in the presence of basic fibroblast growth factor (bFGF) more effectively. At a high concentration PSK suppresses tube formation of HUVECs and their adhesion to extracellular matrix proteins. Administration of PSK could reduce the bFGF-induced angiogenesis in an in vivo rat cornea assay. These data suggest that PSK binds to bFGF and interferes with its signal transduction to inhibit the proliferation of HUVECs, resulting in the suppression of angiogenesis (Wada et al., 2002). Quantitative analysis of microcorrosion casting of the tumor tissue in the study using the S180 tumor-bearing mouse model shows that the polysaccharopeptide, PSP, isolated from the edible mushroom *C. versicolor* has antiangiogenesis properties. The expression of VEGF in these tumors is suppressed (Ho et al., 2004). Moreover, the EPSF of a cultivated *C. sinensis* (Cs) inhibits tumor growth in the mice (C57BL/6). The levels of expression of c-Myc, c-Fos, and VEGF in the lungs and livers of the EPSF-treated mice are significantly lower than those of the untreated mice (Yang et al., 2005).

Polysaccharide peptide (Gl-PP) isolated from *G. lucidum* has antitumor effects in mice and potential antiangiogenesis. Gl-PP treatment of HUVECs could induce cell apoptosis directly and lead to a reduction of Bcl-2 antiapoptotic protein expression and an increase of Bax proapoptotic protein expression of HUVECs. It is thus suggested that Gl-PP may directly inhibit vascular endothelial cell proliferation or indirectly decrease VEGF expression of tumor cells (Cao and Lin, 2006).

In an investigation on antiangiogenic activities of several medicinal fungi, including *Antrodia cinnamomea, Antrodia malicola, Antrodia xantha, Antrodiella liebmannii, Agaricus murrill*, and *Rigidoporus ulmarius*, on VEGF-induced tube formation in endothelial cells (ECs), Chen et al. (2005) reported that polysaccharides isolated from *A. xantha* and *R. ulmarius* produce greater inhibition of endothelial tube formation compared to those from other fungi and polysaccharides isolated from *A. xantha* and *R. ulmarius* provide greater antiangiogenesis than those from commercialized *A. murrill* (Brazilian mushroom) and *A. cinnamomea*. Polysaccharides isolated from *A. cinnamomea* inhibit cyclin D1 expression through inhibition of VEGF receptor signaling, leading to the suppression of angiogenesis. *Antrodia cinnamomea* polysaccharides also block VEGF-induced migration and capillary-like tube formation of ECs on Matrigel (Cheng et al., 2005).

5.4 STRUCTURE AND ANTITUMOR ACTIVITY RELATIONSHIP OF POLYSACCHARIDES

Polysaccharides having antitumor action differ greatly in their chemical composition, branching configuration, helical conformation, and other physical properties. Antitumor activity is exhibited in a wide range of glycans extending from homopolymers to highly complex heteropolymers. Although it is difficult to correlate the structure and antitumor activity of complex polysaccharides, some possible relationships can be inferred. It has been reported that most of the antitumor polysaccharides such as lentinan and schizophyllan show the same basic β-glucan structure with different types of glycosidic linkages. Therefore it is obvious that some structural features such as $(1 \rightarrow 3)$-β linkages in the main chain of the glucan and further β-1,6 branch points are needed for antitumor action (Franz, 1989; Bohn and BeMiller, 1995). It has been suggested that the β-glucans containing mainly $(1 \rightarrow 6)$ linkages have less activities, and polysaccharides with high molecular weight appear to be more effective than those with low molecular weight (Jong et al., 1991; Sakagami et al., 1991; Surenjav et al., 2006). However, obvious variations of antitumor polysaccharides are also noted. There are antitumor polysaccharides with other chemical structures, such as hetero-β-glucan (Mizuno et al., 1995a), heteroglycan (Zhuang and Mizuno, 1995), β-glucan–protein (Kawagishi et al., 1990a), α-manno-β-glucan (Mizuno et al., 1995a), α-glucan–protein (Mizuno et al., 1995a), and heteroglycan–protein complexes (Zhuang et al., 1993; Liu et al., 1996b; Mizuno et al., 1996). For example, PSK and PSP are a β-glucan–protein whereas PSPC from *Tricholoma* species are a heteroglycan–protein complex (Liu et al., 1996b; Mizuno et al., 1996). Therefore, it is difficult to identify which polysaccharide structure is essential for antitumor action. It has been shown that the molecular mass, the degree of branching, conformation, and chemical modification of antitumor polysaccharides significantly affect their antitumor and immunomodulatory activities (Bohn and BeMiller, 1995; Okazaki et al., 1995; Ohno et al., 1995; Surenjav et al., 2006).

5.4.1 Effect of Molecular Mass

A $(1 \rightarrow 3)$-β-glucan extracted from the cultured mycelium of *G. frondosa* with various molecular masses obtained by heat treatment for varying times at 150°C manifests changes of biological activities with different molecular masses (Adachi et al., 1990a). The highest molecular mass fraction (800 kDa) always exhibits the strongest antitumor and immunomodulatory activity (Adachi et al., 1990a; Kim et al., 1990). When PSK is separated into four fractions with different molecular masses (F1: <50 kDa; F2: 50–100 kDa; F3: 100–200 kDa; F4: >200 kDa) by successive filtration, the highest molecular mass fraction (>200 kDa) has the most potent immunomodulatory activity (Kim et al., 1990; Sakagami et al., 1990). The antitumor activity of the native triple helical $(1 \rightarrow 3)$-β-D-glucan with high molecular weight and bound protein isolated from *L. edodes* is higher than that of the modified $(1 \rightarrow 3)$-β-D-glucan having only a single flexible chain and low molecular weight (Surenjav et al., 2006). These investigations suggest that a high molecular mass is required for extensive enhancement of immunological and antitumor activities. However, the study by Peng et al. (2005) indicates that the water-soluble polysaccharide with relatively lower weight-averaged molecular mass from the mycelium of *G. tsugae*, which contains a moderate content of galactose and bound protein, is important in the improvement of antitumor activity of polysaccharides. The chemically modified $(1 \rightarrow 3)$-β-D-glucans, such as schizophyllan and lentinan, having a linear wormlike, triple-helical structure with weight-averaged molecular mass less than 50×10^4 g/mol or larger than 110×10^4 g/mol might stimulate the monocytes in vitro more efficiently to secrete TNF-α than the samples with molecular mass in the range $67 - 10^4 - 110 \times 10^4$ g/mol (Falch et al., 2000). However, lentinan and schizophyllan with low molecular weight have the same antitumor activity against sarcoma 180 as those with higher molecular weight (Sasaki et al., 1976; Tabata et al., 1981; Ogawa and Kaburagi, 1982). Therefore, the discrepancy about the effect of molecular mass of polysaccharides on antitumor activity and immunomodulation remains to be clarified.

5.4.2 Impact of Branching Configuration

In general, β-glucans are active antitumor agents if they have mainly linear structure, possessing not excessively long branches. Some data indicate that it is the distribution of the branch units along the backbone chain of $(1 \rightarrow 3)$-β-D-glucan that is responsible for the antitumor activity. For example, pachyman, a branched $(1 \rightarrow 3)$-β-D-glucan obtained from *P. cocos*, is inactive, whereas pachymaran, obtained by debranching pachyman using periodate oxidation and mild hydrolysis, exhibits pronounced antitumor activity (Chihara et al., 1970a). Lentinan (2/5) is a $(1 \rightarrow 3)$-β-D-glucan with two branches for every five D-glucopyranosyl residues (Chihara, 1992). Schizophyllan (1/3) is also a $(1 \rightarrow 3)$-β-D-glucan with one branch for every three D-glucopyranosyl residues (Tabata et al., 1981). The debranched lentinan preparations are more effective against sarcoma 180 than the native lentinan at a dose of 2.0 mg/kg for five days in mice (Sasaki et al., 1976). In addition, when the highly branched $(1 \rightarrow 3)$-β-D-glucan (2/3, two branches for every three D-glucopyranosyl residues), called OL-2, extracted from *Omphalia lapidescens* is modified to Smith-type degradation product with approximately one branch at every 24 main-chain glucosyl units at each C-6 position (number of all main-chain glucosyl units is on average), it still exhibits potent antitumor activity on the solid form of sarcoma 180 in mice and increases the life span of mice treated with the ascites form of sarcoma 180 and MH 134 hepatoma (Saito et al., 1992). The alkali-insoluble, branched $(1 \rightarrow 3)$-β-D-glucan (glucan II), which is a major constituent of the fruiting body of *Auricularia auricula-judae*, shows essentially no inhibitory activity against implanted sarcoma 180 solid tumor in mice. When glucan II, having numerous branches attached, is modified by controlled, periodate oxidation, borohydride reduction, and mild acid hydrolysis, the resulting water-soluble, degraded glucan, having covalently linked polyhydroxy groups attached at the O-6 of the $(1 \rightarrow 3)$-linked D-glucosyl residues, shows potent antitumor activity (Misaki et al., 1981). At the molecular level, it is found that the patterns of cytokine expression between schizophyllan (1/3) from *P. cocos* and OL-2 (2/3) from *O. lapidescens* are also different. The mice administered with OL-2 strongly express IL-1α, IL-1β, and IL-1ra genes in peritoneal exudate cells (macrophages), while schizophyllan only induces the expression of IL-1α. In the genes related to hematopoiesis, OL-2 induces the expression of G-CSF and GM-CSF, but schizophyllan induces the expression of IL-3 (Nemoto et al., 1993, 1994). Therefore, the relationship between the antitumor activity and the branch ratios of β-glucan appear to be quite complicated. The data indicate that the $(1 \rightarrow 3)$-β-D-glucan backbone is essential, and the most active polymers have degrees of branching (DB) between 0.20 and 0.33.

5.4.3 Relationship of Antitumor Activity and Conformation

Conformations of antitumor polysaccharides include single-helix, triple-helix, and random-coiled conformers. A triple-helix conformer is usually more stable than the single-helix conformer, since a certain part of the single-helix conformer is gradually changed to the triple one. Lentinan, schizophyllan, and glucan moieties of PSK all have triple-helix structure (Tsukagoshi et al., 1984; Chihara 1992; Ohno et al., 1995). Lentinan, a $(1 \rightarrow 3)$-β-D-glucan isolated from *L. edodes*, exists as triple-helical conformation in aqueous NaCl and as single flexible chains in dimethyl sulfoxide (DMSO) (Zhang et al., 2002). The antitumor activity of the native triple-helical $(1 \rightarrow 3)$-β-D-glucan with high molecular weight and bound protein isolated from *L. edodes* is higher than that of the modified $(1 \rightarrow 3)$-β-D-glucan having only a single flexible chain and low molecular weight

(Surenjav et al., 2006). Pachyman from *P. cocos* which is a (1 → 3)-β-D-glucan having a single-helix conformer is biologically inactive against tumor growth. However, when pachyman is treated with periodate oxidation and mild hydrolysis, the newly formed conformer, pachymaran or carboxyl-methyl-pachymaran, exhibits pronounced antitumor activity (Chihara et al., 1970a; Kanayama et al., 1986).

Schizophyllan-OH, which has a single-helix structure derived from the alkaline-treated schizophyllan, shows a reduced ability to inhibit tumor growth as compared to the native schizophyllan (Chihara, 1984). The cytokine stimulating activity of (1 → 3)-β-D-glucans is also found related to the triple-helix conformation (Falch et al., 2000). Therefore, the antitumor and immunopharmacological activities of polysaccharides are dependent on the helical conformation, and the conformation dependency may vary according to the assays and analytical methods used (Ohno et al., 1995).The relationship of conformation and antitumor activity of polysaccharides or polysaccharide–protein complexes suggests the existence of a biological system within the host body that recognizes the configurational structure of polysaccharides. The antitumor and immunomodulating properties of branched (1 → 3)-β-D-glucans may depend on their branching pattern, molecular weight, and conformation or higher order structure.

5.4.4 Improvement of Antitumor Activity by Chemical Modifications

To improve the biological activity of antitumor polysaccharides by chemical modification, carboxymethylated (CM), hydroxylated, formylmethylated, aminethylated, and sulfated products have been designed. The linear (1 → 3)-α-glucans from *Amanita muscaria* and *Agrocybe cylindracea* have little antitumor activity, but their CM derivatives show potent antitumor activity against sarcoma 180 in mice (Kiho et al., 1989, 1994; Yoshida et al., 1996). Debranched pachymaran and CM pachynmaran are more effective antitumor (1 → 3)-β-D-glucans, as compared with the natural glucan, pachyman (Chihara et al., 1970a; Kanayama et al., 1986). The administration of hydroxylated schizophyllan could induce the production of higher concentrations of NO and TNF-α by peritoneal exudate cells (macrophages) than that by the native schizophyllan in vivo (Ohno et al., 1995). The antitumor activity of the formylmethylated and aminoethylated derivatives of schizophyllan against sarcoma 180 solid tumor in mice is increased more effectively than that of the native schizophyllan (Usui et al., 1995). A chemically sulfated polysaccharide (S-GAP-P), derived from the water-insoluble glucan of *G. frondosa* mycelia, could inhibit in vitro the proliferation of SGC-7901 tumor cells and induce apoptosis in a dose-dependent manner. In vivo experiments demonstrate that S-GAP-P significantly inhibits the tumor growth and enhances the phagocytosis of peritoneal macrophages in S180-bearing mice (Nie et al., 2006). The sulfated (1 → 3)-α-D-glucan modified from the native glucan of *L. edodes* exhibits potent antiproliferation action on human MCF-7 breast carcinoma cells, whereas its native water-insoluble (1 → 3)-α-D-glucan has only moderate antitumor activity (Zhang and Cheung, 2002). The native water-soluble (1 → 3)-β-D-glucan from *L. edodes*, which contains bound protein and exists as triple-helical conformation with high molecular weight, shows prominent antitumor activity, whereas the modified (1 → 3)-β-D-glucan having only a single flexible chain has significantly decreased effect. The data suggest that the antitumor activity of the polysaccharides may be related to their conformation, molecular weight, and content of the bound protein (Surenjav et al., , 2006). Furthermore, all of the water-insoluble (1 → 3)-α-D-glucans from *L. edodes*, which were O-sulfonated to obtain derivatives, exhibit higher antitumor activity than those of the native glucan. The triple-helical conformation and the effect of O-sulfonation of the polysaccharides seem to play an important role in the improvement of their antitumor activity (Unursaikhan et al., 2006). The antitumor activity of the sulfated derivatives of water-insoluble (1 → 3)-α-D-glucans isolated from the mycelium of *P. cocos* against sarcoma 180 tumor both in vitro and in vivo is significantly higher than that of the native α-D-glucan (Y. L. Lin et al., 2004). It is also worth mentioning that the native (1 → 3)-β-D-glucan isolated from the fresh sclerotium of *P. cocos* has no antitumor activity, whereas the sulfated and carboxymethylated derivatives exhibit significant antitumor activities against S-180 and gastric carcinoma tumor cells (Y. Wang et al., 2004). The data show that good water solubility, relatively high chain stiffness, and moderate molecular mass of the chemically modified derivatives in aqueous solution contribute favorably to the enhancement of antitumor activity (Y. Wang et al., 2004).

In a series of investigations on chemical modifications of glucan fractions isolated from the sclerotia of *P. tuber-regium*, M. Zhang et al. (2003, 2004) reported that seven water-insoluble (1 → 3)-β-D-glucan fractions (TM8-1 to TM8-7) and six hot alkali extracts (HAE-1 to HAE-6) of mushroom (1 → 3)-β-glucans having different molecular mass were carboxymethylated to produce their corresponding water-soluble derivatives (CTM8-1 to CTM8-7 and CMHAE-1 to CMHAE-6, respectively). On the whole, all the carboxymethylated β-glucans have higher water solubility and CMHAE show higher antitumor activity in vivo (sarcoma 180 solid tumor implanted on BALB/c mice) as well as in vitro (HL-60 tumor cell culture) than that of the native HAE β-glucans. The results suggest that the antitumor activity of the carboxymethylated β-glucans may be correlated to its water solubility and relatively extended chain (M. Zhang et al., 2003, 2004). In another study, a water-soluble hyperbranched β-glucan, coded as TM3b, extracted from the same mushroom was fractionated by treatment with chlorosulfonic acid at 35°C to synthesize sulfated β-glucan derivatives. It reveals that both the native TM3b and its sulfated derivatives exist in a spherical chain conformation in NaCl. Furthermore, the native and sulfated TM3b fractions show potent antitumor activities in vivo and in vitro, but the sulfated derivatives exhibit relatively higher in vitro antitumor activity against human hepatic cancer cell line HepG2 than the native TM3b. Water solubility and introduction of sulfate groups appear to be the main factors in enhancing the antitumor activities (Tao et al., 2006). These studies show that the chemical modification of polysaccharides might be an effective approach for the improvement of the biological activities of polysaccharides.

5.5 CONCLUSIONS

In the last few decades, a large number of macrofungi (mushrooms) have been extensively used as a source of medicinal agents and therapeutic adjuvants or health food supplements. There is an increasing interest in identifying effective and safe constituents from medicinal mushrooms for cancer prevention and treatment. Most mushrooms contain biologically active polysaccharides in fruiting bodies, cultured mycelia, sclerotia, and culture filtrates. One of the most promising activities of the polysaccharides or polysaccharide–protein complexes derived from mushrooms is their immunomodulation and anticancer effects. However, their mechanisms of antitumor and immunomodulating actions are not clearly understood. It is widely accepted that antitumor polysaccharides from higher fungi enhance various immune responses in vivo and in vitro and act as BRMs. Their actions are predominantly host mediated. The cell-mediated immunity plays a critical role in the antitumor activity of polysaccharides or polysaccharide–protein complexes which are able to induce production of various immunomodulatory cytokines. The major immunomodulatory effects of these BRMs involve the activation of immunocompetent cells such as monocytes, macrophages, DCs, NK cells, Th cells, Tc cells, and B cells and hematopoietic stem cells and the activation of alternative signaling pathways. Macrophages and DCs may be activated by mushroom polysaccharides to produce various cytokines and NO which affect the response pathways of cell-mediated immunity. Activated DCs and macrophages produce cytokines (e.g. IL-12), that stimulate NK cells to rapidly secrete other cytokines (including IFN-γ, TNF-α, and GM-CSF) and enhance the cytotoxicity of NK cells. Various data support that their mode of action is due to the enhancement and potentiation of cell-mediated immunity through the regulation of immunomodulatory cytokines and the activation of a complement system. Polysaccharides or polysaccharide–protein complexes are considered multicytokine inducers able to induce gene expression of various cytokines and cytokine receptors.

The detailed mechanisms of action of polysaccharide immunomodulators or biological response modifiers are not fully known, but it is generally accepted that they act on different immunocompetent cells which may initiate a cascade of signal transduction pathways that are responsible for the immune responses. The first step of the polysaccharide BRM in the modulation of cellular activity is the recognition of BRM and the binding to immune cell receptors. Some evidence shows that there are PRRs available for the molecular reception of polysaccharide BRMs. It has been suggested that some groups of PRRs can recognize the polysaccharide BRMs, such as complement receptor 3 (CR3 or CD11b/CD18), dectin-1, and toll-like receptors (TLR-2 and TLR-4). These PRRs play an important role as the membrane β-glucan receptors, which are mainly expressed on cell membranes of monocytes, macrophages, neutrophils, NK cells, and DCs. The binding of BRMs to PRRs may initiate a signaling cascade for the immune responses. PRRs can mediate a variety of cellular functions such as cytotoxic and phagocytic responses as they have the ability to bind various ligands, for example, ICAM-1, β-glucan, and others, through various specific signal transduction pathways.

Although the mechanisms of antitumor action of polysaccharides or polysaccharide–protein complexes are generally thought to be due to the enhancement and modulation of the immune system, many of these macromolecules have also been documented to possess direct cytotoxic effects on cancer cells. The polysaccharides or polysaccharide–protein complexes may exert anticancer activity by antiproliferative effect on tumor cells and induction of cell cycle arrest and apoptosis. It is possible that, in some instances, these two types of inhibitory action may be interwoven. Therefore, the possible modes of anticancer action may include both (1) direct cytotoxicity to cancer cells as shown in many in vitro studies and (2) indirect antitumor inhibition through immunomodulation of the body defense system. Other mechanisms of antitumor action include antiangiogenesis, antimetastasis, and antigenotoxicity.

Mushroom polysaccharides with antitumor and immunomodulatory properties have been extensively documented. However, the relationship of structure and anticancer activity of polysaccharides at the cellular and molecular levels is still not well characterized. Future studies should therefore focus on investigation of the relationship between their structure and antitumor activity, elucidation of their antitumor mechanism at the molecular level, improvement of their various biological activities by chemical modifications, and clinical trials on therapeutic efficacy of mushroom polysaccharides. It is noteworthy that a thorough understanding of the recognition of mushroom polysaccharides (particularly β-glucan) by certain receptors on the immune cells and activation of signal transduction pathways will also be a challenge of future immunomodulation research.

REFERENCES

Adachi, Y., Ohno, N., Ohsawa, M., Oikawa, S., and Yadomae, T. (1990a). Change of biological-activities of (1 → 3)-β-D-glucan from *Grifola frondosa* upon molecular-weight reduction by heat-treatment. *Chemical and Pharmaceutical Bulletin*, *38*, 477–481.

Adachi, Y., Ohno, N., Yadomae, T., Suzuki, Y., Ohsawa, M., and Oikawa, S. (1990b). Thermal-denaturation of (1 → 3)-β-D-glucans in neutral aqueous-solution above 130°—Effect on physicochemical properties. *Carbohydrate Research*, *198*, 111–122.

Adachi, Y., Okazaki, M., Ohno, N., and Yadomae, T. (1994). Enhancement of cytokine production by macrophages stimulated with (1 → 3)-β-D-glucan, Grifolan (Grn), isolated from *Grifola frondosa*. *Biological and Pharmaceutical Bulletin*, *17*, 1554–1560.

Angeli, J. P. F., Ribeiro, L. R., Gonzaga, M. L. C., Soares, S. D., Ricardo, M. P. S. N., Tsuboy, M. S., Stidl, R., Knasmueller, S., Linhares, R. E., and Mantovani, M. S. (2006). Protective effects of β-glucan extracted from *Agaricus brasiliensis* against chemically induced DNA damage in human lymphocytes. *Cell Biology and Toxicology*, *22*, 285–291.

Bae, J. S., Jang, K. H., and Jin, H. K. (2006). Effects of natural polysaccharides on the growth and peritoneal carcinomatosis of human gastric adenocarcinoma in a nude mouse model. *Cancer Letters*, *235*, 60–68.

Baek, S. J., Kim, Y. S., Yong, H. M., Chae, J. B., Lee, S. A., Bae, W. C., Park, D. W., Kim, D. Y., Lee, J. W., and Park, S. K. (2002). Antimetastatic effect of proteoglycan isolated from the mycelium of *Ganoderma lucidum* IY009 *in vitro* and *in vivo*. *Yakhak Hoeji*, *46*, 11–17.

Banchereau, J., Briere, F., Caux, C., Davoust, J., Lebecque, S., Liu, Y. T., Pulendran, B., and Palucka, K. (2000). Immunobiology of dendritic cells. *Annual Review of Immunology*, *18*, 767–811.

Bao, X. F., Liu, C. P., Fang, J. N., and Li, X. Y. (2001). Structural and immunological studies of a major polysaccharide from spores of *Ganoderma lucidum* (Fr.) Karst. *Carbohydrate Research*, *332*, 67–74.

Bohn, J. A., and BeMiller, J. N. (1995). $(1 \rightarrow 3)$-β-D-Glucans as biological response modifiers: A review of structure-functional activity relationships. *Carbohydrate Polymers*, *28*, 3–14.

Borchers, A. T., Keen, C. L., and Gershwin, M. E. (2004). Mushrooms, tumors, and immunity: An update. *Experimental Biology and Medicine*, *229*, 393–406.

Camelini, C. M., Maraschin, M., de Mendonca, M. M., Zucco, C., Ferreira, A. G., and Tavares, L. A. (2005). Structural characterization of β-glucans of *Agaricus brasiliensis* in different stages of fruiting body maturity and their use in nutraceutical products. *Biotechnology Letters*, *27*, 1295–1299.

Cao, L. Z., and Lin, Z. B. (2002). Regulation on maturation and function of dendritic cells by *Ganoderma lucidum* polysaccharides. *Immunology Letters*, *83*, 163–169.

Cao, Q. Z. and Lin, Z. B. (2006). *Ganoderma lucidum* polysaccharides peptide inhibits the growth of vascular endothelial cell and the induction of VEGF in human lung cancer cell. *Life Sciences*, *78*, 1457–1463.

Carbonero, E. R., Gracher, A. H. P., Smiderle, F. R., Rosado, F. R., Sassaki, G. L., Gorin, P. A. J., and Iacomini, M. (2006). A β-glucan from the fruit bodies of edible mushrooms *Pleurotus eryngii* and *Pleurotus ostreatoroseus*. *Carbohydrate Polymers*, *66*, 252–257.

Chakraborty, I., Mondal, S., Rout, D., and Islam, S. S. (2006). A water-insoluble $(1 \rightarrow 3)$-β-D-glucan from the alkaline extract of an edible mushroom *Termitomyces eurhizus*. *Carbohydrate Research*, *341*, 2990–2993.

Chan, W. K., Lam, D. T. W., Law, H. K. W., Wong, W. T., Koo, M. W. L., Lau, A. S. Y., Lau, Y. L., and Chan, G. C. F. (2005). *Ganoderma lucidum* mycelium and spore extracts as natural adjuvants for immunotherapy. *Journal of Alternative and Complementary Medicine*, *11*, 1047–1057.

Chang, S. T. and Miles, P. G. (1989). *Edible Mushrooms and Their Cultivation*. Boca Raton, FL: CRC Press.

Chen, C., Parangi, S., Tolentino, M. J., and Folkman, J. (1995). A strategy to discover circulating angiogenesis inhibitors generated by human tumors. *Cancer Research*, *55*, 4230–4233.

Chen, S. C., Lu, M. K., Cheng, J. J., and Wang, D. L. (2005). Antiangiogenic activities of polysaccharides isolated from medicinal fungi. *FEMS Microbiology Letters*, *249*, 247–254.

Chen, Y. Y. and Chang, H. M. (2004). Antiproliferative and differentiating effects of polysaccharide fraction from fu-ling(*Poria cocos*) on human leukemic U937 and HL-60 cells. *Food and Chemical Toxicology*, *42*, 759–769.

Cheng, J. J., Huang, N. K., Chang, T. T., Wang, D. L., and Lu, M. K. (2005). Study for anti-angiogenic activities of polysaccharides isolated from *Antrodia cinnamomea* in endothelial cells. *Life Sciences*, *76*, 3029–3042.

Chihara, G. (1984). Immunopharmacology of lentinan and the glucans. *EOS-Rivista di Immunologia ed Immunofarmacologia*, *4*, 85–96.

Chihara, G. (1992). Immunopharmacology of lentinan, a polysaccharide isolated from *Lentinus edodes*: Its application as a host defense potentiator. *International Journal of Oriental Medicine*, *17*, 57–77.

Chihara, G., Hamuro, J., Maeda, Y., Arai, Y., and Fukuoka, F. (1970a). Antitumour polysaccharide derived chemically from natural glucan (pachyman). *Nature*, *225*, 943–944.

Chihara, G., Hamuro, J., Maeda, Y., Arai, Y., and Fukuoka, F. (1970b). Fractionation and purification of the polysaccharides with marked antitumor activity, especially lentinan, from *Lentinus edodes* (Berk.) Sing. (an edible mushroom). *Cancer Research*, *30*, 2776–2781.

Chihara, G., Maeda, Y. Y., Suga, T., and Hamuro, J. (1989). Lentinan as a host defense potentiator (Hdp). *International Journal of Immunotherapy*, *5*, 145–154.

Cho, E. J., Oh, J. Y., Chang, H. Y., and Yun, J. W. (2006). Production of exopolysaccharides by submerged mycelial culture of a mushroom *Tremella fuciformis*. *Journal of Biotechnology*, *127*, 129–140.

Cho, K. J., and Chung, K. S. (1999). Antitumor and antileukopenic activity of HCA, the protein-polysaccharide fraction of cultured mycelia of *Hebeloma crustuliniforme*. *Yakhak Hoeji*, *43*, 629–634.

Cho, S. M., Koshino, H., Yu, S. H., and Yoo, I. D. (1998). A mannofucogalactan, fomitellan A, with mitogenic effect from fruit bodies of *Fomitella fraxinea* (Imaz.). *Carbohydrate Polymers*, *37*, 13–18.

Cho, S. M., Lee, J., Han, S. B., Kim, H. M., and Yu, S. H. (1995). Immuno-stimulating polysaccharides from the fruiting bodies of *Fomitella fraxinea*: II. Isolation and characterization of hot-water extracted polysaccharides. *Korean Journal of Mycology*, *23*, 340–347.

Cho, S. M., Park, J. S., Kim, K. P., Cha, D. Y., Kim, H. M., and Yoo, I. D. (1999). Chemical features and purification of immunostimulating polysaccharides from the fruit bodies of *Agaricus blazei*. *Korean Journal of Mycology*, *27*, 170–174.

Chow, L. W. C., Lo, C. S. Y., Loo, W. T. Y., Hu, X. C., and Sham, J. S. T. (2003). Polysaccharide peptide mediates apoptosis by up-regulating p21 gene and down-regulating cyclin D-1 gene. *American Journal of Chinese Medicine*, *31*, 1–9.

Chung, H. Y., Cho, Y. J., and Kim, T. (2005). Isolation and characterization of a water-soluble polysaccharide from the mycelia of solid cultured *Agaricus blazei* murill. *Food Science and Biotechnology*, *14*, 259–262.

Collins, L., Zhu, T., Guo, J., Xiao, Z. J., and Chen, C. Y. (2006). *Phellinus linteus* sensitises apoptosis induced by doxorubicin in prostate cancer. *British Journal of Cancer*, *95*, 282–288.

Cui, F. J., Tao, W. Y., Xu, Z. H., Guo, W. J., Xu, H. Y., Ao, Z. H., Jin, J., and Wei, Y. Q. (2007). Structural analysis of anti-tumor heteropolysaccharide GFPS1b from the cultured mycelia of *Grifola frondosa* GF9801. *Bioresource Technology*, *98*, 395–401.

Cui, T. and Chisti, Y. (2003). Polysaccharopeptides of *Coriolus versicolor*: Physiological activity, uses, and production. *Biotechnology Advances*, *21*, 109–122.

Cun, Z., Ito, H., Shimura, K., Sumiya, T., and Kawade, M. (1994). Antitumor activity and immunological property of polysaccharides from the mycelium of liquid-cultured *Grifola frondosa*. *Journal of the Japanese Society for Food Science and Technology, 41*, 724–732.

Czarnecki, R. and Grzybek, J. (1995). Antiinflammatory and vasoprotective activities of polysaccharides isolated from fruit bodies of higher fungi. 1. Polysaccharides from *Trametes Gibbosa* (Pers.: Fr)Fr (Polyporaceae). *Phytotherapy Research, 9*, 123–127.

Dong, Q., Jia, L. M., and Fang, J. N. (2006). A β-D-glucan isolated from the fruiting bodies of *Hericium erinaceus* and its aqueous conformation. *Carbohydrate Research, 341*, 791–795.

Duncan, C. J. G., Pugh, N., Pasco, D. S., and Ross, S. A. (2002). Isolation of a galactomannan that enhances macrophage activation from the edible fungus *Morchella esculenta*. *Journal of Agricultural and Food Chemistry, 50*, 5683–5685.

Falch, B. H., Espevik, T., Ryan, L., and Stokke, B. T. (2000). The cytokine stimulating activity of $(1 \rightarrow 3)$-β-D-glucans is dependent on the triple helix conformation. *Carbohydrate Research, 329*, 587–596.

Fan, L., Soccol, A. T., Pandey, A., and Soccol, C. R. (2007). Effect of nutritional and environmental conditions on the production of exo-polysaccharide of *Agaricus brasiliensis* by submerged fermentation and its antitumor activity. *LWT Food Science and Technology, 40*, 30–35.

Fang, N., Li, Q., Yu, S., Zhang, J., He, L., Ronis, M. J., and Badger, T. M. (2006). Inhibition of growth and induction of apoptosis in human cancer cell lines by an ethyl acetate fraction from shiitake mushrooms. *Journal of Alternative and Complementary Medicine, 12*, 125–132.

Folkman, J. (1995). Angiogenesis in cancer, vascular, rheumatoid and other disease. *Nature Medicine, 1*, 27–31.

Franz, G. (1989). Polysaccharides in pharmacy—Current applications and future concepts. *Planta Medica, 6*, 493–497.

Fujii, T., Maeda, H., Suzuki, F., and Ishida, N. (1978). Isolation and characterization of a new antitumor polysaccharide, KS-2, extracted from culture mycelia of *Lentinus edodes*. *Journal of Antibiotics, 31*, 1079–1090.

Fujimiya, Y., Suzuki, Y., Oshiman, K., Kobori, H., Moriguchi, K., Nakashima, H., Matumoto, Y., Takahara, S., Ebina, T., and Katakura, R. (1998). Selective tumoricidal effect of soluble proteoglucan extracted from the basidiomycete, *Agaricus blazei* Murill, mediated via natural killer cell activation and apoptosis. *Cancer Immunology Immunotherapy, 46*, 147–159.

Fukuda, K., Uematsu, T., Hamada, A., Akiya, S., Komatsu, N., and Okubo, S. (1975). Polysaccharide from *Lampteromyces japonicus*. *Chemical and Pharmaceutical Bulletin, 23*, 1955–1959.

Fullerton, S. A., Samadi, A. A., Tortorelis, D. G., Choudhury, M. S., Mallouh, C., Tazaki, H., and Konno, S. (2000). Induction of apoptosis in human prostatic cancer cells with β-glucan (Maitake mushroom polysaccharide). *Molecular Urology, 4*, 7–13.

Gao, Q., Jiang, R., Chen, H., Jensen, E., and Seljelid, R. (1996). Characterization and cytokine stimulating activities of heteroglycans from *Tremella fuciformis*. *Planta Medica, 62*, 297–302.

Gao, Y., Chan, E., and Zhou, S. (2004). Immunomodulating activities of Ganoderma, a mushroom with medicinal properties. *Food Reviews International, 20*, 123–161.

Gao, Y., Dai, X., Chen, G., Ye, J., and Zhou, S. (2003). A randomized, placebo-controlled, multicenter study of *Ganoderma lucidum* (W. Curt.: Fr.) LLoyd (Aphyllophoromycetideae) polysaccharides (Ganopoly®) in patients with advanced lung cancer. *International Journal of Medicinal Mushrooms*, *5*, 369–381.

Gao, Y., Gao, H., Chan, E., Tang, W., Xu, A., Yang, H., Huang, M., Lan, J., Li, X., Duan, W., Xu, C., and Zhou, S. (2005). Antitumor activity and underlying mechanisms of ganopoly, the refined polysaccharides extracted from *Ganoderma lucidum*, in mice. *Immunological Investigations*, *34*, 171–198.

Gasiorowski, K., Brokos, B., Lamer-Zarawska, E., and Trocha-Grimshaw, J. (1993). Polysaccharides from *Laetiporus sulphureus* (Basidiomycetes) II. Evaluation of immunostimulative and antitumour activity. *Bulletin of the Polish Academy of Sciences, Biological Sciences*, *41*, 347–352.

Gonzaga, M. L. C., Ricardo, N., Heatley, F., and Soares, S. D. (2005). Isolation and characterization of polysaccharides from *Agaricus blazei* Murill. *Carbohydrate Polymers*, *60*, 43–49.

Grzybek, J., Zgorniak-Nowolsielska, I., and Kohlmunzer, S. (1990). Antitumor and cytotoxic activity of tylopilan, a fungal polysaccharide from *Tylopilus felleus* fruit bodies. *Planta Medica*, *56*, 670–671.

Guterres, Z. D. R., Mantovani, M. S., da Eira, A. F., Ribeiro, L. R., and Jordao, B. Q. (2005). Genotoxic and antigenotoxic effects of organic extracts of mushroom *Agaricus blazei* Murrill on V79 cells. *Genetics and Molecular Biology*, *28*, 458–463.

Gutierrez, A., Prieto, A., and Martinez, A. T. (1996). Structural characterization of extracellular polysaccharides produced by fungi from the genus *Pleurotus*. *Carbohydrate Research*, *281*, 143–154.

Han, M. D., Jeong, H., Lee, J. W., Back, S. J., Kim, S. U., and Yoon, K. H. (1995). The composition and bioactivities of Ganoderan by mycelial fractionation of *Ganoderma lucidum* IY009. *Korean Journal of Mycology*, *23*, 285–297.

Hanahan, D. and Folkman, J. (1996). Patterns and emerging mechanisms of the angiogenic switch during tumorigenesis. *Cell*, *86*, 353–364.

Hara, C., Kumazawa, Y., Inagaki, K., Kaneko, M., Kiho, T., and Ukai, S. (1991). Mitogenic and colony-stimulating factor-inducing activities of polysaccharide fractions from the fruit bodies of *Dictyophora indusiata* Fisch. *Chemical and Pharmaceutical Bulletin*, *39*, 1615–1616.

Harada, M., Matsunaga, K., Oguchi, Y., Iijima, H., Tamada, K., Abe, K., Takenoyama, M., Ito, O., Kimura, G., and Nomoto, K. (1997). Oral administration of PSK can improve the impaired anti-tumor CD4(+) T-cell response in gut-associated lymphoid tissue (GALT) of specific-pathogen-free mice. *International Journal of Cancer*, *70*, 362–372.

Harada, T., Kawaminami, H., Miura, N. N., Adachi, Y., Nakajima, M., Yadomae, T., and Ohno, N. (2006). Comparison of the immunomodulating activities of $(1 \rightarrow 3)$-β-glucan fractions from the culinary-medicinal mushroom *Sparassis crispa* Wulf.: Fr. (Aphyllophoromycetideae). *International Journal of Medicinal Mushrooms*, *8*, 231–244.

Harada, T., Miura, N., Adachi, Y., Nakajima, M., Yadomae, T., and Ohno, N. (2002). Effect of SCG, $(1 \rightarrow 3)$-β-D-glucan from *Sparassis crispa* on the hematopoietic response in cyclophosphamide induced leukopenic mice. *Biological and Pharmaceutical Bulletin*, *25*, 931–939.

Harada, T., Miura, N. N., Adachi, Y., Nakajima, M., Yadomae, T., and Ohno, N. (2004). Granulocyte-macrophage colony-stimulating factor (GM-CSF) regulates cytokine induction by (1 → 3)-β-D-glucan SCG in DBA/2 mice in vitro. *Journal of Interferon and Cytokine Research, 24*, 478–489.

Hattori, T. S., Komatsu, N., Shichijo, S., and Itoh, K. (2004). Protein-bound polysaccharide K induced apoptosis of the human Burkitt lymphoma cell line, Namalwa. *Biomedicine and Pharmacotherapy*, *58*, 226–230.

Hawksworth, D. L. (2001). Mushrooms: The extent of the unexplored potential. *International Journal of Medicinal Mushrooms, 3*, 333–340.

Ho, C. Y., Kim, C. F., Leung, K. N., Fung, K. P., Tse, T. F., Chan, H., and Lau, C. B. S. (2005). Differential anti-tumor activity of *Coriolus versicolor* (Yunzhi) extract through p53- and/or Bcl-2-dependent apoptotic pathway in human breast cancer cells. *Cancer Biology and Therapy*, *4*, 638–644.

Ho, J. C. K., Konerding, M. A., Gaumann, A., Groth, M., and Liu, W. (2004). Fungal polysaccharopeptide inhibits tumor angiogenesis and tumor growth in mice. *Life Sciences*, *75*, 1343–1356.

Hong, J. H. and Choi, Y. H. (2007). Physico-chemical properties of protein-bound polysaccharide from *Agaricus blazei* Murill prepared by ultrafiltration and spray drying process. *International Journal of Food Science and Technology, 42*, 1–8.

Hong, K. J., Dunn, D. M., Shen, C. L., and Pence, B. C. (2004). Effects of *Ganoderma lucidum* on apoptotic and anti-inflammatory function in HT-29 human colonic carcinoma cells. *Phytotherapy Research*, *18*, 768–770.

Hoshi, H., Yagi, Y., Iijima, H., Matsunaga, K., Ishihara, Y., and Yasuhara, T. (2005). Isolation and characterization of a novel immunomodulatory α-glucan-protein complex from the mycelium of *Tricholoma matsutake* in basidiomycetes. *Journal of Agricultural and Food Chemistry, 53*, 8948–8956.

Hsu, M. J., Lee, S. S., Lee, S. T., and Lin, W. W. (2003). Signaling mechanisms of enhanced neutrophil phagocytosis and chemotaxis by the polysaccharide purified from *Ganoderma lucidum. British Journal of Pharmacology, 139*, 289–298.

Hua, K. F., Hsu, H. Y., Chao, L. K., Chen, S. T., Yang, W. B., Hsu, J., and Wong, C. H. (2007). *Ganoderma lucidum* polysaccharides enhance CD14 endocytosis of LPS and promote TLR4 signal transduction of cytokine expression. *Journal of Cellular Physiology, 212*, 537–550.

Hui, K. P. Y., Sit, W. H., and Wan, J. M. F. (2005). Induction of S phase cell arrest and caspase activation by polysaccharide peptide isolated from *Coriolus versicolor* enhanced the cell cycle dependent activity and apoptotic cell death of doxorubicin and etoposide, but not cytarabine in HL-60 cells. *Oncology Reports*, *14*, 145–155.

Ikekawa, T., Ikeda, Y., Yoshioka, Y., Nakanishi, K., Yokoyama, E., and Yamazaki, E. (1982). Studies on antitumor polysaccharides of *Flammulina velutipes* (Curt. ex Fr.) Sing. II. The structure of EA3 and further purification of EA5. *Journal of Pharmacobio-dynamics, 5*, 576–581.

Ikekawa, T., Saitoh, H., Feng, W. J., Zhang, H. L., Li, L. F., and Matsuzawa, T. (1992). Antitumor-activity of hypsizigus-marmoreus. 1. Antitumor-activity of extracts and polysaccharides. *Chemical and Pharmaceutical Bulletin, 40*, 1954–1957.

Ikekawa, T., Uehara, N., Maeda, Y., Nakanish, M., and Fukuoka, F. (1969). Antitumor activity of aqueous extracts of edible mushrooms. *Cancer Research*, *29*, 734–735.

Ishibashi, K., Miura, N. N., Adachi, Y., Ohno, N., and Yadomae, T. (2001). Relationship between solubility of grifolan, a fungal $(1 \rightarrow 3)$-β-D-glucan, and production of tumor necrosis factor by macrophages in vitro. *Bioscience, Biotechnology, and Biochemistry, 65*, 1993–2000.

Ito, H., Shimura, K., Itoh, H., and Kawade, M. (1997). Antitumor effects of a new polysaccharide-protein complex (ATOM) prepared from *Agaricus blazei* (Iwade strain 101) "Himematsutake" and its mechanisms in tumor-bearing mice. *Anticancer Research, 17*, 277–284.

Ito, H., Sugiura, M., and Miyazaki, T. (1976). Antitumor polysaccharide fraction from culture filtrate of *Fomes fomentarius*. *Chemical and Pharmaceutical Bulletin, 24*, 2575.

Itoh, H., Ito, H., Amano, H., and Noda, H. (1994). Inhibitory action of a $(1 \rightarrow 6)$-β-D-glucan-protein complex (FIII-2-b) isolated from *Agaricus blazei* Murill ("Himematsutake") on Meth A fibrosarcoma-bearing mice and its antitumor mechanism. *Japanese Journal of Pharmacology, 66*, 265–271.

Jia, L. M., Liu, L., Dong, Q., and Fang, J. N. (2004). Structural investigation of a novel rhamnoglucogalactan isolated from the fruiting bodies of the fungus *Hericium erinaceus*. *Carbohydrate Research, 339*, 2667–2671.

Jiang, J. H., Slivova, V., Harvey, K., Valachovicova, T., and Sliva, D. (2004a). *Ganoderma lucidum* suppresses growth of breast cancer cells through the inhibition of Akt/NF-kappa B signaling. *Nutrition and Cancer—an International Journal, 49*, 209–216.

Jiang, J. H., Slivova, V., Valachovicova, T., Harvey, K., and Sliva, D. (2004b). *Ganoderma lucidum* inhibits proliferation and induces apoptosis in human prostate cancer cells PC-3. *International Journal of Oncology, 24*, 1093–1099.

Jiang, Y., Wang, H., Lu, L., and Tian Geng, Y. (2005). Chemistry of polysaccharide Lzps-1 from *Ganoderma lucidum* spore and anti-tumor activity of its total polysaccharides. *Yaoxue Xuebao, 40*, 347–350.

Jin, M., Jeon, H., Jung, H. J., Kim, B., Shin, S. S., Choi, J. J., Lee, J. K., Kang, C. Y., and Kim, S. (2003a). Enhancement of repopulation and hematopoiesis of bone marrow cells in irradiated mice by oral administration of PG101, a water-soluble extract from *Lentinus lepideus*. *Experimental Biology and Medicine, 228*, 759–766.

Jin, M., Jung, H. J., Choi, J. J., Jeon, H., Oh, J. H., Kim, B., Shin, S. S., Lee, J. K., Yoon, K., and Kim, S. (2003b). Activation of selective transcription factors and cytokines by water-soluble extract from *Lentinus lepideus*. *Experimental Biology and Medicine, 228*, 749–758.

Jin, Y., Zhang, L. N., Zhang, M., Chen, L., Cheung, P. C. K., Ooi, V. E. C., and Lin, Y. L. (2003c). Antitumor activities of heteropolysaccharides of *Poria cocos* mycelia from different strains and culture media. *Carbohydrate Research, 338*, 1517–1521.

Jong, S. C., Birmingham, J. M., and Pai, S. H. (1991). Immunomodulatory substances of fungal origin. *Eos-Rivista di Immunologia ed Immunofarmacologia, 11*, 115–122.

Kanayama, H., Adachi, N., Fukai, Y., Takeuchi, I., and Togami, M. (1986). Studies on the antitumor active polysaccharides from the mycelia of *Poria cocos* wolf. 2. Structural-analysis of antitumor polysaccharide h-11. *Yakugaku Zasshi—Journal of the Pharmaceutical Society of Japan, 106*, 206–211.

Kanazawa, M., Mori, Y., Yoshihara, K., Iwadate, M., Suzuki, S., Endoh, Y., Ohki, S., Takita, K., Sekikawa, K., and Takenoshita, S. (2004). Effect of PSK on the maturation of dendritic cells derived from human peripheral blood monocytes. *Immunology Letters, 91*, 229–238.

Kanazawa, M., Yoshihara, K., Abe, H., Iwadate, M., Watanabe, K., Suzuki, S., Endoh, Y., Takita, K., Sekikawa, K., Takenoshita, S., Ogata, T., and Ohto, H. (2005). Effects of PSK on T and dendritic cells differentiation in gastric or colorectal cancer patients. *Anticancer Research*, *25*, 443–449.

Kato, M., Hirose, K., Saito, Y., Izutani, R., Noguchi, J., Kuroda, D., Nomura, H., Nishimatsu, S., and Ohoyanagi, H. (1995). Induction of gene-expression for immunomodulating cytokines in peripheral-blood mononuclear-cells in response to orally-administered PSK, an immunomodulating protein-bound polysaccharide. *Cancer Immunology Immunotherapy*, *40*, 152–156.

Kawagishi, H., Kanao, T., Inagaki, R., Mizuno, T., Shimura, K., Ito, H., Hagiwara, T., and Nakamura, T. (1990). Formolysis of a potent antitumor (1 → 6)-β-D-glucan protein complex from *Agaricus blazei* fruiting bodies and antitumor-activity of the resulting products. *Carbohydrate Polymers*, *12*, 393–403.

Kerekgyarto, C., Virag, L., Tanko, L., Chihara, G., and Fachet, J. (1996). Strain differences in the cytotoxic activity and TNF production of murine macrophages stimulated by lentinan. *International Journal of Immunopharmacology*, *18*, 347–353.

Kiho, T., Katsuragawa, M., Nagai, K., Ukai, S., and Haga, M. (1992a). Structure and antitumor activity of a branched (1 → 3)-β-D-glucan from the alkaline extract of *Amanita muscaria*. *Carbohydrate Research*, *224*, 237–243.

Kiho, T., Nagai, Y. S. K., and Sakushima, M. (1992b). Polysaccharides in fungi: XXIX. Structural features of two antitumor polysaccharides from the fruiting bodies of *Armillariella tabescens*. *Chemical and Pharmaceutical Bulletin (Tokyo)*, *40*, 2212–2214.

Kiho, T., Yoshida, I., Katsuragawa, M., Sakushima, M., and Usui, S. (1994). Polysaccharides in fungi: XXXIV. A polysaccharide from the fruiting bodies of *Amanita muscaria* and the antitumor activity of its carboxymethylated product. *Biological and Pharmaceutical Bulletin*, *17*, 1460–1462.

Kiho, T., Yoshida, I., Nagai, K., Ukai, S., and Hara, C. (1989). Polysaccharides in fungi. 13. (1 → 3)-α-D-Glucan from an alkaline extract of *Agrocybe cylindracea,* and antitumor-activity of its O-(carboxymethyl)ated derivatives. *Carbohydrate Research*, *189*, 273–279.

Kim, F., Sakagami, H., Tanuma, S. I., and Konno, K. (1990). Stimulation of interferon-γ-induced human myelogenous leukemic cell differentiation by high molecular weight Psk subfraction. *Anticancer Research*, *10*, 55–58.

Kim, G. Y., Choi, G. S., Lee, S. H., and Park, Y. M. (2004a). Acidic polysaccharide isolated from *Phellinus linteus* enhances through the up-regulation of nitric oxide and tumor necrosis factor-alpha from peritoneal macrophages. *Journal of Ethnopharmacology*, *95*, 69–76.

Kim, G. Y., Han, M. G., Song, Y. S., Shin, B. C., Shin, Y. I., Lee, H. J., Moon, D. O., Lee, C. M., Kwak, J. Y., Bae, Y. S., Lee, J. D., and Park, Y. M. (2004b). Proteoglycan isolated from *Phellinus linteus* induces toll-like receptors 2- and 4-mediated maturation of murine dendritic cells via activation of ERK, p38, and NF-kappa B. *Biological and Pharmaceutical Bulletin*, *27*, 1656–1662.

Kim, G. Y., Lee, J. Y., Lee, J. O., Ryu, C. H., Choi, B. T., Jeong, Y. K., Lee, K. W., Jeong, S. C., and Choi, Y. H. (2006). Partial characterization and immunostimulatory effect of a novel polysaccharide-protein complex extracted from *Phellinus linteus*. *Bioscience, Biotechnology, and Biochemistry*, *70*, 1218–1226.

Kim, G. Y., Oh, W. K., Shin, B. C., Shin, Y. I., Park, Y. C., Ahn, S. C., Lee, J. D., Bae, Y. S., Kwak, J. Y., and Park, Y. M. (2004c). Proteoglycan isolated from *Phellinus linteus* inhibits tumor growth through mechanisms leading to an activation of $CD11c^+$ $CD8^+$ DC and type I helper T cell-dominant immune state. *FEBS Letters*, *576*, 391–400.

Kim, G. Y., Oh, Y. H., and Park, Y. M. (2003). Acidic polysaccharide isolated from *Phellinus linteus* induces nitric oxide-mediated tumoricidal activity of macrophages through protein tyrosine kinase and protein kinase C. *Biochemical and Biophysical Research Communications*, *309*, 399–407.

Kim, H. M., Han, S. B., Oh, G. T., Kim, Y. H., Hong, D. H., Hong, N. D., and Yoo, I. D. (1996). Stimulation of humoral and cell mediated immunity by polysaccharide from mushroom *Phellinus linteus*. *International Journal of Immunopharmacology*, *18*, 295–303.

Kim, H. S., Kacew, S., and Lee, B. M. (1999). *In vitro* chemopreventive effects of plant polysaccharides (*Aloe barbadensis* Miller, *Lentinus edodes, Ganoderma lucidum* and *Coriolus versicolor*). *Carcinogenesis*, *20*, 1637–1640.

Kim, S. H., Kim, J. S., Jin, M. R., Kim, H. W., Choi, E. C., and Kim, B. K. (1993). Studies on antitumor components of *Collybia confluens*. *Korean Journal of Pharmacognosy*, *24*, 267–281.

Kim, Y. O., Han, S. B., Lee, H. W., Ahn, H. J., Yoon, Y. D., Jung, J. K., Kim, H. M., and Shin, C. S. (2005). Immuno-stimulating effect of the endo-polysaccharide produced by submerged culture of *Inonotus obliquus*. *Life Sciences*, *77*, 2438–2456.

Kim, Y. O., Park, H. W., Kim, J. H., Lee, J. Y., Moon, S. H., and Shin, C. S. (2006). Anti-cancer effect and structural characterization of endo-polysaccharide from cultivated mycelia of *Inonotus obliquus*. *Life Sciences*, *79*, 72–80.

Kishida, E., Sone, Y., and Misaki, A. (1989). Purification of an antitumor-active, branched $(1 \rightarrow 3)$-β-D-glucan from *Volvariella volvacea*, and elucidation of its fine structure. *Carbohydrate Research, 193*, 227–239.

Kitamura, S., Hori, T., Kurita, K., Takeo, K., Hara, C., Itoh, W., Tabata, K., Elgsaeter, A., and Stokke, B. T. (1994). An antitumor, branched $(1 \rightarrow 3)$-β-D-Glucan from a water extract of fruiting bodies of *Cryptoporus volvatus*. *Carbohydrate Research*, *263* (1), 111–121.

Kobayashi, H., Matsunaga, K., and Fujii, M. (1993). PSK as a chemopreventive agent (review). *Cancer Epidemiology Biomarkers and Prevention*, *2*, 271–276.

Kobayashi, H., Matsunaga, K., and Oguchi, Y. (1995). Antimetastatic effects of PSK (Krestin), a protein-bound polysaccharide obtained from basidiomycetes: An overview. *Cancer Epidemiology Biomarkers and Prevention*, *4*, 275–281.

Kobayashi, H., Yoshida, R., Kanada, Y., Fukuda, Y., Yagyu, T., Inagaki, K., Kondo, T., Kurita, N., Suzuki, M., Kanayama, N., and Terao, T. (2005). Suppressing effects of daily oral supplementation of β-glucan extracted from *Agaricus blazei* Murill on spontaneous and peritoneal disseminated metastasis in mouse model. *Journal of Cancer Research and Clinical Oncology*, *131*, 527–538.

Kodama, N., Asakawa, A., Inui, A., Masuda, Y., and Nanba, H. (2005a). Enhancement of cytotoxicity of NK cells by D-fraction, a polysaccharide from *Grifola frondosa*. *Oncology Reports*, *13*, 497–502.

Kodama, N., Komuta, K., and Nanba, H. (2003). Effect of maitake (*Grifola frondosa)* D-fraction on the activation of NK cells in cancer patients. *Journal of Medicinal Food*, *6*, 371–377.

Kodama, N., Komuta, K., Sakai, N., and Nanba, H. (2002). Effects of D-fraction, a polysaccharide from *Grifola frondosa* on tumor growth involve activation of NK cells. *Biological and Pharmaceutical Bulletin*, *25*, 1647–1650.

Kodama, N., Murata, Y., Asakawa, A., Inui, A., Hayashi, M., Sakai, N., and Nanba, H. (2005b). Maitake D-fraction enhances antitumor effects and reduces immunosuppression by mitomycin-C in tumor-bearing mice. *Nutrition*, *21*, 624–629.

Kodama, N., Murata, Y., and Nanba, H. (2004). Administration of a polysaccharide from *Grifola frondosa* stimulates immune function of normal mice. *Journal of Medicinal Food*, *7*, 141–145.

Kohlmunzer, S., Grzybek, J., and Wegiel, J. (1990). Biologically active polysaccharides from *Tylopilus felleus*. *Planta Medica*, *56*, 675.

Komatsu, N., Okubo, S., Kikumoto, S., Kimura, K., Saito, G., and Sakai, S. (1969). Host mediated antitumor action of schizophyllan a glucan produced by *Schizophyllum commune*. *Gan*, *60*, 137–144.

Konno, S., Aynehchi, S., Dolin, D.-J., Schwartz, A.-M., Choudhury, M.-S., and Tazaki, H. (2002). Anticancer and hypoglycemic effects of polysaccharides in edible and medicinal maitake mushroom (*Grifola frondosa* (Dicks.: Fr.) S. F. Gray). *International Journal of Medicinal Mushrooms*, *4*, 185–195.

Ladanyi, A., Timar, J., and Lapis, K. (1993). Effect of lentinan on macrophage cytotoxicity against metastatic tumor cells. *Cancer Immunology and Immunotherapy*, *36*, 123–126.

Lau, C. B. S., Ho, C. Y., Kim, C. F., Leung, K. N., Fung, K. P., Tse, T. F., Chan, H. H. L., and Chow, M. S. S. (2004). Cytotoxic activities of *Coriolus versicolor* (Yunzhi) extract on human leukemia and lymphoma cells by induction of apoptosis. *Life Sciences*, *75*, 797–808.

Lavi, I., Friesem, D., Geresh, S., Hadar, Y., and Schwartz, B. (2006). An aqueous polysaccharide extract from the edible mushroom *Pleurotus ostreatus* induces anti-proliferative and pro-apoptotic effects on HT-29 colon cancer cells. *Cancer Letters*, *244*, 61–70.

Lee, H. J., Lee, H. J., Lim, E. S., Ahn, K. S., Shim, B. S., Kim, H. M., Gong, S. J., Kim, D. K., and Kim, S. H. (2005). Cambodian *Phellinus linteus* inhibits experimental metastasis of melanoma cells in mice via regulation of urokinase type plasminogen activator. *Biological and Pharmaceutical Bulletin*, *28*, 27–31.

Lee, S. S., Lee, P. L., Chen, C. F., Wang, S. Y., and Chen, K. Y. (2003). Antitumor effects of polysaccharides of *Ganoderma lucidum* (Curt.: Fr.) P. Karst. (Ling Zhi, Reishi mushroom) (Aphyllophoromycetideae). *International Journal of Medicinal Mushrooms*, *5*, 1–16.

Leung, M. Y. K., and Fung, K. P. (1997). The isolation and characterization of an immunomodulatory and anti-tumor polysaccharide preparation from *Flammulina velutipes*. *Immunopharmacology*, *35*, 255–263.

Leung, M. Y. K., Liu, C., Koon, J. C. M., and Fung, K. P. (2006). Polysaccharide biological response modifiers. *Immunology Letters*, *105*, 101–114.

Li, G., Kim, D. H., Kim, T. D., Park, B. J., Park, H. D., Park, J. I. L., Na, M. K., Kim, H. C., Hong, N. D., Lim, K., Hwang, B. D., and Yoon, W. H. (2004). Protein-bound polysaccharide from *Phellinus linteus* induces G2/M phase arrest and apoptosis in SW480 human colon cancer cells. *Cancer Letters*, *216*, 175–181.

Lim, J. M., Joo, J. H., Kim, H. O., Kim, H. M., Kim, S. W., Hwang, H. J., and Yun, J. W. (2005). Structural analysis and molecular characterization of exopolysaccharides produced by submerged mycelial culture of *Collybia maculata* TG-1. *Carbohydrate Polymers*, *61*, 296–303.

Lin, H., Cheung, S. W. Y., Nesin, M., Cassileth, B. R., and Cunningham-Rundles, S. (2007). Enhancement of umbilical cord blood cell hematopoiesis by maitake β-glucan is mediated by granulocyte colony-stimulating factor production. *Clinical and Vaccine Immunology*, *14*, 21–27.

Lin, H., She, Y. H., Cassileth, B. R., Sirotnak, F., and Rundles, S. C. (2004). Maitake β-glucan MD-fraction enhances bone marrow colony formation and reduces doxorubicin toxicity in vitro. *International Immunopharmacology*, *4*, 91–99.

Lin, Y. L., Lee, S. S., Hou, S. M., and Chiang, B. L. (2006). Polysaccharide purified from *Ganoderma lucidum* induces gene expression changes in human dendritic cells and promotes T helper 1 immune response in BALB/c mice. *Molecular Pharmacology*, *70*, 637–644.

Lin, Y. L., Liang, Y. C., Lee, S. S., and Chiang, B. L. (2005). Polysaccharide purified from *Ganoderma lucidum* induced activation and maturation of human monocyte-derived dendritic cells by the NF-kappaB and p38 mitogen-activated protein kinase pathways. *Journal of Leukocyte Biology*, *78*, 533–543.

Lin, Y. L., Zhang, L. N., Chen, L., Jin, Y., Zeng, F. B., Jin, J., Wan, B., and Cheung, P. C. K. (2004). Molecular mass and antitumor activities of sulfated derivatives of α-glucan from *Poria cocos* mycelia. *International Journal of Biological Macromolecules*, *34*, 289–294.

Lin, Z. B. (2005). Cellular and molecular mechanisms of immuno-modulation by *Ganoderma lucidum*. *Journal of Pharmacological Sciences*, *99*, 144–153.

Liu, F., Fung, M. C., Ooi, V. E. C., and Chang, S. T. (1996a). Induction in the mouse of gene expression of immunomodulating cytokines by mushroom polysaccharide-protein complexes. *Life Sciences*, *58*, 1795–1803.

Liu, F. and Ooi, V. E. C. (1995). Anti-tumour components of the culture filtrates from *Tricholom* a sp. *World Journal of Microbiology and Biotechnology*, *11*, 486–490.

Liu, F., Ooi, V. E. C., Liu, W. K., and Chang, S. T. (1996b). Immunomodulation and antitumor activity of polysaccharide-protein complex from the culture filtrates of a local edible mushroom, *Tricholoma lobayense*. *General Pharmacology*, *27*, 621–624.

Liu, J. C., Su, F. Q., and Sun, J. N. (2006). Enhancing effect and mechanism of *Agaricus blazei* Murrill copolysaccharides on cyclophosphamide. *Zhongguo Linchuang Kangfu*, *10*, 118–121.

Liu, J. J., Huang, T. S., Hsu, M. L., Chen, C. C., Lin, W. S., Lu, F. J., and Chang, W. H. (2004). Antitumor effects of the partially purified polysaccharides from *Antrodia camphorata* and the mechanism of its action. *Toxicology and Applied Pharmacology*, *201*, 186–193.

Liu, M. Q., Li, J. Z., Kong, F. Z., Lin, J. Y., and Gao, Y. (1998). Induction of immunomodulating cytokines by a new polysaccharide-peptide complex from culture mycelia of *Lentinus edodes*. *Immunopharmacology*, *40*, 187–198.

Liu, W. K., Ng, T. B., Sze, S. F., and Tsui, K. W. (1993). Activation of peritoneal macrophages by polysaccharopeptide from the mushroom, *Coriolus versicolor*. *Immunopharmacology*, *26*, 139–146.

Lowe, E., Rice, P., Ha, T. Z., Li, C. F., Kelley, J., Ensley, H., Lopez-Perez, J., Kalbfleisch, J., Lowman, D., Margl, P., Browder, W., and Williams, D. (2001). A $(1 \rightarrow 3)$-β-D-linked heptasaccharide is the unit ligand for glucan pattern recognition receptors on human monocytes. *Microbes and Infection*, *3*, 789–797.

Lull, C., Wichers, H. J., and Savelkoul, H. F. (2005). Antiinflammatory and immunomodulating properties of fungal metabolites. *Mediators Inflammation*, *2005*, 63–80.

Ma, Y., Mizuno, T., and Ito, H. (1991). Studies on the host-mediated antitumor polysaccharides. 15. Antitumor-activity of some polysaccharides isolated from a Chinese mushroom, Huangmo, the fruiting body of *Hohenbuehelia serotina*. *Agricultural and Biological Chemistry*, *55*, 2701–2710.

Maeda, Y. Y., Watanabe, S. T., Chihara, C., and Rokutanda, M. (1988). Denaturation and renaturation of a (1 → 6)-(1 → 3)-β-glucan, Lentinan, associated with expression of T-cell-mediated responses. *Cancer Research*, *48*, 671–675.

Maruyama, H. and Ikekawa, T. (2005). Combination therapy of transplanted Meth-A fibrosarcoma in BALB/c mice with protein-bound polysaccharide EA6 isolated from enokitake mushroom *Flammulina velutipes* (W. Curt.: Fr.) Singer and surgical excision. *International Journal of Medicinal Mushrooms*, *7*, 213–220.

Miller, J. S. (2002). Biology of natural killer cells in cancer and infection. *Cancer Investigation*, *20*, 405–419.

Misaki, A. and Kakuta, M. (1995). Kikurage (tree-ear) and shirokikurage (white jelly-leaf)—*Auricularia auricula* and *Tremella fuciformis*. *Food Reviews International*, *11*, 211–218.

Misaki, A., Kakuta, M., Sasaki, T., Tanaka, M., and Miyaji, H. (1981). Studies on interrelation of structure and antitumor effects of polysaccharides: Antitumor action of periodate-modified, branched (1 → 3)-β-D-glucan of *Auricularia auricula judae*, and other polysaccharides containing (1 → 3)-β glycosidic linkages. *Carbohydrate Research*, *92*, 115–129.

Mitamura, T., Sakamoto, S., Suzuki, S., Yoshimura, S., Maemura, M., and Kudo, H. (2000). Effects of lentinan on colorectal carcinogenesis in mice with ulcerative colitis. *Oncology Reports*, *7*, 599–601.

Mizuno, M., Morimoto, M., Minato, K., and Tsuchida, H. (1998). Polysaccharides from *Agaricus blazei* stimulate lymphocyte T-cell subsets in mice. *Bioscience, Biotechnology, and Biochemistry*, *62*, 434–437.

Mizuno, M., Minato, K., Ito, H., Kawade, M., Terai, H., and Tsuchida, H. (1999). Anti-tumor polysaccharide from the mycelium of liquid-cultured *Agaricus blazei* mill. *Biochemistry and Molecular Biology International*, *47*, 707–714.

Mizuno, M., Shiomi, Y., Minato, K., Kawakami, S., Ashida, H., and Tsuchida, H. (2000). Fucogalactan isolated from *Sarcodon aspratus* elicits release of tumor necrosis factor-α and nitric oxide from murine macrophages. *Immunopharmacology*, *46*, 113–121.

Mizuno, T. (1995). Sarunokoshikake: *Polyporaecea* fungi—kofukisarunokoshikake, *Ganoderma applanatum* and tsugasarunokoshikake, *Fomitopsis pinicola*. *Food Reviews International*, *11*, 129–133.

Mizuno, T. and Zhuang, C. (1995). Maitake, *Grifola frondosa*—pharmacological effects. *Food Reviews International*, *11*, 135–149.

Mizuno, T., Ohsawa, K., Hagiwara, N., and Kuboyama, R. (1986). Studies on the host-mediated antitumor polysaccharides. 9. Fractionation and characterization of antitumor polysaccharides from maitake, *Grifola frondosa*. *Agricultural and Biological Chemistry*, *50*, 1679–1688.

Mizuno, T., Hagiwara, T., Nakamura, T., Ito, H., Shimura, K., Sumiya, T., and Asakura, A. (1990a). Studies on the host-mediated antitumor polysaccharides. 13. Antitumor-activity and some properties of water-soluble polysaccharides from himematsutake, the fruiting body of *Agaricus blazei* Murill. *Agricultural and Biological Chemistry*, *54*, 2889–2896.

Mizuno, T., Inagaki, R., Kanao, T., Hagiwara, T., Nakamura, T., Ito, H., Shimura, K., Sumiya, T., and Asakura, A. (1990b). Antitumor activity and some properties of water-insoluble hetero-glycans from himematsutake the fruiting body of *Agaricus blazei* Murill. *Agricultural and Biological Chemistry*, *54*, 2897–2906.

Mizuno, T., Ando, M., Sugie, R., Ito, H., Shimura, K., Sumiya, T., and Matsuura, A. (1992a). Studies on the host-mediated antitumor polysaccharides. 16. Antitumor-activity of some polysaccharides isolated from an edible mushroom, ningyotake, the fruiting body and the cultured mycelium of *Polyporus confluens*. *Bioscience, Biotechnology, and Biochemistry*, *56*, 34–41.

Mizuno, T., Wasa, T., Ito, H., Suzuki, C., and Ukai, N. (1992b). Antitumor active polysaccharides isolated from the fruiting body of *Hericium erinaceum*, an edible and medicinal mushroom called yamabushitake or houtou. *Bioscience, Biotechnology, and Biochemistry*, *56*, 347–348.

Mizuno, T., Kinoshita, T., Zhuang, C., Ito, H., and Mayuzumi, Y. (1995a). Antitumor active heteroglycans from niohshimeji mushroom, *Tricholoma giganteum*. *Bioscience, Biotechnology, and Biochemistry*, *59*, 568–571.

Mizuno, T., Saito, H., Nishitoba, T., and Kawagishi, H. (1995b). Antitumor active substances from mushrooms. *Food Reviews International*, *11*, 23–61.

Mizuno, T., Wang, G. Y., Zhang, J., Kawagishi, H., Nishitoba, T., and Li, J. X. (1995c). Reishi, *Ganoderma lucidum* and *Ganoderma tsugae*—Bioactive substances and medicinal effects. *Food Reviews International*, *11*, 151–166.

Mizuno, T., Yeohlui, P., Kinoshita, T., Zhuang, C., Ito, H., and Mayuzumi, Y. (1996). Antitumor activity and chemical modification of polysaccharides from niohshimeji mushroom, *Tricholoma giganteum*. *Bioscience, Biotechnology, and Biochemistry*, *60*, 30–33.

Mizuno, T., Zhuang, C., Abe, K., Okmoto, H., Kiho, T., Ukai, N., S., L., and Meijer, L. (1999). Antitumor and hypoglycemic activities of polysaccharides from the sclerotia and mycelia of *Inonotus obliquus* (Pers.: Fr.) Pil. (Aphyllophoromycetideae). *International Journal of Medicinal Mushrooms*, *1*, 301–316.

Moradali, M. F., Mostafavi, H., Ghods, S., and Hedjaroude, G. A. (2007). Immunomodulating and anticancer agents in the realm of macromycetes fungi (macrofungi). *International Immunopharmacology*, *7*, 701–724.

Mueller, A., Raptis, J., Rice, P. J., Kalbfleisch, J. H., Stout, R. D., Ensley, H. E., Browder, W., and Williams, D. L. (2000). The influence of glucan polymer structure and solution conformation on binding to (1 → 3)-β-D-glucan receptors in a human monocyte-like cell line. *Glycobiology*, *10*, 339–346.

Nemoto, J., Ohno, N., Saito, K., Adachi, Y., and Yadomae, T. (1993). Expression of interleukin-1 family messenger-Rnas by a highly branched (1 → 3)-β-D-glucan, Ol-2. *Biological and Pharmaceutical Bulletin*, *16*, 1046–1050.

Nemoto, J., Ohno, N., Saito, K., Adachi, Y., and Yadomae, T. (1994). Analysis of cytokine messenger-Rnas induced by the administration of a highly branched (1 → 3)-β-D-glucan, Ol-2. *Biological and Pharmaceutical Bulletin*, *17*, 948–954.

Nie, X., Shi, B., Ding, Y., and Tao, W. (2006). Preparation of a chemically sulfated polysaccharide derived from *Grifola frondosa* and its potential biological activities. *International Journal of Biological Macromolecules*, *39*, 228–233.

Ogawa, T. and Kaburagi, T. (1982). Glucan synthesis. 2. Synthesis of a branched D-glucotetraose, the repeating unit of the extracellular polysaccharides of *Grifola umbellate, Sclerotinia libertiana, Porodisculus pendulus*, and *Schizophyllum commune* Fries. *Carbohydrate Research, 103*, 53–64.

Ogihara, T., Iinuma, H., and Okinaga, K. (2004). Usefulness of immunomodulators for maturation of dendritic cells. *International Journal of Oncology, 25*, 453–459.

Oh, G. S., Lee, M. S., Pae, H. O., Kwon, J., Lee, S. S., Jeong, J. G., Shin, M. K., Kwon, T. O., and Chung, H. T. (2006). Effects of oral administration of *Phellinus linteus* on the production of Th1- and Th2-type cytokines in mice. *Immunopharmacology and Immunotoxicology, 28*, 281–293.

Ohkuma, T., Otagiri, K., Ikekawa, T., and Tanaka, S. (1982). Augmentation of antitumor activity by combined cryo-destruction of sarcoma 180 and protein-bound polysaccharide, EA6, isolated from *Flammulina velutipes* (Curt. ex Fr.) Sing. in ICR mice. *Journal of Pharmacobio-dynamics, 5*, 439–444.

Ohmori, T., Tamura, K., Takaoka, H., Sawai, T., Kawanishi, G., Yanahira, S., Tsuru, S., and Nomoto, K. (1988a). Component analysis of protein-bound polysaccharide (SN-C) from *Cordyceps ophioglossoides* and its effects on syngeneic murine tumors. *Chemical and Pharmaceutical Bulletin, 36*, 4505–4511.

Ohmori, T., Tamura, K., Wakaiki, A., Kawanishi, G., Tsuru, S., Yadomae, T., and Nomoto, K. (1988b). Dissociation of a glucan fraction (CO-1) from protein-bound polysaccharide of *Cordyceps ophioglossoides* and analysis of its antitumor effect. *Chemical and Pharmaceutical Bulletin, 36*, 4512–4518.

Ohno, N., Furukawa, M., Miura, N. N., Adachi, Y., Motoi, M., and Yadomae, T. (2001). Antitumor β glucan from the cultured fruit body of *Agaricus blazei. Biological and Pharmaceutical Bulletin, 24*, 820–828.

Ohno, N., Miura, N. N., Chiba, N., Adachi, Y., and Yadomae, T. (1995). Comparison of the immunopharmacological activities of triple and single-helical schizophyllan in mice. *Biological and Pharmaceutical Bulletin, 18*, 1242–1247.

Ohno, N., Miura, N. N., Nakajima, M., and Yadomae, T. (2000). Antitumor 1,3-β-glucan from cultured fruit body of *Sparassis crispa. Biological and Pharmaceutical Bulletin, 23*, 866–872.

Ohno, N., Saito, K., Nemoto, J., Kaneko, S., Adachi, Y., Nishijima, M., Miyazaki, T., and Yadomae, T. (1993). Immunopharmacological characterization of a highly branched fungal $(1 \rightarrow 3)$-β-D-glucan, OL-2, isolated from *Omphalia lapidescens. Biological and Pharmaceutical Bulletin, 16*, 414–419.

Okazaki, M., Adachi, Y., Ohno, N., and Yadomae, T. (1995). Structure activity relationship of $(1 \rightarrow 3)$-β-D-glucans in the induction of cytokine production from macrophages, in vitro. *Biological and Pharmaceutical Bulletin, 18*, 1320–1327.

Okuzawa, M., Shinohara, H., Kobayashi, T., Iwamoto, M., Toyoda, M., and Tanigawa, N. (2002). PSK, a protein-bound polysaccharide, overcomes defective maturation of dendritic cells exposed to tumor-derived factors in vitro. *International Journal of Oncology, 20*, 1189–1195.

Ooi, L.-S. M., Ooi, V. E. C., and Fung, M. C. (2002). Induction of gene expression of immunomodulatory cytokines in the mouse by a polysaccharide from *Ganoderma lucidum* (Curt.: Fr.) P. Karst. (Aphyllophoromycetideae). *International Journal of Medicinal Mushrooms, 4*, 27–35.

Ooi, V. E. and Liu, F. (2000). Immunomodulation and anti-cancer activity of polysaccharide-protein complexes. *Current Medicinal Chemistry, 7*, 715–729.

Ooi, V. E. C. and Liu, F. (1999). A review of pharmacological activities of mushroom polysaccharides. *International Journal of Medicinal Mushrooms, 1*, 195–206.

Pacheco-Sanchez, M., Boutin, Y., Angers, P., Gosselin, A., and Tweddell, R. J. (2006). A bioactive (1 → 3)-, (1 → 4)-β-D-glucan from *Collybia dryophila* and other mushrooms. *Mycologia, 98*, 180–185.

Pacheco-Sanchez, M., Boutin, Y., Angers, P., Gosselin, A., and Tweddell, R. J. (2007). Inhibitory effect of CDP, a polysaccharide extracted from the mushroom *Collybia dryophila*, on nitric oxide synthase expression and nitric oxide production in macrophages. *European Journal of Pharmacology, 555*, 61–66.

Palleschi, A., Bocchinfuso, G., Coviello, T., and Alhaique, F. (2005). Molecular dynamics investigations of the polysaccharide scleroglucan: First study on the triple helix structure. *Carbohydrate Research, 340*, 2154–2162.

Park, S. K., Kim, G. Y., Lim, J. Y., Kwak, J. Y., Bae, Y. S., Lee, J. D., Oh, Y. H., Ahn, S. C., and Park, Y. M. (2003). Acidic polysaccharides isolated from *Phellinus linteus* induce phenotypic and functional maturation of murine dendritic cells. *Biochemical and Biophysical Research Communications, 312*, 449–458.

Peng, Y., Zhang, L., Zeng, F., and Kennedy, J. F. (2005). Structure and antitumor activities of the water-soluble polysaccharides from *Ganoderma tsugae* mycelium. *Carbohydrate Polymers, 59*, 385–392.

Pramanik, M., Mondal, S., Chakraborty, I., Rout, D., and Islam, S. S. (2005). Structural investigation of a polysaccharide (Fr. II) isolated from the aqueous extract of an edible mushroom, *Pleurotus sajor-caju*. *Carbohydrate Research, 340*, 629–636.

Ross, G. D., Cain, J. A., Myones, B. L., Newman, S. L., and Lachmann, P. J. (1987). Specificity of membrane complement receptor type 3 (CR3) for β-glucans. *Complement, 4*, 61–74.

Rout, D., Mondal, S., Chakraborty, I., Pramanik, M., and Islam, S. S. (2004). Structural characterisation of an immunomodulating polysaccharide isolated from aqueous extract of *Pleurotus florida* fruit-bodies. *Medicinal Chemistry Research, 13*, 509–517.

Rout, D., Mondal, S., Chakraborty, I., Pramanik, M., and Islam, S. S. (2005). Chemical analysis of a new (1 → 3)-, (1 → 6)-branched glucan from an edible mushroom, *Pleurotus florida*. *Carbohydrate Research, 340*, 2533–2539.

Ruan, Z., Su, J., Dai, H., and Wu, M. (2005). Characterization and immunomodulating activities of polysaccharide from *Lentinus edodes*. *International Immunopharmacology, 5*, 811–820.

Saito, K., Nishijima, M., and Miyazaki, T. (1989). Studies on fungal polysaccharides 35. Structural-analysis of an acidic polysaccharide from *Ganoderma lucidum*. *Chemical and Pharmaceutical Bulletin, 37*, 3134–3136.

Saito, K., Nishijima, M., Ohno, N., Yadomae, T., and Miyazaki, T. (1992). Studies on fungal polysaccharide 38. Structure and antitumor activity of the less-branched derivatives of an alkali-soluble glucan isolated from *Omphalia lapidescens*. *Chemical and Pharmaceutical Bulletin, 40*, 261–263.

Sakagami, H., Aoki, T., Simpson, A., and Tanuma, S. I. (1991). Induction of immunopotentiation activity by a protein-bound polysaccharide, Psk. *Anticancer Research, 11*, 993–999.

Sakagami, H., Kim, F., and Konno, K. (1990). Stimulation of human peripheral-blood polymorphonuclear cell iodination by Psk subfractions. *Anticancer Research*, *10*, 697–702.

Sanzen, I., Imanishi, N., Takamatsu, N., Konosu, S., Mantani, N., Terasawa, K., Tazawa, K., Odaira, Y., Watanabe, M., Takeyama, M., and Ochiai, H. (2001). Nitric oxide-mediated antitumor activity induced by the extract from *Grifola frondosa* (maitake mushroom) in a macrophage cell line, RAW264.7. *Journal of Experimental and Clinical Cancer Research*, *20*, 591–597.

Sarangi, I., Ghosh, D., Bhutia, S. K., Mallick, S. K., and Maiti, T. K. (2006). Anti-tumor and immunomodulating effects of *Pleurotus ostreatus* mycelia-derived proteoglycans. *International Immunopharmacology*, *6*, 1287–1297.

Sasaki, T., Takasuka, N., Chihara, G., and Maeda, Y. Y. (1976). Antitumor activity of degraded products of lentinan—Its correlation with molecular-weight. *Gann*, *67*, 191–195.

Shao, B. M., Dai, H., Xu, W., Lin, Z. B., and Gao, X. M. (2004). Immune receptors for polysaccharides from *Ganoderma lucidum*. *Biochemical and Biophysical Research Communications*, *323*, 133–141.

Sharon, N. and Lis, H. (1993). Carbohydrates in cell recognition. *Scientific American*, *268*, 82–89.

Shida, M. and Sakai, N. (2004). Heterogalactans obtained from some typical edible mushrooms. *Journal of the Japanese Society for Food Science and Technology-Nippon Shokuhin Kagaku Kogaku Kaishi*, *51*, 559–562.

Smiderle, F. R., Carbonero, E. R., Mellinger, C. G., Sassaki, G. L., Gorin, P. A. J., and Iacomini, M. (2006). Structural characterization of a polysaccharide and a β-glucan isolated from the edible mushroom *Flammulina velutipes*. *Phytochemistry*, *67*, 2189–2196.

Song, K. S., Cho, S. M., Lee, J., Kim, H. M., Han, S. B., Ko, K. S., and Yoo, I. D. (1995). B-lymphocyte-stimulating polysaccharide from mushroom *Phellinus linteus*. *Chemical and Pharmaceutical Bulletin*, *43*, 2105–2108.

Sorimachi, K., Akimoto, K., Ikehara, Y., Inafuku, K., Okubo, A., and Yamazaki, S. (2001). Secretion of TNF-alpha, IL-8 and nitric oxide by macrophages activated with *Agaricus blazei* Murill fractions in vitro. *Cell Structure and Function*, *26*, 103–108.

Suga, T., Shiio, T., Maeda, Y. Y., and Chihara, G. (1984). Antitumor activity of lentinan in murine syngeneic and autochthonous hosts and its suppressive effect on 3-methylcholanthrene-induced carcinogenesis. *Cancer Research*, *44*, 5132–5137.

Suga, T., Yoshihama, T., Tsuchiya, Y., Shiio, T., Maeda, Y. Y., and Chihara, G. (1989). Prevention of tumour metastasis and recurrence of DBA/2.MC.CS-1, DBA/2.MC.CS-T fibrosarcoma, MH-134 hepatoma and other murine tumours using lentinan. *International Journal of Immunotherapy*, *5*, 187–193.

Sullivan, R., Smith, J. E., and Rowan, N. J. (2006). Medicinal mushrooms and cancer therapy—Translating a traditional practice into Western medicine. *Perspectives in Biology and Medicine*, *49*, 159–170.

Surenjav, U., Zhang, L. N., Xu, X. J., Zhang, M., Cheung, P. C. K., and Zeng, F. B. (2005). Structure, molecular weight and bioactivities of $(1 \rightarrow 3)$-β-D-glucans and its sulfated derivatives from four kinds of *Lentinus edodes*. *Chinese Journal of Polymer Science*, *23*, 327–336.

Surenjav, U., Zhang, L., Xu, X., Zhang, X., and Zeng, F. (2006). Effects of molecular structure on antitumor activities of $(1 \rightarrow 3)$-β-D-glucans from different *Lentinus edodes*. *Carbohydrate Polymers*, *63*, 97–104.

Tabata, K., Ito, W., Kojima, T., Kawabata, S., and Misaki, A. (1981). Ultrasonic degradation of schizophyllan, an anti-tumor polysaccharide produced by *Schizophyllum commune* Fries. *Carbohydrate Research*, *89*, 121–135.

Tani, M., Tanimura, H., Yamaue, H., Tsunoda, T., Iwahashi, M., Noguchi, K., Tamai, M., Hotta, T., and Mizobata, S. (1993). Augmentation of lymphokine-activated killer-cell activity by lentinan. *Anticancer Research*, *13*, 1773–1776.

Tao, Y., Zhang, L., and Cheung, P. C. K. (2006). Physicochemical properties and antitumor activities of water-soluble native and sulfated hyperbranched mushroom polysaccharides. *Carbohydrate Research*, *341*, 2261–2269.

Tao, Y. Z. and Zhang, L. (2006). Determination of molecular size and shape of hyperbranched polysaccharide in solution. *Biopolymers*, *83*, 414–423.

Taylor, P. R., Brown, G. D., Reid, D. M., Willment, J. A., Martinez-Pomares, L., Gordon, S., and Wong, S. Y. C. (2002). The β-glucan receptor, dectin-1, is predominantly expressed on the surface of cells of the monocyte/macrophage and neutrophil lineages. *Journal of Immunology*, *169*, 3876–3882.

Thornton, B. P., Vetvicka, V., Pitman, M., Goldman, R. C., and Ross, G. D. (1996). Analysis of the sugar specificity and molecular location of the β-glucan-binding lectin site of complement receptor type 3 (CD11b/CD18). *Journal of Immunology*, *156*, 1235–1246.

Tsukagoshi, S., Hashimoto, Y., Fujii, G., Kobayashi, H., Nomoto, K., and Orita, K. (1984). Krestin (Psk). *Cancer Treatment Reviews*, *11*, 131–155.

Ukawa, Y., Ito, H., and Hisamatsu, M. (2000). Antitumor effects of $(1 \rightarrow 3)$-β-D-glucan and $(1 \rightarrow 6)$-β-D-glucan purified from newly cultivated mushroom, Hatakeshimeji (*Lyophyllum decastes* Sing.). *Journal of Bioscience and Bioengineering*, *90*, 98–104.

Unursaikhan, S., Xu, X., Zeng, F., and Zhang, L. (2006). Antitumor activities of O-sulfonated derivatives of $(1 \rightarrow 3)$-α-D-glucan from different *Lentinus edodes*. *Bioscience, Biotechnology, and Biochemistry*, *70*, 38–46.

Usui, S., Tomono, Y., Sakai, M., Kiho, T., and Ukai, S. (1995). Preparation and antitumor activities of β-$(1 \rightarrow 6)$ branched $(1 \rightarrow 3)$-β-D-glucan derivatives. *Biological and Pharmaceutical Bulletin*, *18*, 1630–1636.

Vinogradov, E., Petersen, B. O., Duus, J. O., and Wasser, S. (2004). The structure of the glucuronoxylomannan produced by culinary-medicinal yellow brain mushroom (*Tremella mesenterica* Ritz.: Fr., Heterobasidiomycetes) grown as one cell biomass in submerged culture. *Carbohydrate Research*, *339*, 1483–1489.

Vivier, E., Nunes, J. A., and Vely, F. (2004). Natural killer cell signaling pathways. *Science*, *306*, 1517–1519.

Wada, T., Wakamatsu, Y., Bannai, K., Kato, M., Oguchi, Y., Matsunaga, K., Ando, T., and Nomoto, K. (2002). Suppression mechanism of angiogenesis by PSK. *Annals of Cancer Research and Therapy*, *10*, 93–106.

Wang, G. Y., Zhang, J., Mizuno, T., Zhuang, C., Ito, H., Mayuzumi, H., Okamoto, H., and Li, J. (1993). Antitumor active polysaccharides from the Chinese mushroom Songshan Lingzhi, the fruiting body of *Ganoderma tsugae*. *Bioscience, Biotechnology, and Biochemistry*, *57*, 894–900.

Wang, H. X., Ng, T. B., Liu, W. K., Ooi, V. E. C., and Chang, S. T. (1996a). Polysaccharide-peptide complexes from the cultured mycelia of the mushroom *Coriolus versicolor* and their culture medium activate mouse lymphocytes and macrophages. *International Journal of Biochemistry and Cell Biology*, *28*, 601–607.

Wang, H. X., Ng, T. B., Ooi, V. E. C., Liu, W. K., and Chang, S. T. (1996b). A polysaccharide-peptide complex from cultured mycelia of the mushroom *Tricholoma mongolicum* with immunoenhancing and antitumor activities. *Biochemistry and Cell Biology*, *74*, 95–100.

Wang, S. Y., Hsu, M. L., Hsu, H. C., Tzeng, C. H., Lee, S. S., Shiao, M. S., and Ho, C. K. (1997). The anti-tumor effect of *Ganoderma lucidum* is mediated by cytokines released from activated macrophages and T lymphocytes. *International Journal of Cancer*, *70*, 699–705.

Wang, Y., Zhang, L., Li, Y., Hou, X., and Zeng, F. (2004). Correlation of structure to antitumor activities of five derivatives of a β-glucan from *Poria cocos* sclerotium. *Carbohydrate Research*, *339*, 2567–2574.

Wang, Z. J., Luo, D. H., and Liang, Z. Y. (2004). Structure of polysaccharides from the fruiting body of *Hericium erinaceus* Pers. *Carbohydrate Polymers*, *57*, 241–247.

Wasser, S. P. (2002). Medicinal mushrooms as a source of antitumor and immunomodulating polysaccharides. *Applied Microbiology and Biotechnology*, *60*, 258–274.

Wasser, S. P., Tan, K. K., and Elisashvili, V. I. (2002). Hypoglycemic, interferonogenous, and immunomodulatory activity of Tremellastin from the submerged culture of *Tremella mesenterica* Retz.: Fr. (Heterobasidiomycetes). *International Journal of Medicinal Mushrooms*, *4*, 215–227.

Wong, S. M., Wong, K. K., Chiu, L. C. M., and Cheung, P. C. K. (2007). Non-starch polysaccharides from different developmental stages of *Pleurotus tuber-regium* inhibited the growth of human acute promyelocytic leukemia HL-60 cells by cell-cycle arrest and/or apoptotic induction. *Carbohydrate Polymers*, *68*, 206–217.

Yalin, W., Ishurd, O., Cuirong, S., and Yuanjiang, P. (2005). Structure analysis and antitumor activity of $(1 \rightarrow 3)$-β-D-glucans (Cordyglucans) from the mycelia of *Cordyceps sinensis*. *Planta Medica*, *71*, 381–384.

Yang, J., Zhang, W., Shi, P., Chen, J., Han, X., and Wang, Y. (2005). Effects of exopolysaccharide fraction (EPSF) from a cultivated *Cordyceps sinensis* fungus on c-Myc, c-Fos, and VEGF expression in B16 melanoma-bearing mice. *Pathology Research and Practice*, *201*, 745–750.

Yang, Q. Y., Jong, S. C., Li, X. Y., Zhou, J. X., Chen, R. T., and Xu, L. Z. (1992). Antitumor and immunomodulating activities of the polysaccharide-peptide (Psp) of *Coriolus versicolor*. *EOS-Rivista di Immunologia ed Immunofarmacologia*, *12*, 29–34.

Yap, A. T., and Ng, M. L. (2003). Immunopotentiating properties of lentinan $(1 \rightarrow 3)$-β-D-glucan extracted from culinary-medicinal shiitake mushroom *Lentinus edodes* (Berk.) Singer (Agaricomycetideae). *International Journal of Medicinal Mushrooms*, *5*, 339–358.

Yokoyama, W. M., Kim, S. J., and French, A. R. (2004). The dynamic life of natural killer cells. *Annual Review of Immunology*, *22*, 405–429.

Yoshida, I., Kiho, T., Usui, S., Sakushima, M., and Ukai, S. (1996). Polysaccharides in fungi. 37. Immunomodulating activities of carboxymethylated derivatives of linear $(1 \rightarrow 3)$-α-D-glucans extracted from the fruiting bodies of *Agrocybe cylindracea* and *Amanita muscaria*. *Biological and Pharmaceutical Bulletin*, *19*, 114–121.

Yoshikawa, R., Yanagi, H., Hashimoto-Tamaoki, T., Morinaga, T., Nakano, Y., Noda, M., Fujiwara, Y., Okamura, H., and Yamamura, T. (2004). Gene expression in response to anti-tumour intervention by polysaccharide-K (PSK) in colorectal carcinoma cells. *Oncology Reports*, *12*, 1287–1293.

Yoshioka, Y., Tabeta, R., Saito, H., Uehara, N., and Fukuoka, F. (1985). Antitumor polysaccharides from *P. ostreatus* (Fr.) Quel.: Isolation and structure of a β-glucan. *Carbohydrate Research, 140*, 93–100.

Yui, T., Ogawa, K., Kakuta, M., and Misaki, A. (1995). Chain Conformation of a glucurono-xylo-mannan isolated from fruit body of *Tremella fuciformis* Berk. *Journal of Carbohydrate Chemistry, 14*, 255–263.

Zaidman, B. Z., Yassin, M., Mahajna, J., and Wasser, S. P. (2005). Medicinal mushroom modulators of molecular targets as cancer therapeutics. *Applied Microbiology and Biotechnology, 67*, 453–468.

Zekovic, D. B., Kwiatkowski, S., Vrvic, M. M., Jakovljevic, D., and Moran, C. A. (2005). Natural and modified $(1 \rightarrow 3)$-β-D-glucans in health promotion and disease alleviation. *Critical Reviews in Biotechnology, 25*, 205–230.

Zhang, A. Q., Zhang, J. S., Tang, Q. U., Jia, W., Yang, Y., Liu, Y. F., Fan, J. M., and Pan, Y. J. (2006). Structural elucidation of a novel fucogalactan that contains 3-*O*-methyl rhamnose isolated from the fruiting bodies of the fungus, *Hericium erinaceus*. *Carbohydrate Research, 341*, 645–649.

Zhang, J., Wang, G., Li, H., Zhuang, C., Ito, H., Mayuzumi, H., Okamoto, H., and Li, J. (1994a). Antitumor active protein-containing glycans from the Chinese mushroom Songshan Lingzhi, *Ganoderma tsugae* mycelium. *Bioscience, Biotechnology, and Biochemistry, 58*, 1202–1205.

Zhang, J., Wang, G., Li, H., Zhuang, C., Ito, H., Suzuki, C., Okamoto, H., and Li, J. (1994b). Antitumor polysaccharides from a Chinese mushroom, "Yuhuangmo," the fruiting body of *Pleurotus citrinopileatus*. *Bioscience, Biotechnology, and Biochemistry, 58*, 1195–1201.

Zhang, J. S., Tang, Q. J., Zimmerman-Kordmann, M., Reutter, W., and Fan, H. (2002). Activation of B lymphocytes by GLIS, a bioactive proteoglycan from *Ganoderma lucidum*. *Life Sciences, 71*, 623–638.

Zhang, L., Li, X. L., Zhou, Q., Zhang, X. F., and Chen, R. Q. (2002). Transition from triple helix to coil of lentinan in solution measured by SEC, viscometry, and C-13 NMR. *Polymer Journal, 34*, 443–449.

Zhang, M., Zhang, L., and Cheung, P. C. (2003). Molecular mass and chain conformation of carboxymethylated derivatives of β-glucan from sclerotia of *Pleurotus tuber-regium*. *Biopolymers, 68*, 150–159.

Zhang, M., Cheung, P. C. K., Zhang, L., Chiu, C. M., and Ooi, V. E. C. (2004). Carboxymethylated β-glucans from mushroom sclerotium of *Pleurotus tuber-regium* as novel water-soluble anti-tumor agent. *Carbohydrate Polymers, 57*, 319–325.

Zhang, M., Cheung, P. C. K., Chiu, L. C. M., Wong, E. Y. L., and Ooi, V. E. C. (2006a). Cell-cycle arrest and apoptosis induction in human breast carcinoma MCF-7 cells by carboxymethylated β-glucan from the mushroom sclerotia of *Pleurotus tuber-regium*. *Carbohydrate Polymers, 66*, 455–462.

Zhang, M., Chiu, L. C. M., Cheung, P. C. K., and Ooi, V. E. C. (2006b). Growth-inhibitory effects of a β-glucan from the mycelium of *Poria cocos* on human breast carcinoma MCF-7 cells: Cell-cycle arrest and apoptosis induction. *Oncology Reports, 15*, 637–643.

Zhang, M., Cui, S. W., Cheung, P. C. K., and Wang, Q. (2007). Polysaccharides from mushrooms: A review on their isolation process, structural characteristics and antitumor activity. *Trends in Food Science and Technology, 18*, 4–19.

Zhang, P. and Cheung, P. C. K. (2002). Evaluation of sulfated *Lentinus edodes* α-(1 $\rightarrow$ 3)-D-glucan as a potential antitumor agent. *Bioscience, Biotechnology, and Biochemistry*, *66*, 1052–1056.

Zhang, W. Y., Wang, Y., and Hou, Y. Y. (2004). Effects of Chinese medicinal fungus water extract on tumor metastasis and some parameters of immune function. *International Immunopharmacology*, *4*, 461–468.

Zhang, W. Y., Yang, J. Y., Chen, J. P., Hou, Y. Y., and Han, X. D. (2005). Immunomodulatory and antitumour effects of an exopolysaccharide fraction from cultivated *Cordyceps sinensis* (Chinese caterpillar fungus) on tumour-bearing mice. *Biotechnology and Applied Biochemistry*, *42*, 9–15.

Zhu, T., Guo, J., Collins, L., Kelly, J., Xiao, Z. J., Kim, S. H., and Chen, C. Y. (2007). *Phellinus linteus* activates different pathways to induce apoptosis in prostate cancer cells. *British Journal of Cancer*, *96*, 583–590.

Zhuang, C. and Mizuno, T. (1995). Niohshimeji, *Tricholoma giganteum* —Antitumor-active heteroglycans. *Food Reviews International*, *11*, 197–202.

Zhuang, C., Mizuno, T., Shimada, A., Ito, H., Suzuki, C., Mayuzumi, Y., Okamoto, H., Ma, Y., and Li, J. X. (1993). Antitumor protein-containing polysaccharides from a Chinese mushroom Fengweigu or Houbitake, *Pleurotus sajor-caju* (Fr) Sings. *Bioscience, Biotechnology, and Biochemistry*, *57*, 901–906.

Zhuang, C., Mizuno, T., Ito, H., Shimura, K., Sumiya, T., and Kawade, M. (1994a). Antitumor-activity and immunological property of polysaccharides from the mycelium of liquid cultured *Grifola frondosa*. *Journal of the Japanese Society for Food Science and Technology-Nippon Shokuhin Kagaku Kogaku Kaishi*, *41*, 724–732.

Zhuang, C., Mizuno, T., Ito, H., Shimura, K., Sumiya, T., and Kawade, M. (1994b). Fractionation and antitumor-activity of polysaccharides from *Grifola frondosa* mycelium. *Bioscience, Biotechnology, and Biochemistry*, *58*, 185–188.

Zimmerman, J. W., Lindermuth, J., Fish, P. A., Palace, G. P., Stevenson, T. T., and Demong, D. E. (1998). A novel carbohydrate-glycosphingolipid interaction between a β-(1 $\rightarrow$ 3)-glucan immunomodulator, PGG-glucan, and lactosylceramide of human leukocytes. *Journal of Biological Chemistry*, *273*, 22014–22020.

CHAPTER 6

Regulatory Issues of Mushrooms as Functional Foods and Dietary Supplements: Safety and Efficacy

Solomon P. Wasser and Eden Akavia

International Center of Biotechnology and Biodiversity of Fungi, Institute of Evolution, University of Haifa, Haifa, Israel

CONTENTS

6.1 INTRODUCTION

In the second half of the twentieth century, mushroom-producing technologies grew enormously. In 2004, the value of world mushroom production was estimated at about U.S. $40 billion, which is almost the same as the value of coffee production (Wasser et al., 2004; Chang and Miles, 2004; Chang, 2006).

Many pharmaceutical substances with potent and unique valuable properties are used worldwide. Unique anticancer preparations have been developed based on certain components such as the polysaccharides lentinan [from *Lentinus edodes*

Mushrooms as Functional Foods, Edited by Peter C. K. Cheung

(Berk.) Singer], krestin [from *Trametes versicolor* (L.:Fr.) Lloyd], maitake D-fraction [from *Grifola frondosa* (Dicks.: Fr.) S. F. Gray], schizophyllan (from *Schizophyllum commune* Fr.: Fr.), and befungin [from *Inonotus obliquus* (Pers.: Fr.) Pilat] (Mizuno, 1999; Chang, 1999, 2001; Wasser and Weis, 1999a, b; Stamets, 2000, 2002; Wasser et al., 2000a, b, 2001, 2004; Wasser, 2002; Chang and Buswell, 2003; Chang and Miles, 2004).

Most mushroom-derived preparations and substances find use not as pharmaceuticals ("real" medicines) but rather as a novel class of products by a variety of names: dietary supplements (DSs), functional foods, nutraceuticals, nutriceuticals, phytochemicals, mycochemicals, biochemopreventives, and designer foods (Chang and Buswell, 1999, 2003; Zeisel, 1999; Wasser et al., 2000a, b, 2001, 2004; Chang and Miles, 2004; Chang, 2006). These terms vary in meaning from country to country, as does the regulation of these products. In the United States, the formal definition under the Dietary Supplements Health and Education Act (DSHEA) (1994) exists only for DSs. "Functional foods" has no legal status or general acceptance but is used as a definition in the dietetic profession. "Nutraceuticals" and "nutriceuticals" are also accepted definitions in the nutrition/science community but are not embodied in law or regulation. "Medical foods" are regulated by the U.S. Food and Drug Administration (FDA) Office of Special Nutritionals on a case-by-case basis as enteral foods or foods for sick infants (Smith et al., 1996, 2000).

DSs are ingredients obtained from foods, plants, and mushrooms (fungi) that are taken, without further modification, separately from foods for their presumed health-enhancing benefits (McNamara, 1999). The term DS includes one or more of certain dietary ingredients—a vitamin, mineral, amino acid, herb, or other botanical—or it is a dietary substance used to supplement the diet by increasing the total dietary intake and is intended for ingestion in the form of a capsule, powder, softgel, or gelcap and is not represented as a conventional food or as a sole item of a meal or a diet (DSHEA, Public Law 103–417, 1994). Supplements are easy to add to the daily diet and are often the first steps consumers undertake toward greater nutritional awareness and the adoption of a healthy lifestyle. The DS pyramid has been developed recently, which, just like the food pyramid, has its own rules.

At the same time, the increased interest in traditional remedies for various physiological disorders and the recognition of numerous biological activity of mushroom products have led to the coining of the term "mushroom nutriceuticals" (Chang and Buswell, 1996–2003; Chang, 1999, 2006; Wasser et al., 2000a, b, 2001, 2004; Chang and Miles, 2004), not to be confused with nutraceuticals, functional foods, and pharmaceuticals. A mushroom nutriceutical is a refined, or partially refined, extract or dried biomass from either mycelium or the fruiting body of a mushroom, which is consumed in the form of capsules or tablets as a DS (not a food) and has potentially therapeutic applications. Regular intake may enhance the immune response of the human body, thereby increasing resistance to disease and in some cases causing regression of the disease state. Thus, acting as immunopotentiators, mushroom preparations modify host biological responses (Hobbs, 1995;

Wasser and Weis, 1999a, b; Wasser, 2002; Wasser et al., 2004; Chang, 2006) and are also known as biological response modifiers (BRMs).

The term DS (nutriceutical) is broader than BRM and better describes the qualities of this class of substances. Another shortcoming of the term "biological response modifiers" in this context is that many of the mushroom preparations possess particular physiological effects, such as lowering blood cholesterol or hepatoprotectivie activity (Hobbs, 1995; Chang and Buswell, 1999, 2003; Wasser and Weis, 1999a, b; Stamets, 2000, 2002).

Several types of culinary–medicinal products are available on the market today:

1. Artificially cultivated fruit body powders, hot-water or alcohol extracts of these, or the same extract concentrates and their mixtures
2. Dried and pulverized preparations of the combined substrate, mycelium, and mushroom primordia after inoculation of edible semisolid medium (usually grains)
3. Biomass or extracts from mycelium harvested from submerged liquid culture grown in a fermentation tank
4. Naturally growing, dried mushroom fruiting bodies in the form of capsules or tablets

There is no doubt that medicinal mushroom-based products can serve as superior DSs. The problem is that mushroom-based DSs are so diverse, and there are currently no standard protocols for guaranteeing their product quality and critical testing. There is a serious need for improved quality and legal control, which are essential both to increase and maintain consumers' confidence and to meet current and future standards set by regulatory authorities (Chang and Buswell, 1999, 2003; Wasser et al., 2000a, b, 2001, 2004; Chang and Miles, 2004; Chang, 2006).

It should be mentioned that most suppliers and distributors of mushroom DSs provide very little and highly variable information on the source of their materials, ways of preparation, and composition. The field of mushroom DSs today is very far from unification and standardization (Wasser et al., 2000a, b; Chang and Buswell, 2003).

One of the difficulties facing this area is the lack of international consistency in the regulatory management of DSs per se. In some countries regulation is effectively based on a three-category system, that is, foods, medicines, and DSs (or alternate term). In other countries, regulations pertaining to DSs (or other) generally sit under a broader (two-category) legislative system for either foods or medicines. By way of example of the potential confusion inherent in some of these systems, the New Zealand Dietary Supplements Regulations (NZDSR) regulate products that are predominantly therapeutic products in Australia, yet the NZDSR place them under the Food Act (1981) in New Zealand. Countries that have instituted a three-category approach include New Zealand, the United States, Canada (proposed system), Europe, India, and Japan whereas countries such as Australia and the United Kingdom use a two-category system, that is, foods and medicines (under therapeutic products) [Australia New Zealand Food Authority (ANZFA), 2002].

6.2 LEGAL AND REGULATORY ISSUES OF INTRODUCING AND CONTROLLING DIETARY SUPPLEMENTS FROM MEDICINAL MUSHROOMS IN DIFFERENT COUNTRIES

6.2.1 World Health Organization Guidelines

In 1991, the World Health Organization (WHO) published its *Guidelines for the Assessment of Herbal Medicines*. This document was created over a period of five years (1986–1991) at several international conferences of drug regulatory authorities and is based on the WHO's recognition that 80% of the world's population in developing countries relies on traditional medicine; "a major part of the traditional therapies involves the use of active constituents of plant extracts"; and "considerable growth has occurred in popular, official, and commercial interest in the use of natural products."

The objective of the WHO guidelines "is to define basic criteria for the evaluation of quality, safety, and efficacy" of all herbal (as usual, including mushrooms among the herbs) medicines. "As a general rule in this assessment, traditional experience means that long-term use as well as the medical, historical, and ethnological background of those products shall be taken into account" (WHO, 1991). Depending on each country's situation, "the definition of long-term use may vary, but would be at least several decades. . . . Prolonged and apparently uneventful use of a substance usually offers testimony of its safety."

According to Alkerele (1992), "safety should be overriding criterion in the selection of herbal medicines for use in health service systems." The WHO guidelines call for various assessments of quality, safety, efficacy, and intended use of herbal medicines. The guidelines call for reference to pharmacopeia monographs, if they exist. If none exist, a manufacturer applying for marketing licenses or registration should supply a monograph with the same components as an official pharmacopeia. All procedures should correspond to good manufacturing practices (GMPs), and there needs to be standard testing of the product in its final packaging form. The assessment should also make a distinction between old and new combination products (Wasser et al., 2000a, b, 2004).

With regard to safety, "a guiding principle should be that if the product has been traditionally used without demonstrated harm, no specific restrictive regulatory action should be undertaken unless new evidence demands a revised risk-benefit assessment" (Alkerele, 1992).

Finally, consumer product information is recommended, including a quantitative list of active ingredients, dosage, dosage form, indications, mode of administration, duration of use, any major adverse effects, contraindications, warnings, and so on (Wasser et al., 2000a, b, 2004; Chang and Buswell, 2003; Chang, 2006).

6.2.2 Codex Alimentarius

The Codex Alimentarius Commission implements the joint Food and Agriculture Organization (FAO, 2005) and WHO Food Standards Programme, the purpose of which is to protect the health of consumers and to ensure fair practices in the food trade. The Codex Alimentarius (Latin, meaning food law or code) is a collection of internationally adopted food standards presented in a uniform manner. It also includes provisions of an advisory nature in the form of codes of practice, guidelines, and other recommended measures to assist in achieving the purposes of the Codex Alimentarius. The commission has expressed the view that codes of practice might provide useful checklists of requirements for national food control or enforcement authorities. The publication of the Codex Alimentarius is intended to guide and promote the elaboration and establishment of definitions and requirements for foods, to assist in their harmonization, and, in doing so, to facilitate international trade (FAO, 2005).

The twenty-eighth session of the Codex Alimentarius Commission was held during July 2005. This event has been the subject of considerable controversy, in part because many member countries regulate substances as therapeutic goods or pharmaceuticals and not as foods (if they were not foods, they would be excluded from the Codex Alimentarius). The Codex seeks not to ban supplements but to subject them to labeling and composition requirements. Some groups have pointed to greater concerns related to restrictions on dietary supplement ingredients in Europe via the European Food Supplements Directive (which utilizes approved lists of ingredients and ingredient forms) and potentially restrictive dosage limits to be based on a Codex model via the FAO and WHO Nutrient Risk Assessment Project. The Codex regulations have been rejected by the FDA in the United States but accepted by the European Union (EU) trading bloc.

6.2.3 United States

Nearly all Americans have benefited from natural nutritional supplements of some kind in their lives. Whether taking a multivitamin or eating breakfast cereal fortified with nutrients, people all over the world live healthier lives by supplementing their diets. Surveys estimate that more than half the U.S. population, or more than 100 million Americans, use DSs such as vitamins, minerals, and herbs as a safe and natural way to maintain good health and supplement inadequate diets (Natural Products Association, 2007).

There are approximately 10,000 natural product stores in the United States. In 2004, The U.S. nutrition industry reached sales of \$68.6 billion. The industry is broken down into the following categories (in billions): functional foods \$24.5, supplements \$20.3, natural/organic foods \$18.4, and natural personal care \$5.5.

Natural/organic foods and natural personal care products are the fastest growing segment of the industry, at a rate of 10% per year, taking 35% of the market. In the United States, virtually all facets of DS manufacturing, labeling, and advertising are covered by extensive regulations issued and enforced by the Center for Food Safety and Applied Nutrition (CFSAN, 2005a) and the U.S. Federal Trade Commission (FTC, 2001).

For decades the FDA regulated DSs as foods to ensure that they were safe and wholesome and that their labeling was truthful and not misleading. An important facet of ensuring safety was the FDA's evaluation of the safety of all new ingredients, including those used in DSs, under the 1958 Food Additive Amendments to the Federal Food, Drug, and Cosmetic Act (Zeisel, 1999). GRAS is an acronym for the phrase "generally recognized as safe." Under Sections 201(s) and 409 of the Federal Food, Drug, and Cosmetic Act, any substance that is intentionally added to food is a food additive, that is, subject to premarket review and approval by FDA, unless the substance is generally recognized, among qualified experts, as having been adequately shown to be safe under the conditions of its intended use or unless the use of the substance is otherwise excluded from the definition of a food additive. For example, substances whose use meets the definition of a pesticide, a dietary ingredient of a dietary supplement, a color additive, a new animal drug, or a substance approved for such use prior to September 6, 1958, are excluded from the definition of food additive. Sections 201(s) and 409 were enacted in 1958 as part of the Food Additives Amendment to the act. While it is impracticable to list all ingredients whose use is GRAS, the FDA published a partial list of food ingredients whose use is GRAS to aid the industry's understanding of what did not require approval. The use of a food substance may be GRAS either through scientific procedures or, for a substance used in food before 1958, through experience based on common use in food. Interestingly, the 4-hydroxymethylbenzenediazonium (HMBD) ion has been detected in the commonly cultivated and consumed mushroom *Agaricus bisporus* at levels near 0.6 ppm (Ross et al., 1982). HMBD is a fungal metabolite of agaritine, a naturally occurring substance found in *Agaricus* species. HMBD has been implicated in causing DNA strand breaks at 1.2 mg per 70-kg person (Hiramoto et al., 1995). The calculated safe dose is less than 4 g mushroom per day or one meal every 100 days. However, since mushrooms are GRAS, HMBD is exempt in this food.

However, with passage of the DSHEA, Congress amended the Food, Drug, and Cosmetic Act to include several provisions that apply only to DSs and dietary ingredients of DSs. As a result of these provisions, dietary ingredients used in DSs are no longer subject to the premarket safety evaluations required for other new food ingredients or for new uses of old food ingredients. However, they must satisfy several other safety provisions. The Council for Responsible Nutrition (CRN, 2002) outlines different regulations as they apply to foods, DSs, and drugs when reviewing products.

The DSHEA acknowledges that millions of consumers believe that DSs may help to augment daily diets and provide health benefits. The intent of Congress in enacting the DSHEA was to meet the concerns of consumers and manufacturers to help ensure that safe and appropriately labeled products are available to those who want to use them. In the findings associated with the DSHEA, Congress stated that there may be a positive relationship between sound dietary practice and good health and that, although further scientific research is needed, there may be a connection between dietary supplement use, reduced health care expenses, and disease prevention.

One of the most important matters of regulation, in which the FDA has invested much effort in recent years, is the labeling of DS products. The new nutrition-labeling format for DSs includes a supplement facts box, which lists vitamins and minerals always in the same order. The new labels show the amount of each nutrient in terms of metric quantities (e.g., 60 mg of vitamin C) and in terms of percentages of the daily value (DV), if one has been established. All ingredients in the product must be listed, either within the fact box or in a separate list of ingredients below the fact box. For extracts, additional information may be provided including the solvent used and the concentration of the extract (Wasser et al., 2000a, b, 2004).

The final rules for structure–function (SF) claims for DSs are given under the DSHEA. In April 1988, the FDA published proposed regulations that included a highly controversial redefinition of the word "disease." Many industry and consumer groups viewed this as an attempt to restrict the scope of the SF claims under the DSHEA. Since then, the FDA has received 235,000 comments from the public about this measure: 213,000 as form letters circulated by consumer and trade groups and 22,000 as individual letters from consumers, members of industry, and other interested parties (Wasser et al., 2000a, b, 2004; Blumenthal, 2001).

Formerly, the definition of disease was "any deviation from, impairment of, or interruption of the normal structure of function." Now, the FDA uses the definition "damage to an organ, structure, or system..." By this alternation of the definition, the FDA automatically reduces the range or number of health claims for DSs.

On January 6, 2000, the FDA issued its final regulations on SF claims for DSs under the DSHEA of 1994 (65 FR 1000–1050). The full text is available on the Internet (FDA, 2000). In these rules, a significant shift is made in relation to SF claims for over-the-counter (OTC) drugs including DSs. In particular, the FDA has enlarged the range of SF claims by agreeing with the comments by the American Herbal Products Association (AHPA) that some claims are not disease claims but, instead, are claims that deal with the SF of the body. Thus, DSs will be able to make claims for antacid, digestive aid, short-term laxative, and other uses previously prohibited to DSs (Blumenthal, 2001).

Many examples of conditions and statements for which SF claims are allowed under the DSHEA are worked out by these regulations. They include many that are important for mushroom-derived DSs, such as "immune system function," "maintenance of cholesterol levels that are already within the normal range," "tonic," and "as part of the diet to maintain blood sugar levels." Many other examples remain disease claims, such as "anti-inflammatory," "lowers cholesterol," and "controls blood sugar in persons with insufficient insulin." Another important regulation states that "the name of a product should not contain the name or recognizable portion of the name of the disease" (Wasser et al., 2000a, b, 2004).

Thus, the FDA regulates all key steps of DS development and application in the United States. The seemingly important drawback of the FDA empowerment is that consumers have to be injured prior to banning the product on the shelf. However, if the FDA has indications that a DS is unsafe, it can issue a warning that might have an impact on the consumer.

In April 2005, the CFSAN/Office of Nutritional Products, Labeling, and Dietary Supplements released the Dietary Supplement Labeling Guide. The guide applies to both dietary supplement ingredients and finished dietary supplement products manufactured or produced in the United States as well as those produced in other countries. The guide covers all aspects of dietary supplement labeling, including requirements related to statements of identity, net quantity listing supplement facts panels, ingredient listing, health and SF claims, and premarket notifications for new dietary ingredients. There are several types of claims (CFSAN, 2005a, b):

(a) An authorized health claim is an "explicit or implied characterization of a relationship between a substance and a disease or a health-related condition." According to the FDA, a health claim "describes the effect a substance has on reducing the risk of or preventing a disease, e.g., 'calcium may reduce the risk of osteoporosis.'" This type of claim requires significant scientific agreement and must be authorized by the FDA.

(b) A qualified health claim is supported by less scientific evidence than an authorized health claim. The FDA requires that qualified claims be accompanied by a disclaimer that explains the level of the scientific evidence supporting the relationship.

(c) A SF claim describes the role of a substance intended to maintain the structure or function of the body. Structure–function claims do not require preapproval by the FDA. Structure–function claims are permitted if such statement is truthful and not misleading, including a required disclaimer, and the marketer notifies the FDA of the claim no later than 30 days after the first marketing of the product.

When the FDA learns of a supplement claim it believes violates the provisions of the Food, Drug and Cosmetic Act or regulations promulgated by the act, it sends a letter to the supplement manufacturer or marketer advising them of the questionable claims, the potential violations raised by those claims, and a request that the claims be withdrawn. The FDA posts these letters in a database on the Internet that is searchable by date, month, and type of violation. Responses to these letters are also posted when received.

The guide is intended to address questions about dietary supplement labeling issues and reflects the FDA's current position on these issues. However, it is not legally binding or precedent setting but rather is merely an interpretation of the applicable statutes and regulations to educate the industry on how the FDA believes compliance can be met (CFSAN, 2005b).

Several examples of actions the FDA has taken against violators of the health claims regulations are available. For example, in December 2001, the FDA's New York district office recommended detention without physical examination for Essence of Mushrooms capsules, 400 mg. The product, manufactured by Windsor Health Products, Kowloon, Hong Kong, was shipped as vitamins via Federal Express. However, FDA examination found accompanying labeling promoting the product for treatment of cancer. In addition, the labeling also identified the manufacturer's website, which was found to be promoting the Essence of Mushrooms as an alternative therapy for cancer. The FDA refused entry of the product.

In 2005, a company named Vitapurity, headquartered in the United States, made therapeutic claims on its website regarding Miracle Mushroom Blend. The company established that this product was a drug because it was intended for use in the cure, mitigation, treatment, or prevention of diseases. The marketing of this product with these claims violates the Federal Food, Drug, and Cosmetic Act. The claims were as follows:

- "Reishi Mushroom was comparable to hydrocortisone and aspirin in its ability to reduce inflammation."
- "When more than 2,000 patients with chronic bronchitis were given Reishi Mushroom Extract, 60 to 90 percent of these patients showed a marked improvement in health."
- "Effective in treating conditions such as stomach ulcers and high blood pressure."
- "The Shiitake Mushroom is . . . an aid in the prevention of cerebral hemorrhagic strokes."
- "Shiitake Mushrooms are promoted to fight the development and progression of cancer and AIDS. . . . These mushrooms are also said to help prevent heart disease by lowering cholesterol levels in the blood and treating infections."
- "Research in Japan shows that the mushroom itself can lower blood pressure in those with hypertension."
- "Giving Shiitake to patients with probable pre AIDS improved the patients [sic] general conditions and improved their immune status. . . . This agent may prove to be effective in the suppression of the AIDS condition."
- "The compounds contained in Maitake have the capacity to . . . inhibit tumor growth."
- "People with Type 2 Diabetes may also benefit from Maitake."
- "Maitake Mushrooms may have the ability to . . . induce apoptosis (cell death) in cancer cells."
- "Mushroom extracts may improve overall survival and quality of life for cancer patients."

The FDA warned Vitapurity the product was not GRAS and effective for the above-referenced conditions and, therefore, the product is a "new drug" and may not be legally marketed in the United States.

A similar warning was issued to Health Food Emporium in Idaho, which made similar claims regarding Garden of Life RM-10. The product contains a combination of 10 organic medicinal mushrooms which "aid the immune system to fight Viral Disorders, Allergies, Asthma, Inflammatory Bowel Disease, Hepatitis, Cancer, Candida Yeast Overgrowth, Diabetes, Parasitic Infections, Multiple Sclerosis, Psoriasis, Eczema, Aids, Rheumatoid Arthritis, Lupus, etc." Today the company lists the following statement on its website: "Statements on this website have not been evaluated by the Food and Drug Administration. These products are not intended to diagnose, treat, cure, or prevent any disease, but rather are dietary supplements intended solely for nutritional use."

In October 2006 the FDA announced a public hearing on the regulation of certain conventional foods that companies are marketing as "functional foods." The purpose of the hearing was for the agency to share its current regulatory framework and rationale regarding the safety evaluation and labeling of these foods and to solicit information and comments from interested persons on how the FDA should regulate these foods under the agency's existing legal authority [U.S. Department of Health and Human Services (DHHS), 2006].

In December 2006, the Senate passed the Dietary Supplement and Nonprescription Drug Consumer Protection Act, dubbed the AER bill (S.3546.ES). The legislation will amend the Federal Food, Drug and Cosmetic Act to require the reporting of "serious" adverse events for both OTC drugs and dietary supplements to the FDA. President Bush signed the act on December 22, 2006. The bipartisan legislation includes several provisions that were instrumental in earning the Natural Products Association's support. Under this bill, companies would be required to include contact information on their products' labels for consumers to use in reporting adverse events. They would further be required to notify the FDA of any serious adverse event reports received within 15 business days. According to the bill, the term "adverse event" means any health-related event associated with the use of a nonprescription drug that is adverse, including:

(a) An event occurring from an overdose of the drug, whether accidental or intentional
(b) An event occurring from abuse of the drug
(c) An event occurring from withdrawal from the drug
(d) Any failure of expected pharmacological action of the drug

The term "serious adverse event" is an adverse event that results in death, a life-threatening experience, in-patient hospitalization, a persistent or significant disability or incapacity, or a congenital anomaly or birth defect or requires, based on reasonable medical judgment, a medical or surgical intervention to prevent one of the above outcomes. Among these key provisions are those requiring that only serious adverse events, not just any complaint, be reported, exempting retail stores from reporting directly to the FDA and preempting a potential patchwork of state laws on the issue.

6.2.4 European Union

In Europe, the DSs are known mostly as "food supplements." "Health food" is a marketing term rather than a legal term, normally denoting a product sold in specialty stores (Smith et al., 1996; Wasser et al., 2004). In addition, the term DS is used by the EU for products regulated as Foods for Particular Nutritional Uses (PARNUTS).

Consumer safety is Europe's primary concern. A good example of the statement is that individual countries have already banned several supplements available in the United States. European countries follow the precautionary principle, which means that when they suspect a product may cause harm, they do not wait for proof before they take action against it. For example, the FDA warned American consumers in March 2002 that supplements containing kava, promoted as a relaxant, may cause liver injury—sometimes so severe that a transplant is required, sometimes even fatal, based on the case of a 45-year-old woman using kava who suddenly required a liver transplant as well as reports of 25 similar cases in Europe. However, while the FDA issued only a warning, several European countries, including Germany and France, banned the sale of kava (Hecht, 2003).

As the European countries vary in the ways they regulate DSs, the European Commission (EC) aims to provide a uniform standard of quality and safety for all member states. The EC divided DSs into two groups for the purpose of regulation: vitamins/minerals and herbals. The groups are treated separately.

Directive 2002/46/EC, which harmonized EC legislation on vitamins and minerals as food supplements, was published in the *Official Journal of the EC* (L183/51) on June 10, 2002 (Directive 2002/46/EC). The directive defined the term "food supplements" and set out labeling requirements. The directive did not immediately outlaw any products already on the market.

The set standards enable a product approved in one country to be sold in other member countries. Also, everything in the EC directive follows from the basic principle that, under normal circumstances, a balanced diet can provide the necessary nutrients for development and health. For instance, the directive forbids manufacturers to state or even suggest the contrary; a manufacturer may not state that a supplement is a substitute for a varied diet or that it can prevent, treat, or cure a disease. Furthermore, rather than setting only general guidelines, the directive lists exactly which vitamins and minerals may be sold in member countries and in what form (Hecht, 2003).

Despite the directive's apparent advantages for the consumer, there are some disadvantages. It will tighten regulation in some countries but loosen it in others that previously had stricter laws. It also allows each country to decide whether it will require government notification from manufacturers when a new supplement comes on the market.

The EC is developing a list of alternative herbal substances. It includes details about each substance: what it is to be used for, the strength and daily dose, how it is to be administered, and possible drug interactions or adverse effects. During 2005, the Committee on Herbal Medicinal Products (HMPC) adopted regulatory guidance documents, which include guidelines on the documentation to be submitted for inclusion in the list of herbal substances, preparations, and combinations thereof (EMEA/HMPC/107399/2005). The guideline is in effect now, and the various documents are updated from time to time (HMPC 2006).

The EU relates to herbal DSs in the same way it does to conventional medications. European countries vary in the degree to which they use herbals and how carefully they regulate them. Germany, Austria, Switzerland, and France lean toward a stricter approach, while the Netherlands and the United Kingdom have traditionally been more lenient. The same products sold as medicine in Germany are sold as food supplements in the Netherlands (Hecht, 2003).

Germany was in the vanguard of the move toward stricter herbal regulation for all EU countries, resulting in Directive 2001/83/EC. Although the directive does not relate to homeopathic medicines, it does cover all herbals used in the treatment or prevention of disease or as modifiers of physiological functions, including inducing relaxation. Therefore, under the directive, "natural" will no longer automatically mean "good." Manufacturers are not allowed to claim that a product is safe and effective simply because it is natural (Hecht, 2003). EC legislation does not refer to mushroom products directly.

The dietary supplements industry in Europe strongly opposed the food supplements directive (2002/46/EC). A large number of consumers throughout Europe, including over one million in the United Kingdom, and many doctors and scientists have signed petitions against what are viewed by the petitioners as unjustified restrictions of consumer choice. The European Court of Justice ruled on July 12, 2005, that the directive is valid although the court's own advocate general advised that the declaration was invalid under EU law [Court of Justice of the European Communities (CURIA), 2005].

At the end of 2006, Regulation (EC) No 1924/2006 on the use of nutrition and health claims for foods was adopted by the Council and Parliament. This regulation made standard rules for the use of health or nutritional claims (such as "low fat," "high fibre," and "helps lower cholesterol") on foodstuffs based on nutrient profiles. The Health Claims Regulation ensures that any claim made on a food label in the EU is clear, accurate, and substantiated. In doing so, it enables consumers to make informed and meaningful choices when it comes to food and drinks. This also contributes to a higher level of human health protection, as it ties in with the EC campaign for healthier lifestyle choices by allowing citizens to know exactly what they are consuming. The regulation also aims to ensure fair competition and promote and protect innovation in the area of food. Only products offering genuine health or nutritional benefits will be allowed to refer to them on their labels (European Commission, 2006).

6.2.5 Canada

In Canada, the regulatory framework governing foods, as in other major jurisdictions, is evolving. Food research, product and process innovation, and change in consumer behavior are all outpacing the adaptation of regulation to new market realities. Among these realities is growing consumer awareness of nutrition and interest in health promotion. Increasingly, this awareness is manifested through consumption of particular foods and dietary supplements believed to contribute to good health and, in some cases, to hold therapeutic value in the treatment or prevention of specific afflictions or diseases (Smith et al., 1996; Wasser et al., 2004).

Many of these food products are becoming commonly known as nutraceuticals, or "functional foods." The regulatory environment governing functional foods is so restrictive that the development of a functional foods industry or even functional food products in Canada will be severely impaired, if not entirely precluded (Smith et al., 1996; Wasser et al., 2004).

The *Food and Drugs Act* (F&DA) is the primary piece of legislation governing the safety and quality of food sold in Canada. Its scope includes: food labeling, advertising and claims; food standards and compositional requirements; fortification; foods for special dietary uses; food additives; chemical and microbial hazards; veterinary drug residues; packaging material and pesticides. Created in 1953, the founding premise of the F&DA is to protect the public from adulterated food, drink, and drugs and their associated health effects. That same orientation continues today, and any amendments to the Act must be aligned with this premise.

Health Canada is an agency similar to the U.S. Food and Drug Administration. It is Health Canada's responsibility to establish the detailed set of regulations and guidelines that accompany the F&DA to protect the consumer from being misled about the characteristics of food products. However, the Canadian Food Inspection Agency (CFIA) is responsible for enforcing the F&DA regulations.

The F&DA specifically forbids the labeling or advertising of food in a manner that is false, misleading, deceptive, or likely to create an erroneous impression regarding its character, value, quantity, composition, merit, or safety. Section 3 makes it an offence to advertise or sell a food to the general public as a treatment, preventative, or cure for any disease referred to in Schedule A of the FD&A. Heart disease, hypertension, and cancer are examples of diseases listed in Schedule A. Section 3 and Schedule A of the F&DA prohibits diet-related health messages from being used on packaging and in advertising.

In an effort to ensure that consumers are provided with consistent information on the foods they buy, regulations introduced on January 1, 2003 require mandatory nutrition labeling. The nutrition labeling requirements to the Food and Drug Regulations provided a three year transition period, making the Nutrition Facts table mandatory for most prepackaged foods since December 12, 2005 (CFIA, 2003)

The framework for the authorization of health claims for foods in Canada distinguishes between generic claims and product-specific claims. Generic claims are associated with nutrients, other food components, foods or food groups that contribute to a dietary pattern of eating associated with a reduction in risk for a disease. Product-specific claims are associated with a specific food which has demonstrated a measurable health benefit beyond normal body function, growth, development, or maintenance of good health. The five authorized health claims are considered generic claims (Kennedy 2006).

Currently there are three types of generic claims that are associated with nutrients, other food components, foods, or food groups that contribute to a dietary pattern of eating associated with a reduction in risk for a disease. The types are: nutrient content claims, biological role claims, and diet-related health claims. Nutrition claims typically highlight one nutrient found in a food. To ensure nutrition claims comply with the F&DA requirements (i.e., truthful and not misleading) Health Canada has established specific rules regarding the language, format, and compositional criteria for each of these three types of claims. Based on these generic types, Health Canada authorized five health claims, all of them applicable to mushrooms:

1. "A healthy diet containing foods high in potassium and low in sodium may reduce the risk of high blood pressure, a risk factor for stroke and heart disease."
2. "A healthy diet with adequate calcium and vitamin D, and regular physical activity, help to achieve strong bones and may reduce the risk of osteoporosis."
3. "A healthy diet low in saturated and trans fats may reduce the risk of heart disease."
4. "A healthy diet rich in a variety of vegetables and fruit may help reduce the risk of some types of cancer."
5. "Won't cause cavities" or "Does not promote tooth decay" or "Does not promote dental caries" or "Noncariogenic."

Currently Health Canada is embarking on a major renewal of its health protection legislation and has proposed the creation of a new Canada Health Protection Act to replace a number of pieces of legislation, including the Food and Drugs Act. The goal is to integrate them into a single piece of legislation that has an overall general policy direction. Until new legislation replaces the F&DA, diet and health claims for foods must be made within the existing legislative framework (Kennedy 2006).

6.2.6 Australia and New Zealand

The use of natural products in Australia has been widespread for many years, as evidenced by the presence of an active organization of herbalists formed in 1920 (Blumenthal, 1999). In 1985, the Australian Parliament established the Working Party on Natural and Nutrition Supplements to review the quality, safety, efficacy, and labeling of herbs and related products, including mushrooms, for appropriate regulation under the Therapeutic Goods Act of 1990 (Wasser et al., 2000a, b, 2004).

The 1990 South Australian Working Party on Natural and Nutritional Supplements report dealt with approximately 1144 herbs and "therapeutic substances," dividing them into three groups, each having a recommended level of control and labeling, "appropriate to their level of toxicity."

As in the EU and United States, food regulations between New Zealand and Australia were awaiting harmonization and standardization. The basic body responsible for this was the ANZFA. The ANZFA's role is to protect the health and safety of people in Australia and New Zealand by maintaining a safe food supply. The ANZFA is a partnership between the Commonwealth government, Australian States and Territories governments, and the New Zealand government. As an independent expert body, the ANZFA is responsible for developing and reviewing food standards for both Australia and New Zealand. The ANZFA makes recommendations to change the food standards to the Australia New Zealand Food Standards Council, a ministerial council made up of commonwealth, state and territory, and New Zealand health ministers. If the council approves the recommendations made by the ANZFA, the food standards are automatically adopted as regulations into the food laws of the Australian States and Territories and New Zealand (ANZFA, 2002).

Specific regulations were applied to novel and special-purpose foods. DSs derived from fungi are included (along with herbs and species) into the category of novel foods. "Novel food" means a nontraditional food (a food that does not have a history of significant human consumption by the broad community in Australia or New Zealand) for which there is insufficient knowledge in the broad community to enable safe use in the form or context in which it is presented, taking into account:

(a) Composition or structure of product
(b) Levels of undesirable substances in product
(c) Known potential for adverse effects in humans
(d) Traditional preparation and cooking methods
(e) Patterns and levels of consumption of product

Traditional foods of a particular community may be considered novel if they are made available to a new or wider community without adequate information regarding applicable presentation and use. Novel foods can take on many forms, and not all are natural substances. There is an increase in the number and variety of novel foods on the market as a result of technological developments, trade opportunities, scientific advances, and increasing ethnic diversity of the population (ANZFA, 2002)

The New Zealand Food Safety Authority (NZFSA) was established in July 2002 to improve the effectiveness of New Zealand's food safety system by coordinating and harmonizing food safety efforts. Specifically, New Zealand wanted to address inconsistencies between the methods used in the Ministry of Agriculture and Forestry's export food safety program and the Ministry of Health's domestic food safety program. The NZFSA has farm-to-table responsibilities—from primary production through processing to retailers, importing, and exporting as well as responsibility for consumer education [U.S. General Accountability Office (GAO), 2005].

The NZFSA has two main areas of focus: to protect and promote public health and safety through the administration of food-related legislation and to facilitate access to markets for New Zealand food products and related products (State Services Commission, 2006).

6.2.7 Japan

In Japan, food with health claims (FHC) refers to foods that comply with the specifications and standards established by the Minister of Health, Labor and Welfare

(MHLW) and are labeled with certain nutritional or health functions. These foods are categorized into two groups, according to differences in purpose and function: (a) foods for specified health uses (FOSHU), that is, foods officially approved to claim their physiological effects on the human body, and (b) foods with nutrient function claims (FNFC), that is, foods that are labeled with the functions of nutritional ingredients (vitamins and minerals).

Under Japan's Nutrition Improvement Act of 1992 the regulatory system of FOSHU was developed to delineate functional foods from the pharmaceutical regulations. The Japanese government defines FOSHU as "foods that are expected to have certain health benefits, and have been licensed to bear a label claiming that a person using them for a specified health use may expect to obtain the health use through the consumption thereof" (Shimizu, 2003). The classification or list has no status outside Japan. As of January 8, 2004, FOSHU totaled 402 items. Among them, 91 products have been approved since the beginning of 2003. As of 2005, FOSHU account for over $6 billion with over 500 approved products. The total market value including non-FOSHU functional foods and DSs is three times higher. FOSHU refer to foods containing ingredients with functions for health and officially approved to claim its physiological effects on the human body. FOSHU are intended to be consumed for the maintenance/promotion of health or special health uses by people who wish to control health conditions, including blood pressure or blood cholesterol. In order to sell a food as FOSHU, the assessment for the safety of the food and effectiveness of the functions for health is required, and the claim must be approved by the MHLW.

Requirements for FOSHU approval include the following:

Effectiveness on the human body is clearly proven.

Absence of any safety issues (animal toxicity tests, confirmation of effects in the cases of excess intake, etc.).

Use of nutritionally appropriate ingredients (e.g., no excessive use of salt, etc.).

Guarantee of compatibility with product specifications by the time of consumption.

Established quality control methods, such as specifications of products and ingredients, processes, and methods of analysis.

In addition to "regular" FOSHU, new types of FOSHU were introduced to facilitate applicants for FOSHU approvals:

- *Qualified FOSHU* Food with health function which is not substantiated on scientific evidence that meets the level of FOSHU or food with certain effectiveness but without established mechanism of the effective element for the function will be approved as qualified FOSHU.
- *Standardized FOSHU* Standards and specifications are established for foods with sufficient FOSHU approvals and accumulation of scientific evidence. Standardized FOSHU are approved when they meet the standards and specifications.

- *Reduction of Disease Risk FOSHU* This claim is permitted when reduction of disease risk is clinically and nutritionally established in an ingredient. FOSHU regulated products are typically novel products and not value-added general-purpose foods. The majority of products are specifically focused on prevention of disease and maintenance of health status rather than the direct treatment of disease states. Health agencies of the Japanese government develop and maintain the regulatory framework that covers products that are to be regulated under the FOSHU system.
- *Foods for Special Dietary Uses (FOSDU)* These refer to foods that are approved/permitted to display that the food is appropriate for specified dietary use. There are five categories of FOSDU: (1) formulas for pregnant or lactating women; (2) infant formulas; (3) foods for the elderly with difficulty in masticating or swallowing; (4) medical foods for the ill; and (5) FOSHU.
- *Prohibition of Exaggerated and Misleading Claims (under Health Promotion Law)* Any claims related to health or function made on functional foods must be relevant and substantiated by scientific evidence. Advertisements are used in different media to promote food sales. Some are advertised to convey positive effects on health maintenance and promotion not necessarily having scientific evidence on the claim. When these advertisements are not regulated and uncontrolled, consumers who believed the claim might miss an opportunity for an adequate medical consultation, resulting in adverse affects on health. Under Paragraph 2, Article 32 of the Food Promotion Law, exaggerated and misleading claims are prohibited.

6.2.8 Israel

Hand in hand with the public's growing interest in health care, there has been an increasing demand for natural health products considered both safe and medically effective. However, on the one hand, many such products have not been shown to meet efficacy and safety criteria and, therefore, cannot be registered as pharmaceuticals. On the other hand, it is quite clear that some products do have pharmacological activity and are being used for therapeutic or preventive effects. In Israel, the marketing rules for food or DSs prevent their manufacturers from claiming medicinal/healing properties that the product might have and allow only limited health statements. As great demand for these products has created tremendous notoriety, medicinal indications have been attributed to products whose quality, efficacy, and safety have not been examined and proven according to accepted medical criteria (Lavy et al., 2000; Wasser et al., 2004).

In October 2006 the Ministry of Health, Department of Food and Dietary Services updated the list of mushroom species allowed as food and as DSs in Israel. The change was necessary because the ministry wanted to align its policy with those of other countries for products for which knowledge regarding safety was accumulated. The list includes mushrooms on the FAO list, but only those labeled as food or edible. Mushrooms on the list labeled as medicinal are not permitted. Nineteen additional mushrooms are permitted and listed in a separate file. An importer or cultivator may request the use of new species in writing.

The FAO list of mushrooms is annexed in the book *Wild Edible Fungi: A Global Overview of Their Use and Importance to People* (Boa, 2004). Wild edible fungi are collected for food and to earn money in more than 80 countries. There is a huge diversity of different types, from truffles to milk-caps, chanterelles to termite mushrooms, with more than 1100 species recorded during the preparation of this book. A small group of species are of economic importance in terms of exports, but the wider significance of wild edible fungi lies with their extensive subsistence uses in developing countries. They provide a notable contribution to diet in central and southern Africa during the months of the year when the supply of food is often perilously low. Elsewhere they are a valued and valuable addition to diets of rural people.

6.3 SAFETY AND DIVERSITY OF DIETARY SUPPLEMENT TYPES FROM CULINARY–MEDICINAL MUSHROOMS

Safety of the substances considered is a central feature of any regulation measure. The substances might have obvious or hidden beneficial actions or exploit a placebo effect as long as the ideas of health are attached to them in the public's mind. However, their safety should be verified and proven as thoroughly as possible.

Drugs that affect body functions such as immune response, blood pressure, diuresis, and others are called pharmacodynamic agents. This is the way DSs are evaluated. Pharmacodynamics is the spectrum of biological responses produced by an intervention at a given time. The presence of therapeutic pharmacodynamic activity implies a lack of safety at a sufficiently high dose. This means that a mushroom preparation, like any other pharmacodynamic agent, cannot have pharmacological action without toxicological action. A completely safe agent would be without any activity whatsoever (Huxtable, 1999; Chang and Buswell, 1999, 2003; Wasser et al., 2000a, b, 2004; Chang and Miles, 2004; Chang, 2006).

It is commonly believed that many botanicals and mushrooms can be considered safe because of their long history of usage. The shiitake mushroom, for instance, was described in classical Chinese medicine (Shen Nong Ben Cao Jin, Compendium of Material Medica of the East Han dynasty) as long as 2000 years ago (Mizuno, 1999).

However, safety to a pharmacologist is a relative concept which is very different from the public notion of safety as an absolute concept. Here, we outline the reasons why we cannot take the safety of all mushroom-derived DSs for granted simply because they were used for many centuries in traditional human cultures:

1. Considering the historical perspective, "safety" in traditional terms is very different from that in modern times. First, mortality patterns of developed societies today are very different from those of traditional ones. Life expectancy in the United States today is 76 years or more (Fries and Crapo, 1981), but in medieval China it was 20 years less (these figures do not reflect

a maximum life span, but only a median age at death). Death itself (the likelihood of dying at a certain age) had a different kinetics in society. Second, traditional users rarely had the means to evaluate long-term or chronic toxicity of the agents. However, we do have cautionary instances of plants that have been used medicinally for centuries and recently proved to carry delayed toxic effects. One such example is comfrey (*Symphytum officinale* L.), containing pyrrolizidine, alkaloids, which are acutely hepatoxic in high doses or chronically hepatoxic in low doses (Huxtable, 1992). Nevertheless, comfrey has been used since classic times as a blood purifier, for enlarged glands, for female debility, and many more symptoms (Santillo, 1993). An example of an ill effect is the mushroom *Paxillus involutus* (Batsch.: Fr.) Fr. It was widely used in Europe, and it is still taken as a food in many regions there. *Paxillus* syndrome was described only in 1971 (Schmidt et al., 1971); in fact, this syndrome is an immunohemolytic anemia. A patient who has eaten *Paxillus* over a long period (sometimes years) on occasion may develop *Paxillus* syndrome in a short time period of 1 or 2 hours after consumption. An antigen of still-unknown structure stimulates a severe immune shock resulting in such symptoms as diarrhea, subicterus, oliguria, anuria, hemoglobinuria, and renal pain (Flammer, 1980, 1983; Lefevre, 1982; Bresinsky and Besl, 1990; Wasser et al., 2000a, b, 2004).

2. Many supposedly traditional mushroom products are now marketed in ways markedly different from those in the past. Today, larger amounts may typically be taken, or the material is used more frequently, or it is consumed in the form of enriched extracts, and it may be taken simultaneously with synthetic drugs. The user of shiitake in old China, for instance, could not ingest as much active polysaccharide (lentinan) as a modern user taking it in pure form extracted from shiitake as a DS. Notably, 200 kg of fresh mushrooms is needed for extraction of 31 g of lentinan (Chihara et al., 1970). This heightens the possibility of ill effects from traditionally "safe" mushrooms.
3. Also, many mushrooms or mushroom preparations traditionally taken as treatments for specific conditions are now often marketed for use as prophylactic agents. The idea of DSs, in many cases, implies that they are taken in the absence of any indicated conditions to prevent disturbances of health.
4. Finally, reliance on traditional use as an indication of safety involves a danger, namely, the poor information available to us from antiquity. Huxtable (1999) recently carried out an analysis of historical sources of such information that are in many cases contradictory and vague. Information on acute toxicity may be located on the same page as the recommendations for use. The identities of the plants are sometimes dubious. Herbalists copied extensively from each other over thousands of years. The descriptions of symptoms are often vague and general (e.g., bloodstream disturbances or liver malfunction) (Wasser et al., 2000a, b, 2004).

Besides the problems of direct toxicity of some individual nutrients, nutrient supplementation can cause problems related to nutrient imbalances or adverse interactions with medications. Many problems associated with high doses of a single nutrient may reflect interactions resulting in a relative deficiency of another nutrient. Many examples are available from studies of DSs that have been used for a long time and are more common on the market then those derived from mushrooms. High amounts of calcium inhibit absorption of iron (Cook et al., 1991). Folic acid can mask hematological signs of vitamin B_{12} deficiency, which, if untreated, can result in irreversible neurological damage. Folic acid can also interact adversely with anticonvulsant medications (Food and Nutrition Board, 1989). Abundant zinc supplementation can reduce copper status, impair immune responses, and decrease high-density lipoprotein cholesterol levels..

Because we lack many strict data on the safety of many mushroom preparations, we could only make estimates on the basis of available scientific knowledge. The clear advantages of using mushroom-based DSs with regard to safety (as opposed to herbal preparations) are the following:

1. The overwhelming majority of mushrooms used for production of DSs are cultivated commercially (and not gathered in the wild). This guarantees proper identification and pure, unadulterated products. In many cases it also means genetic uniformity.
2. Mushrooms are easily propagated vegetatively and thus keep to one clone. The mycelium can be stored for a long time, and the genetic and biochemical consistency may be checked after considerable time.
3. The main advantage might be the fact that many mushrooms are capable of growing in the form of mycelial biomass in submerged cultures (Buchalo, 1988; Chang and Buswell, 1996, 2003; Pointing et al., 2000; Wasser et al., 2000a, b, 2004; Wasser and Reshetnikov, 2002a, b, c; Chang and Miles, 2004; Chang, 2006).

Chang (2006) suggests adopting the five "G" guidelines to enhance quality mushroom products. Although the text was originally oriented toward mushroom fruit bodies, it is still valid for other forms of mushroom products such as submerged culturing:

1. *GLP (Good Laboratory Practice)* A known mushroom strain must be used; the source and nature of the strain culture must be clearly documented and should be properly maintained and preserved without contamination and degeneration.
2. *GAP (Good Agriculture Practice)* Growth conditions must be known; the substrate should be free of heavy metals and composed of consistent levels of ingredients; the environmental conditions should include unpolluted air and a good sanitary growth area. The product should be harvested at optimal maturity and free of diseases.
3. *GMP (Good Manufacturing Practice)* The parameters for process must be known and maintained. The temperature, duration, and percentage of solvates used in extraction should be constantly monitored.

4. *GPP (Good Production Practice)* The following tests must be conducted: a chemical analysis of the products to determine organic components and heavy metal contents; a microbial analysis to determine if the type and level of microorganisms present are within safe limits; and standardization of the formulation of the products.
5. *GCP (Good Clinical Practice)* Medical practitioners must conduct high-quality clinical trials, including double-blind studies, which should be conducted to ensure standardization and allow appropriate dosage determinations and product formulation for the effective treatment of a particular health problem.

The fifth G is somewhat problematic to DS developers. Once a clinical trial on humans is performed and bioactivity is confirmed, the product acquires the status of unapproved active drug. Until the product is approved, it is banned from the market because it contains an unapproved drug. The process of drug approval is extremely expensive and lasts several years. By the time the product is approved, the manufacturers may lose their business. This problem requires that only large pharmaceutical companies run clinical trials and those large companies do not engage in the research of traditional DSs. Another reason why DS developers do not wish to transform a healthy product into a medical product is because some health organizations, such as the FDA, do not approve a mixture of molecules, as in the case of mycelial or fruit body extracts, but only approve well-defined, patentable molecules. On the one hand, most DS marketers will be just happy enough to sell their mushroom products without printed health claims. They will advertise them in foreign markets that allow health claims on the Internet or with word of mouth. On the other hand, perhaps a whole new industry, or even industries, will arise having greater economic value than those currently producing mushrooms for food.

In 2004, the FDA published *Guidance for Industry: Botanical Drug Products* (FDA, 2004). This guidance explains when a botanical drug may be marketed under an OTC drug monograph and when FDA regulations require approval for marketing of a new drug application (NDA). In addition, this document provides sponsors with guidance on submitting investigational new drug (IND) applications for botanical drug products, including those currently lawfully marketed as foods (including conventional foods and dietary supplements) in the United States. Botanical products are finished, labeled products that contain vegetable matter as ingredients. A botanical product may be a food (including a dietary supplement) or a drug (including a biological drug). The term "botanical" includes plant materials, algae, and macroscopic fungi. It does not include fermentation products such as extracts made from submerged fermentation of mushroom biomass. Also, a botanical drug substance does not include a highly purified or chemically modified substance derived from natural sources. The authors are not aware of any clinical trials in the United States with mushrooms as the botanical drug.

6.4 SUBMERGED CULTURING AS BEST TECHNIQUE FOR OBTAINING CONSISTENT AND SAFE MUSHROOM PRODUCTS

Today, approximately 80% of mushroom products are taken from fruit bodies either collected in the wild or grown commercially. In both cases, the resulting products are considerably diverse and unpredictable. The quality of mushroom fruit bodies is strongly dependent on substrate composition and properties of its ingredients, and usually these are far from constant. This is explained by the fact that the main components for mushroom production are of available agricultural and forest origin such as cereal straw, corn stakes, horse or chicken manure, and wood sawdust.

The production of many biologically active substances is connected with maturation processes of fruit bodies. Lovastatin was found concentrated primarily in the lamellae and basidiospores, but not in the stipe or cap tissue, and its amount depends on fruit body size and age (Gunde-Cimerman, 1999).

Variability of mushroom fruit body composition is the reason why processing for extraction of polysaccharides from fruit bodies is not considered commercially feasible, as physicochemical properties of the products resulting from these processes are not known or regulated (Ohtsuka et al., 1977).

The cultivation of mushrooms for fruit body production is a long-term process, taking one to several months for the first fruiting bodies to appear, depending on species and substrate. By contrast, the growth of pure mushroom cultures in submerged conditions in a liquid culture medium allows one to accelerate the speed of growth and reduces its duration to several days. Optimization of culture medium composition and the physicochemical conditions of growth allow regulation of mushroom metabolism to obtain a high yield of biomass and large amounts of specific substances of consonant composition (Wasser et al., 2000a, b, 2002, 2004; Wasser and Reshetnikov, 2002a, b, c; Wasser, 2002).

6.5 EXPERIENCES OF SEVEN COUNTRIES IN CONSOLIDATING THEIR FOOD SAFETY SYSTEMS

The GAO examined seven countries in consolidating their food safety systems: Canada, Denmark, Germany, Ireland, the Netherlands, New Zealand, and the United Kingdom. These countries varied in their approaches and the extent to which they consolidated. However, the countries' approaches were similar in one respect: Each established a single agency to lead food safety management or enforcement of food safety legislation. These countries had two primary reasons for consolidating their food safety systems: public concern about the safety of the food supply and the need to improve program effectiveness and efficiency. The above-mentioned countries faced challenges in (1) deciding whether to place the agency within the existing health or agriculture ministry or establish it as a stand-alone agency while also determining what responsibilities the new agency would have and (2) helping employees adjust to the new agency's culture and support its priorities.

Although none of the countries had analyzed the results of its consolidation, government officials consistently stated that the net effect of their country's consolidation had been or would likely be beneficial. Officials in most countries stated their new food safety agencies incurred consolidation start-up costs. However, in each country, government officials believed that consolidation costs had been or would likely be exceeded by the benefits. These officials and food industry and consumer stakeholders cited significant qualitative improvements in the effectiveness or efficiency of their food safety systems. These improvements include less overlap in inspections, greater clarity in responsibilities, and more consistent or timely enforcement of food safety laws and regulations. In addition to these qualitative benefits, officials from three countries, Canada, Denmark, and the Netherlands, identified areas where they believed financial savings may be achieved as a result of consolidation. For example, in the Netherlands officials said that reduced duplication in food safety inspections would likely result in decreased food safety spending, and they anticipated savings from an expected 25% reduction in administrative and management personnel.

Although the seven reviewed countries are much smaller than the United States, they are also high-income countries where consumers have very high expectations for food safety. Consequently, those countries' experiences in consolidating food safety systems can offer useful information to U.S. policymakers (GAO, 2005).

6.6 SUMMARY

In this chapter, we discussed legal and regulatory issues introducing and controlling DSs from medicinal mushrooms in different countries, including the United States, the European Community, Canada, Australia, New Zealand, Japan, Israel, and guidelines of the WHO. A lot of attention is drawn to the safety and diversity of DS types from culinary–medicinal mushrooms. Most of the mushroom DSs presently in the marketplace are highly diverse and there are currently few standard protocols to ensure product quality. There must be thorough analysis and improved quality and legal control which will, in turn, increase and maintain consumer confidence and achieve the current and future standards set by national regulatory authorities. We hope that these and future regulations will continue to provide and protect stable medicinal mushroom products of reliable quality and that culturing biotechnology will continue to be the best and the most progressive technique for obtaining consistent and safe mushroom products.

REFERENCES

Alkerele, O. (1992). Summary of WHO guidelines for the assessment of herbal medicines. *Fitoterapia*, *62*, 99–110.

Australia New Zealand Food Authority (ANZFA) (2002). Initial assessment report (prepare a new proposal—section 21). Proposal P235. Review of food-type dietary supplements.

Available: www.foodstandards.gov.au/_srcfiles/P235_IA_Report_v2.pdf. Accessed Oct. 16, 2007.

Blumenthal, M. (1999). Regulatory models for approval of botanicals as traditional medicines. In *Botanical Medicine*. Eshkinazi, D., editor. New York: Mary Ann Liebert, pp. 205–212.

Blumenthal, M. (2001). House committee on the government reform holds hearings on dietary supplements. *HerbalGram*, *52*, 55–56.

Boa, E. (2004). *Wild Edible Fungi: A Global Overview of Their Use and Importance to People*. Available: www.fao.org/docrep/007/y5489e/y5489e00.htm. Accessed Oct. 16, 2007.

Bresinsky, A. and Besl, H. (1990). *A Colour Atlas of Poisonous Fungi*. London: Wolfe Publishing.

Buchalo, A. S. (1988). *Higher Basidiomycetes in Pure Culture*. Kiev: Naukova Dumka, p. 133.

Canadian Food Inspection Agency (CFIA) (2003). *Guide to Food Labelling and Advertising*. Available: http://www.inspection.gc.ca/english/fssa/labeti/guide/toce.shtml. Accessed June 10, 2008.

Center for Food Safety and Applied Nutrition (CFSAN) (2005a). Food and Drug Administration. Dietary supplements: Overview. Available: http://www.cfsan.fda.gov/~dms/supplmnt.html. Accessed Apr. 14, 2005.

Center for Food Safety and Applied Nutrition (CFSAN) (2005b). Food and Drug Administration. Guidance for industry: A dietary supplement labeling guide. Available: http://www.cfsan.fda.gov/~dms/supplmnt.html. Accessed Oct. 16, 2007.

Chang, S. T. (1999). World production of edible and medicinal mushrooms in 1997 with emphasis on *Lentinus edodes* (Berk.) Sing. in China. *International Journal of Medicinal Mushrooms*, *1*, 291–301.

Chang, S. T. (2001). A 40-year journey through bioconversion of lignocellulosic wastes to mushrooms and dietary supplements. *International Journal of Medicinal Mushrooms*, *3*, 299–311.

Chang, S. T. (2006). The need for scientific validation of culinary-medicinal mushroom products. *International Journal of Medicinal Mushrooms*, *8*, 187–195.

Chang, S. T. and Buswell, J. A. (1996). Mushroom nutriceuticals. *World Journal of Microbiology and Biotechnology*, *12*, 473–476.

Chang, S. T. and Buswell, J. A. (1999). *Ganoderma lucidum* (W.Curt.:Fr.) P. Karst. (Aphyllophoromycetideae)—A mushrooming medicinal mushroom. *International Journal of Medicinal Mushrooms*, *1*, 139–146.

Chang, S.T. and Buswell, J. A. (2003). Medicinal mushrooms—A prominent source of nutriceuticals for the 21st century. *Current Topics in Nutraceutical Research*, *1*, 257–280.

Chang, S. T. and Miles, P. G. (2004). *Mushrooms: Cultivation, Nutritional Value, Medicinal Effect, and Environmental Impact*. Boca Raton, FL: CRC Press, p. 451.

Chihara, G., Hamuro, J., Maeda, Y. Y., Arai, Y., and Fukoka, F. (1970). Fractionation and purification of the polysaccharides with marked antitumor activity, especially lentinan, *Lentinus edodes* (Berk.) Sing. (an edible mushroom). *Cancer Research*, *30*, 2776–2781.

Cook, J. D., Dassenko, S. A., and Whittaker, P. (1991). Calcium supplementation: Effect on iron absorption. *American Journal of Clinical Nutrition*, *53*, 106–111.

Council for Responsible Nutrition (CRN) (2002). TABD dietary supplements. A report from the CEOs on their expert group agreements. Available: www.crnusa.org/pdfs/TABD0302final.pdf. Accessed Oct. 16, 2007.

Council for Responsible Nutrition. (2007). Available: http://www.crnusa.org/. Accessed Oct. 16, 2007.

Court of Justice of the European Communities (CURIA) (2005). The court confirms the validity of the community directive on food supplements. Press Release, 66/05. Available: http://curia.europa.eu/en/actu/communiques/cp05/aff/cp050066en.pdf. Accessed Oct. 16, 2007.

EUROPA (2003). Commission proposes common rules for adding vitamins and minerals to foods. Press release, IP/03/1516. Available: europa.eu/rapid/pressReleasesAction.do?reference = IP/03/1516&format = HTML&aged = 0&language = EN&guiLanguage = en. Accessed Oct. 16, 2007.

European Commission (EC) (2006). Food and feed safety. Regulation on health and nutritional claims. Available: http://ec.europa.eu/food/food/labellingnutrition/claims/index_en.htm. Accessed June 10, 2008.

Flammer, R. (1980). *Differentialdiagnose der Pilzvergiftungen*. Stuttgart: Gustav Fischer Verlag.

Flammer, R. (1983). Hamolise bei pilzvergiftungen: Fakten and hypothesen. *Schweizerische Medizinische Wochenschrift*, *113*, 1555–1561.

Food and Agriculture Commission (FAO) (2005). CODEX STAN 1–1985 (Rev. 1–1991), updated 2005. General standard for the labelling of prepackaged foods. FAO: Codex Alimentarius, 2005.

Food and Nutrition Board (1989). *Recommended Dietary Allowances*, 10th ed. Washington: National Academy Press.

Food Standards Australia New Zealand (FSANZ) (2006). FSANZ food standards development work plan. Available: www.foodstandards.gov.au/srcfiles/Workplan_9_November_2006.doc. Accessed Oct. 16, 2007.

Fries, J. F. and Crapo, M. L. (1981). *Vitality and Aging: Implications of the Rectangular Curve*. San Francisco: W. H. Freeman.

Gunde-Cimerman, N. (1999). Medicinal value of the genus *Pleurotus* (Fr.) P. Karst. (Agaricales, Basidiomycetes). *International Journal of Medicinal Mushrooms*, *1*, 69–80.

Hecht, E. (2003). Seller beware: What do tough European supplement laws say about safety in the US? Available: findarticles.com/p/articles/mim0FKA/is_3_65/ai_97823105. Accessed Oct. 16, 2007.

Herbal Medicinal Products Committee (CMPC) (2006). *Human Medicines—Herbal Medicinal Products*. HMPC Status Report, October 2006. Available: http://www.emea.europa.eu/htms/human/hmpc/hmpcstatus.htm. Accessed June 10, 2008.

Hiramoto, K., Kaku, M., Kato, T., and Kikugawa, K. (1995). DNA strand breaking by the carbon-centered radical generated from 4-(hydroxymethyl) benzenediazonium salt, a carcinogen in mushroom *Agaricus bisporus*. *Chemico-Biological Interactions*, *94*, 21–36.

Hobbs, C. (1995). *Medicinal Mushrooms: An Exploration of Tradition, Healing, and Culture*. Santa Cruz: Botanica, p. 251.

Huxtable, R. H. (1992). The toxicology of alkaloids in foods and herbs. In *Handbook of Natural Toxins*, Vol. 7: *Food Poisoning*. Tu, A. T., editor. New York: Marcel Dekker, pp. 237–262.

Huxtable, R. H. (1999). The safety of botanicals: A historic perspective. In *Botanical Medicine*. Eshkinazi, D., editor. New York: Mary Ann Liebert, pp. 87–101.

Kennedy, A. (2006). Nutrition claims in Canada: A primer. Available: http://www4.agr.gc.ca/AAFC-AAC/display-afficher.do?id=1172766094272&lang=e. Accessed June 10, 2008.

Lavy, T., Haran, B., Shemer, J., and Shani, S. (2000). Regulation of natural medicines in Israel and abroad. *Harefuah, 139*, 339–344, 408, 407. Available: www.heilpflanzen-welt.de/science/2000/11341206.htm. Accessed Oct. 16, 2007.

Lefevre, H. (1982). Immunohamolitische anamie nach genuss des kahlen krempling (*Paxillus involutus*). *Deutsche medizinische Wochenschrift, 107*, 1734.

McNamara, H. S. (1999). Dietary supplement legislation. In *Botanical Medicine*. Eshkinazi, D., editor. New York: Mary Ann Liebert, pp. 199–204.

Mizuno, T. (1999). The extraction and development of antitumor-active polysaccharides from medicinal mushrooms in Japan (Review). *International Journal of Medicinal Mushrooms, 1*, 9–31.

Natural Products Association (2007). Natural nutritional supplements historical background. Available: www.naturalproductsassoc.org. Accessed Oct. 16, 2007.

New Zealand Food Safety Authority (NZFSA) (2005). Food safety; Briefing for incoming ministers. Available: www.maf.govt.nz/mafnet/publications/2005-briefing-for-incoming-ministers/food-safety-briefing.pdf. Accessed Oct. 16, 2007.

Ohtsuka, S., Ueno, S., Yoshikumi, C., Hirose, F., Ohmura, Y., Wada, T., Fujii, T., and Takahashi E. (1977). Polysaccharides and method for producing same. U.S. Patent 4,051,314, Sept. 27, 1977.

Pointing, S. B., Jones, E. B. G., and Vrijmoed, L. L. P. (2000). Optimizations of laccase production by *Pycnoporus sanguineus* in submerged liquid culture. *Mycologia, 92*, 139–144.

Ross, A. E., Nagel, D. L., and Toth, B. (1982). Evidence for the occurrence and formation of diazonium ions in the *Agaricus bisporus* mushroom and its extracts. *Journal of Agricultural and Food Chemistry, 30*, 521–525.

Santillo, H. (1993). *Natural Healing with Herbs*. Prescott, AZ: Hohm Press.

Schmidt, J., Hartmann, W., Wurstlin, A., and Deicher, H. (1971). Akutes Nierenversagen durch immunhamolitische Anamie nach Genuss des Kahlen Kremplings (*Paxillus involutus*). *Deutsche Medizinische Wochenschrift, 96*, 1188–1191.

Shimizu, T. (2003). Health claims on functional foods: The Japanese regulations and an international comparison. *Nutrition Research Reviews, 16*, 241–252.

Smith, B. L., Marcotte, M., and Harrison, G. (1996). A comparative analysis of the regulatory framework affecting functional food development and commercialization in Canada, Japan, the European Union and the United States of America. Available: dsp-psd.communication.gc.ca/Collection/WWW96-19E.pdf. Accessed Oct. 16, 2007.

Smith, J. E., Rowan, N. J., and Tan, K. K. (2000). Functional food science and the medicinal mushrooms. *International Journal of Medicinal Mushrooms*, 2, 277–285.

Stamets, P. (2000). *Growing Gourmet and Medicinal Mushrooms*, 3rd ed. Berkeley: Ten Speed, p. 574.

Stamets, P. (2002). *MycoMedicinals: An Informational Treatise on Mushrooms*, 3rd ed. Olympia: MycoMedia Productions, p. 96.

State Services Commission (2006). Machinery of government review of the New Zealand Food Safety Authority. Available: http://www.ssc.govt.nz/display/document.asp?DocID = 5401. Accessed Oct. 16, 2007.

U.S. Department of Health and Human Services (DHHS) (2006). Food and Drug Administration. Conventional foods being marketed as "functional foods"; public hearing; request for comments. Available: www.fda.gov/OHRMS/DOCKETS/98fr/06-8895.htm. Accessed Oct. 16, 2007.

U.S. Federal Trade Commission (FTC) (2001). Dietary supplemets: An advertising guide for industry. Available: www.ftc.gov/bcp/conline/pubs/buspubs/dietsupp.pdf. Accessed Oct. 16, 2007.

U.S. Food and Drug Administration (FDA) (2000). Regulations on statements made for dietary supplements concerning the effect of the product on the structure or function of the body. Food and Drug Administration, 21 CFR part 101, Docket No.98N-0044. Available: www.fda.gov/OHRMS/DOCKETS/98fr/oc99257.pdf. Accessed Oct. 16, 2007.

U.S. Food and Drug Administration (FDA) (2004). *Guidance for Industry: Botanical Drug Products*. Available: www.fda.gov/cder/guidance/4592fnl.pdf. Accessed Oct. 16, 2007.

U.S. Government Accountability Office (GAO) (2005). Food safety; experiences of seven countries in consolidating their food safety systems. GAO Report to Congressional Requesters. Available: www.gao.gov/new.items/d05212.pdf. Accessed Oct. 16, 2007.

Wasser, S. P. (2002). Medicinal mushrooms as a source of antitumor and immunomodulating polysaccharides. *Applied Microbiology and Biotechnology*, *60*, 258–274.

Wasser, S. P., Didukh, M.Y., and Nevo, E. (2004). Dietary supplements from culinary-medicinal mushrooms: A variety of regulations and safety concerns for the 21st century. *International Journal of Medicinal Mushrooms*, *6*, 231–248.

Wasser, S. P., Nevo, E., Sokolov, D., Reshetnikov, S., and Timor-Tismenetsky, M. (2000a). Dietary supplements from medicinal mushrooms: Diversity of types and variety of regulations. *International Journal of Medicinal Mushrooms*, *2*, 1–19.

Wasser, S. P., Nevo, E., Sokolov, D., and Timor-Tismenetsky, M. (2000b). The regulation of dietary supplements from medicinal mushrooms. In *Science and Cultivation of Edible Fungi*. Van Griensven, L. J. L. D., editor. Rotterdam: A. A. Balkema, pp. 789–801.

Wasser, S. P. and Reshetnikov, S. V. (2002a). Process for producing, method and composition of glucuronoxylomannan as a nutriceutical agent from higher Basidiomycetes mushrooms. U.S. Patent 6,383,799 B1, May 7, 2002.

Wasser, S. P. and Reshetnikov, S. V. (2002b). For higher basidiomycetes mushrooms grown (as one cell biomass) in submerged cultures. U.S. Patent 6,362,397 B1, Mar. 26, 2002.

Wasser, S. P. and Reshetnikov, S. V. (2002c). Process for producing, methods and composition of cholesterol lowering agents from higher Basidiomycetes mushrooms. U.S. Patent 6,372,462 B2, Apr. 16, 2002.

Wasser, S. P., Reshetvikov, S. V., Solomko, E. F., Buchalo, A. S., and Nevo, E. (2002). For higher Basidiomycetes mushrooms grown as biomass in submerged culture. U.S. Patent 6,372,964 B1, Apr. 16, 2002.

Wasser, S. P., Sokolov, D., Nevo, E., Reshetnikov, S. and Timor-Tismenetsky, M. (2001). Dietary supplements from medicinal mushrooms: How we are going to ensure their quality and safety. *International Journal of Medicinal Mushrooms*, *3*, 94.

Wasser, S. P. and Weis, A. L. (1999a). Medicinal properties of substances occurring in higher basidiomycetes mushrooms: Current perspectives (Review). *International Journal of Medicinal Mushrooms*, *1*, 31–62.

Wasser, S. P. and Weis, A. L. (1999b). Therapeutic effects of substances occurring in higher Basidiomycetes mushrooms: A modern perspective. *Critical Reviews in Immunology*, *19*, 65–96.

World Health Organization (WHO) (1982). Evaluation of certain food additives and contaminants. 26th Report of the Joint FAO/WHO Expert Committee of Food Additives. Technical Report Series No. 683. Geneva: WHO.

World Health Organization (WHO) (1991). World Health Organization Programme on Traditional Medicines. *Guidelines for the Assessment of Herbal Medicines*. Geneva: WHO Reprinted in *HerbalGram*, *28*, 17–20, 1993. Available: http://content.herbalgram.org/wholefoodsmarket/herbalgram/articleview.asp?a=275.

Zeisel, S. (1999). Regulation of "nutraceuticals." *Science*, *285*, 1853–1855.

INDEX

Mushrooms as Functional Foods, Edited by Peter C. K. Cheung

* 9 7 8 0 4 7 0 0 5 4 0 6 2 *